KV-337-689

World Energy Issues and Policies

Proceedings of the First Oxford Energy Seminar
(September 1979)

EDITED BY
ROBERT MABRO

OXFORD UNIVERSITY PRESS
for THE OXFORD ENERGY SEMINAR
St Catherine's College
1980

Oxford University Press, Walton Street, Oxford OX2 6DP
OXFORD LONDON GLASGOW
NEW YORK TORONTO MELBOURNE WELLINGTON
KUALA LUMPUR SINGAPORE JAKARTA HONG KONG TOKYO
DELHI BOMBAY CALCUTTA MADRAS KARACHI
NAIROBI DAR ES SALAAM CAPE TOWN

Published in the United States by
Oxford University Press, New York

British Library Cataloguing in Publication Data

Oxford Energy Seminar, 1st, 1979

World energy.
1. Power resources - Congresses
I. Title II. Mabro, Robert III. St. Catherine's College, Oxford
333.7 HD9502. A2
ISBN 0-19-920119-6

Typeset by Anne Joshua Associates
and printed in Great Britain by
Billing and Sons Ltd., Guildford and Worcester

CONTRIBUTORS

HE Humberto Calderon Berti	Minister of Energy & Mines, Venezuela
HE Ali Khalifa al Sabah	Minister of Oil, Kuwait
Mr Nordine Aït-Laoussine	Executive Vice President, Sonatrach, Algeria
Dr Ali Attiga	Secretary General, Organization of the Arab Petroleum Exporting Countries
Mr Peter W. Camp	Manager, Nuclear Energy Strategic Planning, General Electric, US
Dr Fadhil al-Chalabi	Deputy Secretary General, Organization of the Petroleum Exporting Countries
Dr George Corm	Professor of Economics, Consultant and Financial Advisor, Lebanon
Dr Richard Eden	Director, Energy Group, University of Cambridge, UK
Mr John Foster	Senior Economic Advisor, Petro-Canada
Professor John Goodenough	Professor of Inorganic Chemistry, Oxford University, UK
Professor Wolf Häfele	Director, Interntional Institute for Applied Systems Analysis, Austria
Mr Jack E. Hartshorn	Consultant, Switzerland
Mr Ali Jaidah	Managing Director, Qatar General Petroleum Corporation
Mr James T. Jensen	Chairman, Jensen Associates, US
Mr T. Philip Jones	Deputy Under Secretary, Department of Energy, UK
Mr Michael Kaser	Professorial Fellow, St Antony's College, Oxford University, UK
Mr Robert Mabro	Fellow, St Antony's College, Oxford University, UK
Professor W. Murgatroyd	Professor of Engineering, Imperial College, London University, UK

Mr René Ortiz	Secretary General, Organization of the Petroleum Exporting Countries
Mr Michael J. Parker	Chief Economist, National Coal Board, UK
Mr Francisco R. Parra	Managing Director, Petroleos de Venezuela (SA), UK
†Mr C.C. Pocock	Late Chairman of the Committee of Managing Directors, Royal Dutch/Shell, UK
Mr Alberto Quiros	President, Maraven, Venezuela
The Staff	OPEC Special Fund
Mr Thomas Stauffer	Centre for Middle Eastern Studies, Harvard University, US
Mr Aziz al Watari	Assistant Secretary General, Organization of the Arab Petroleum Exporting Countries
Mr George Williams	Director, UK Offshore Operations Association Ltd.
Mr Robert G. Wilson	Manager, Energy Policy, Exxon, US
Mr Minos Zombanakis	Chairman, Blyth Eastman Dillon & Co. International

The Oxford Energy Seminar forms part of the eduation and research that St. Catherine's College sponsors and accommodates during University vacation. It is a fully residential, educational conference designed for government officials, industrialists, managers and other professionals engaged in the field of energy. Its objectives are: to enhance the professional qualifications of participants; to improve the understanding of forces shaping the environment in which future investment and policy decisions will be made; to provide an opportunity for close contact and privileged debate between participants from petroleum exporting and importing countries.

The Oxford Energy Seminar is co-sponsored by OPEC and OAPEC. It ensures a balance among speakers and participants between nationals of petroleum exporting and importing countries.

The first Oxford Energy Seminar was held on 3rd–14th September 1979 and involved fifty-three participants and forty speakers. The Seminar is organized by a Board of Management consisting of the Secretary General of OPEC, the Secretary General of OAPEC, Mr. Wilfrid Knapp and Mr. Robert Mabro.

All the views expressed in this book are those of the respective authors, and do not necessarily reflect those of the governments, companies or institutions to which they belong. Nor do they necessarily reflect the views of the management of the Oxford Energy Seminar or of its sponsors.

CONTENTS

Part III: ECONOMIC, INSTITUTIONAL AND POLITICAL STRUCTURES

Part IV: WORLD ENERGY OUTLOOK

Part V: ENERGY ISSUES AND POLICIES: OECD COUNTRIES

Part VI: ENERGY ISSUES AND POLICIES: THE SECOND AND THIRD WORLD

Part VII: ENERGY ISSUES AND POLICIES: OIL-EXPORTING COUNTRIES

ABBREVIATIONS, SYMBOLS AND UNITS

AGR	Advanced gas-cooled reactor
b	Barrel
B	Billion
Bcf	Billion cubic feet
b/d	Barrel per day
BNOC	British National Oil Company
boe	Barrel of oil equivalent
Btu	British thermal unit
BTX	Benzene, Toluene, Xylenes
BWR	Boiling water reactor
c	Velocity of light
CA	Communist area
CFP	Compagnie Française des Pétroles
CIA	Central Intelligence Agency (US)
CIEC	Conference on International Economic Co-operation
c.i.f.	Cost, insurance and freight
DAC	Development Assistance Committee (OECD)
DMT	Dimethyl terephthalate
e	Electronic charge
E	Energy (Physics)
EDC	Ethylene dichloride
EEC	European Economic Communities
ENI	Ente Nazionale Idrocarburi
FAO	Food and Agriculture Organization
f.o.b.	Free on board
GDP	Gross domestic product
GDR	German Democratic Republic
GNP	Gross national product
GRC	Gas Future Requirements Committee (US)
GRP	Gross regional product
GW	Gigawatt
GWe	Gigawatt (electricity)
h	Planck's constant
HDPE	High density polyethylene
HTGR	High temperature gas-cooled reactor
I	Intensity (electricity)

IBRD	International Bank for Reconstruction and Development, World Bank
IEA	International Energy Agency
IGAT	Iran gas trunkline
IIASA	International Institute for Applied Systems Analysis (Austria)
INFCE	International Fuel Cycle Evaluation
K	Kelvin (thermodynamic temperature)
Kcal	Kilocalories
Kgoe	Kilogram of oil equivalent
KWe, kWe	Kilowatt (electricity)
KWh, kWh	Kilowatt hour
λ	Wave length
LCC	Large crude carrier
LDC	Less developed countries
LDPE	Low density polyethylene
LMFBR	Liquid metal fast-breeder reactor
LNG	Liquefied natural gas
LOFT	Loss of fluid test
LP	Linear programming
LPG	Liquefied petroleum gas
LWR	Light water reactor
M	Million
M b/d	Million barrels per day
Mcf	Million cubic feet
Mcf/d	Million cubic feet per day
MNC	Multinational (oil) corporation
MSA	Most seriously affected countries
Mtce	Million tons of coal equivalent
Mtoe	Million tons of oil equivalent
MVC	Monovinyl chloride
MW	Megawatt
NGL	Natural gas liquids
NIOC	National Iranian Oil Company
NOC	National Oil Company
NOPEC	Non-OPEC developing countries
NPC	National People's Congress (China)
O	Oxygen
OAPEC	Organization of Arab Petroleum Exporting Countries
ODA	Official development aid
OECD	Organization for Economic Co-operation and Development
OIDC	Oil-importing developing countries
OPEC	Organization of the Petroleum Exporting Countries
OTEC	Ocean thermal energy conversion
P	Power (physics)
PP	Polypropylene
p.a.	Per annum
PIW	*Petroleum Intelligence Weekly*

PJ	Petajoule
Pu	Plutonium
PVC	Polyvinyl chloride
PWR	Pressurised water reactor
R&D	Research and development
RD and D	Research, development and demonstration
SBM	Seabed moors
SBR	Styrene butadiene rubber
t	Ton, tonne
T	Temperature
tce	Tons coal equivalent
tcf	Trillion cubic feet
Th	Thorium
TMI	Three Mile Island
TPA	Terephthalic acid
TW	Terawatt
TWe	Terawatt (electricity)
TW yr	Terawatt year
U	Uranium
UAE	United Arab Emirates
UKCS	United Kingdom continental shelf
UKOOA	United Kingdom Offshore Operators Association Ltd.
UNCTAD	United Nations Conference for Trade and Development
UNIDO	United Nations Industrial Development Organization
UV	Ultra violet
VCM	Vinylchloride monomer
WOCA	World outside Communist areas

INTRODUCTION

Robert Mabro

The papers collected in this book were presented at the Oxford Energy Seminar, held at St. Catherine's College in September 1979. The Seminar provided a restricted number of participants with a privileged opportunity and a structured framework for a debate on the present problems and prospects of world energy. One distinctive feature of the Seminar was the balanced composition between nationals of oil-consuming and oil-producing countries. Another feature was the involvement of some 40 leading personalities — ministers, senior civil servants, industrialists and academics — as chairmen of sessions, speakers and panelists. The statements made at the conference were authoritative; and the dialogue, established in the formal sessions and continued late in the night in the College bar, was constructive.

The purpose of this book is to present to a wider audience a set of papers which survey a broad range of complex energy issues, the papers which formed the backbone of the Seminar debates. The book embraces a variety of views, and it is hoped that their publication side by side will enable the reader to gain a better understanding of the positions and objectives of parties from a different background, of approaches and interests which he may not share. In some way, the book, like the Seminar, may thus contribute to the bridging of gaps between the perceptions and attitudes of oil consumers and producers, and improve the chances of success of any future formal dialogue which Governments sooner or later will have to initiate.

The objective may thus be construed as political. But I have no apologies to offer, however disturbing or distasteful the thought of politics may be to many fellow academicians. The energy problem is, in the final analysis, political. Its solution — a solution on which the health of the world economy and the welfare of society in the next thirty years so crucially depend — is not hindered by technological difficulties, nor by a lack of ideas on how to design appropriate economic policies. The technology required for the development of alternative sources of energy to oil is largely available; and engineers and scientists when provided with adequate resources rarely fail to produce new technological answers. Further, good energy policies can and have been conceived on paper. The shambles in which energy policy finds itself in the major industrialised countries is almost entirely attributable to powerful conflicts of interests between various constituencies in these countries which Governments find difficult to reconcile. Working for a solution of the energy crisis is, for all parties concerned, both domestically in each country and internationally between developed and developing countries, a matter of political will, a will that may be enhanced by improvements in the understanding of the energy problem.

The book is structured in seven parts. Parts I to III review recent developments and current problems of energy. Parts IV to VII consider the future – the prospects and the problems likely to challenge decision-makers in the medium and long term – and focus more explicitly on issues of policy.

The economic concepts of supply, demand and institutional structure provide a convenient framework for Parts I to III. In Parts IV to VII speculation about the future is followed by papers on the particular policy problems of four sets of nations: industrialized, Communist, non-oil producing, and oil-exporting nations of the third world. These distinctions, however, should not be stretched too far. There is a strong unity of themes in the book; and most, if not all, of the papers involve incursions into the future and significant discussions of policy. The reader should note, for example, that most chapters of Part I (mainly Chapters 2, 4, and 5 to 7), and all chapters of Part II, pose the energy issues faced by the industrialized, oil-importing world, namely those of alternative sources of supply and conservation in use.

Gaps are inevitable in such collections of papers. The reader will probably recognize that the attempt to achieve comprehensive coverage has gone very far. Yet, I would have liked to include a paper on the environment to balance the views presented in Chapter 6, which deals with nuclear energy. Further, a paper by a senior official from a major OECD country, preferably the US, would have been welcome. But many official speakers at the Seminar elected, and were indeed encouraged, to talk off the record. What was gained then in candour and substance more than compensate for what is lost in this book.

The long discussions which followed the presentation of the papers are not published here, simply because they were not minuted. Privilege is essential for a frank dialogue. However, I have attempted in Chapter 29 to summarise the conclusions of the debate and to identify areas where agreement was reached and issues on which differences of opinion remained irreductible.

This book has not been heavily edited, but a measure of formal consistency in the presentation has been sought. A different editorial policy would have involved long publication lags, and may have robbed the papers of their original flavour. In a field where opinions, misunderstandings, and prejudices, alas, tend to prevail, it is essential to respect the integrity of authoritative statements, both in substance and in style. And in any sincere attempt at understanding somebody else's position much can be learned from the way in which the different point of view is put.

Having briefly introduced the objectives, structure and features of this publication I would like to turn to the very agreeable task of recording expressions of gratitude. The contributors to the book, who kindly and promptly authorised the publication of their papers or prepared revised versions, have made the Editor's task easy and pleasurable. Wilfrid Knapp provided well-humoured support throughout the preparation of the book, as he did for long months before when we jointly organized the Seminar. Judy Mabro and Ann Davison did valuable editorial work. Acknowledgements are due to the Editors of *Petroleum Intelligence Weekly* and of *Futures* for permission to reproduce in this book papers originating or relating to the Seminar which they managed to publish before us.

Finally I would like personally to thank Mr. Ali Attiga, Mr. René Ortiz, the speakers and the participants of the Seminar. They made it all worthwhile. Without their encouragement and moral support I would have not been able to marshall the energy needed to put together this book.

Part I

ENERGY SUPPLY

1 OPEC OIL: RECENT DEVELOPMENTS AND PROBLEMS OF SUPPLIES

Francisco R. Parra

The subject assigned to me for this paper concerns recent developments in the OPEC area and the supply scene. I shall therefore give a brief outline of OPEC's present and future activities. I must emphasise that I can only give you the view of an outsider, since I have no official or formal connection with the Organization. However, as a director of a national oil company in one of its member countries, I am constrained to follow its activities with interest.

I am assuming that there is a general agreement among observers of the energy scene on the following interpretation of the difficulties that lie ahead in keeping the world fuelled:

(a) The world, after enjoying two lush decades of low-cost energy from the Middle East, North Africa, Venezuela and elsewhere, is now facing a problem of transition. The underlying reality of this transition is the approaching physical shortage of low-cost oil, compounded by the reluctance of its owners to sell off their reserves with undue haste. The dimensions of the transition are unexpectedly large in terms of cost – incremental energy resources on a significant and expanding scale may average 20 to 30 times as much as the average cost of Middle East oil, a magnitude which is far from being fully accepted by consumers.

(b) Major obstacles (cost apart) to making this transition are as follows. First, the reluctance of the consumer to further foul his own nest by turning to fuels which are thought to pollute more than oil and gas; or by developing oil and gas resources in his own backyard. The environmentalists are particularly strong in the United States, where just about any project can be blocked for an unconscionable time on environmental grounds, but they are also growing in strength (particularly the anti-nuclear lobby) in Japan and Western Europe; and since cannibalism is still frowned upon in these countries, you will not be able to eat the environmentalists when you are cold, hungry and unemployed. Being against the environment is neither politically practical nor sane, but the extremist who seeks to block all development and energy-related projects almost on principle is the enemy within, constituting a threat to US national security for those that see continued US reliance on oil imports that way. It seems, therefore, that the price of energy will have to increase, not only by the greater technical cost of production, but also by an amount which permits it to be produced in a way which causes minimal additional pollution.

Secondly, the reluctance of governments to take sufficiently strong action, particularly on prices. This has been due partly to the great uncertainty among experts of all hues and colours over the likely future course of oil supplies and prices from the OPEC area. It was only a year ago that several authoritative sources were saying aloud that the real energy crisis had been postponed until 1990 at least and perhaps cancelled altogether. With ammunition like this, the finance minister seeking to damp inflation is unlikely to lose out to the energy minister who is thought to have been crying wolf since 1973. Government perceptions are changing in the wake of the Iranian crisis, but the public at large, upon whose support official programmes must ultimately depend, has yet to be convinced that there is not at least a large element of skulduggery in the whole thing.

Thirdly, the high cost of incremental energy sources combined with the relatively undeveloped state of technology in nuclear fusion, solar energy, etc. There is, of course, no single figure for the cost of alternatives – it is a matter of scale on which development is to take place, and where – but they all have one thing in common: they are massively expensive and investment-intensive. A price of $20 per barrel is now considered expensive, but the cost of producing hydrocarbons from (high-priced) coal has yet to be seen. Incidentally, those who envisage a massive expansion of coal usage in the international coal trade by the year 2000 will also have to start considering some interesting geo-political problems that will arise from dependence on coal imports.

(c) The politics of the Middle East pose a constant threat to supply. It has long been a matter for concern that energy supplies are threatened by disruption stemming from political crises, particularly in the Middle East. But it is now more widely felt that there is no way this threat can be materially attenuated until a *modus vivendi* acceptable to most Arab states can be found for the rights of the Palestinian people. And as Iran has so graphically shown us, internal instability in any major oil exporter can produce supply problems which have a lasting impact.

How then do recent developments in the OPEC area fit into this context, which is in effect determining the way in which governments of producing and exporting countries react, and the direction in which they orient their energy policies? Of the many aspects of current trends in the OPEC area, I think four are of particular interest at present and I shall confine my remarks to these. These four are: (1) Production Limitations, (2) Restructuring of the International Oil Trade, (3) Oil Investment Programmes, and (4) The 'Dialogue' between Oil Exporters and Consumers.

Production Limitations

Most of the larger OPEC countries now have production limitations in force. During recent years, these have become increasingly widespread and stringent. They are the source of much misunderstanding among consumer countries. At present, production limitations are in force in Saudi Arabia, Iran, Kuwait, the United Arab Emirates and Venezuela, with a fixed annual production imposed as a ceiling. In Libya and Nigeria, variable limitations have been imposed at times with respect to particular areas or fields. The cumulative effect of such production limitations is to withhold from the market a potential supply the precise quantity of which is not known, but that certainly exceeds 5 M b/d. The overwhelmingly predominant motive for the imposition of these limitations in all of the countries where they exist is to stretch out the country's resources over a period of time which at present appears to be long enough to allow for transition to other forms of economic activity and national income. The OPEC countries, and other large oil exporters, also have their transition problem which, almost by definition, coincides with the problems of transition of the industrialised countries, referred to above. But of course it is of a far greater relative magnitude. The preoccupation of industrialised countries is how to stagger through this transition without losing too much in the way of economic growth in the process. Virtually all energy consumption forecasts published by authoritative sources in the West envisage continued economic growth to the end of the century (or whatever their time horizon is), albeit at a slower rate than previously. The chief worry appears to be whether economic growth can be sustained at a rate high enough to prevent increases in unemployment. But if an oil-exporting country were to produce at full capacity for as long as possible, it would in a span of time which can now be counted as a matter of a few years, encounter decline rates which at first would be small but would rapidly accelerate until production was falling rapidly and alarmingly. You will see this happen right here in the North Sea for production from the presently discovered fields, although its effect may be attenuated by discovery of further fields in newly licensed acreage.

The consequence of decline in an oil-exporting country of the OPEC area would not be limited to struggling to maintain a modest rate of economic growth; it would be a precipitious decline in GNP, presumably accompanied by internal upheavals which would probably aggravate the problem even further by slowing down secondary and tertiary recovery projects (viz Iran today). The transition problems of the oil-importing countries fade into insignificance compared with those which would surely confront any OPEC country depleting its resources as fast as possible before it has something to take their place.

It would be rash for any country to embark on this course. OPEC countries are sometimes thought of as rich countries, but anyone who has travelled through them need not consult the statistics to know otherwise. They are like starving men suddenly faced with an enormous and at first sight apparently endless feast. It does not matter how much they eat, they cannot gain strength and normal weight overnight — but they may make themselves very sick indeed by trying to do so too rapidly. There is an essential difference between the absorptive capacity commonly referred to in the case of revenue-rich exporting countries when world monetary problems are being discussed; and the capacity to invest effectively. In the first case, it is a question of, among other things, balance-of-payments problems and exchange rates; but the absorptive capacity that is truly important for an oil-exporting country is not its capacity to spend money but its capacity to invest effectively in its own economy. There is plenty of evidence to show that this kind of absorptive capacity has not grown anywhere near as fast as the capacity to import consumer goods on a massive scale. For the oil-exporting countries, conservation of resources is not an oil price fixing device; it is a question of economic survival.

Yet many people in consuming countries interpret OPEC's production ceilings as primarily motivated by the desire to maintain and raise prices, and from this they jump to the conclusion that OPEC is a price-fixing cartel. The next step in their logic is that this cartel must somehow be broken. It should be obvious, but is not, that OPEC's price-setting function is separate from the production limitations of its individual member countries. What happens in practice is that prices are set and production then finds its level at those prices. Until recently, such production has mostly been below the aggregate of production limitations. When the reverse occurs, namely demand pushing up against the aggregate of production limitations, OPEC may continue to pretend to set prices, but price is in fact taken out of its hands, and the official government price then includes an element of give-away which may or may not be justified by broader economic and political considerations.

Among the reasons why production limitations in exporting countries have been misunderstood is no doubt the way in which Saudi Arabia has responded to pressures brought upon it from time to time to increase its minimum allowable. At no time have the Saudi increases above its present formal maximum of 8.5 M b/d been in the country's interest, as far as revenue needs or appropriate depletion rates are concerned. They have been a political response to a multiplicity of factors, such as the consequences for the world economy of too great an increase in oil prices over a short period of time, the political situation in the Middle East, an increase in world inflation, etc. But it is difficult in these circumstances to convince the consumer that because Saudi Arabia will sometimes open up its production above maximum allowables that the whole question of conservation of resources through production limitations is not a façade for price fixing. It then appears to be particularly galling to the consumer that so many immediate economic difficulties caused by high oil prices and temporary shortages could be simply relieved by turning the valves in a few countries. But to do so would spell disaster for the producing countries concerned, and would only bring the day of reckoning closer for the consumer countries.

Restructuring of the International Oil Trade

The restructuring of the international oil trade now in the process of completion is one of the fundamental changes which has occurred in the petroleum industry since OPEC member countries took control of investment, output and prices. In principle, the process is quite simple: all sales of crude oil from exporting countries will, when the process has been completed, be made directly by the government entity in charge of international marketing to refiners on a long-term contract basis. The process is not yet complete and no doubt never will be in such a rigid manner. Sales still occur in several countries through traders, brokers and large oil companies acting as intermediaries. The role of these in marketing will be reduced, though probably never entirely eliminated.

In general, also, there has been a desire on the part of OPEC countries to minimise spot sales (but the temptation of high spot prices has not always been resisted) partly because spot sales are associated with brokers and traders making large windfall profits at the expense of producing countries, and partly because any substantial dependence on the spot market increases the country's vulnerability in the event of a down-turn in the market, both with respect to price and volume. However, I do not expect that spot sales will entirely disappear.

There are many consequences to the restructuring of the international oil markets, and I would like to touch on two of them briefly. The first is that it has brought the importing developing countries increasingly into contact with the oil-exporting countries. The major oil companies have been rapidly withdrawing from the so-called third-party crude sales market, an important part of which consisted in sales to national oil companies in the importing developing countries. These countries, usually through their national oil companies, are now buying directly from OPEC countries, who are practically their exclusive suppliers (other net oil-exporting countries ship practically no oil to these LDCs). The elimination of the major oil companies as intermediaries, or 'buffers' as they would term it, is bringing this segment of the oil trade increasingly into the political arena and individual OPEC member countries have, by and large, established a peculiar kind of *ad hoc* relationship with these countries by which they obtain oil on concessional terms through a variety of devices, mostly other than a direct discount off the official government price. In addition, a number of OPEC countries have individually, but in parallel fashion, been assuming responsibility for the supply of oil to these countries, either directly through their sales contracts, or, where the major oil companies are still concerned as intermediaries, by directing their suppliers to continue supplying them at previously established levels.

There seems little doubt that the relationship between oil exporters and LDC importers (which must also be considered against the background of the failed North-South Paris Dialogue, where OPEC attempted to tie discussion of energy issues to broader issues between LDCs and industrialised countries generally) will become more political as time goes by. But it is impossible to predict which group of countries will gain the most from the basic political bargain being struck, namely continued support by the importing developing countries for OPEC in OPEC's relations with industrial countries, in return for some kind of relatively favourable treatment with respect to the supply and/or price of oil.

The second consequence of the restructuring of the international oil markets that I would like to mention here is the increasing rigidity which is being built into the long-term sales contracts concluded by exporting countries. Financially, the terms are getting tougher with respect to credit and method of payment; but of more consequence to the consumer is the increasingly strict adhesion to relatively even and regular liftings over the year of basic contract volumes, and the virtual disappearance of tolerances at the purchaser's option on such volumes. The crude oil purchaser is now usually required to lift the volumes contracted for, on a fairly even schedule throughout the year. This tends to minimise production variations in the supplying country and facilitate the supplier's programming, both with respect to production and finances; but it tends to create a problem of seasonality for the purchaser. In addition, the general rule in the OPEC area is fast becoming that there be little or no tolerance at the purchaser's option to lift more than, or less than, the contracted volume. This provision has the same advantages for the suppliers as the requirement of regular liftings over the year, but again means that there may be some difficulty for the purchaser, who must either buy short and make up on the spot market or risk going long on crude. It should be noted too, that purchasers generally do not have the right to resell crude without the specific permission of the supplying country. In one way or another these conditions will tend to increase the cost of oil to the ultimate consumer and possibly tend to push more of the swing in production into non-OPEC producers.

Oil Investment Programmes

The governments of OPEC member countries are now almost entirely in direct control of investment programmes for the development of their oil resources. Different countries acquired control at different times, but all of them relatively recently. It is only now that the process of formulating investment policies and gearing up to implement them is taking shape. Countries naturally see their development requirements in a different light, usually as a function of their current production-to-reserves ratio. But it is not possible to generalise. This is of course a critically important area for the medium and long-term, since OPEC countries among them not only control the lion's share of world proved reserves, but also much of the best prospective acreage remaining, and immense reserves of natural gas. I therefore regret not being able to be more specific in my comments, but the fact is that OPEC countries' current investment programmes and future investment intentions are not in many cases made public. I can only indicate some of the key issues involved:

(*a*) *The Development of Proved and Probable Reserves.* Countries with large, undeveloped oil reserves in the OPEC area are Saudi Arabia, Iraq, the UAE and (if one counts the Orinoco petroleum belt) Venezuela. Saudi Arabia has a go-slow policy on the development of additional producing capacity from proved reserves, and, as far as I know, has no firm plans to go beyond 12.8 M b/d. Its chief fear, of course, is that if it develops capacity, it will be forced by political circumstances to use it; and the Saudis are probably right. But as the world oil supply picture darkens, political circumstances may also force a change of heart on Saudi Arabia; in addition, recent major discoveries of light crude inland in Saudi Arabia require important investment and policy decisions: should they be developed? And should they affect the current permissible production ratio of 65/35 light/heavy crude, which at present underpins the price-differentials of world crudes and has made the installation of upgrading facilities a matter of urgency in the refining industry?

Little has been published on Iraq, and neither present nor intended future producing capacity is known. The popular figure for current producing capacity is less than 4 M b/d, and to judge from estimates of the number of rigs active in the country (about 30) over the past few years, which have not changed, development is likely to be steady and relatively slow.

The UAE's current investment plans for new capacity revolve mainly around the Upper Zakum project which will eventually add over 1 M b/d of producing capacity to the country. This will be, by Middle East standards, very high cost production; how the circle will be squared *vis-à-vis* production limitations is at present a matter for speculation.

Venezuela has estimated recoverable reserves of oil in the Orinoco petroleum belt of at least 70 billion barrels. There are financial and technical problems associated with its development, but current plans envisage development of over 1 M b/d of capacity by the early 1990s.

(*b*) *Exploration.* Most of the larger OPEC countries have considerable areas of attractive acreage which are far from being fully explored. Algeria, Indonesia, Iran and presumably Iraq have undertaken vigorous exploration programmes over the past few years, although Iran's programme is now faltering. Venezuela is stepping up its exploration (outside the petroleum belt) after years of neglect prior to nationalisation. Other countries, notably Libya and Nigeria, recognise the need for more intensive exploration, but do not yet appear to have a coherent and effective policy.

(*c*) *Other Areas*. I can only mention in passing here investment plans for natural gas (especially in Iran and Algeria), for natural gas liquids (most countries in the Middle East), and export refineries (further changes in the structure of the international oil trade).

The 'Dialogue'

I should like to wind up with a few brief remarks about the 'dialogue' between OPEC and industrial countries on energy issues. OPEC's stated policy is quite clear: there can be no dialogue unless broader economic issues between LDCs and the industrialised countries are also brought in, presumably as part of some sort of package deal. This has been OPEC's position since 1974, throughout the CIEC (Paris) North-South Dialogue, and it was reiterated at OPEC's most recent Ministerial Conference, held a couple of months ago in Geneva.

And yet . . . pressure mounts from various sources in industrialised countries for a dialogue of some sort, leading presumably to an understanding on some things which would include supplies if not the price of oil. There are of course many contacts, at many levels, both political and commercial, between OPEC countries and industrial ones. The proposed dialogue is supposed to lead eventually to something broader and perhaps vaguer, but to encompass informal commitments on supply.

I think, as far as OPEC is concerned, this may be a mirage. There are too many important political, economic and financial matters pending settlement between individual OPEC countries, with quite different interests in the issues, to permit such a development to evolve. And conversely, industrial countries view the various OPEC countries with a discriminating eye, both politically and in the context of their vastly differing resources.

2 CURRENT AND FUTURE OIL ACTIVITIES ON THE EEC's NORTH WEST CONTINENTAL SHELF

George Williams

This paper will attempt to answer the two questions: where have we got to so far on the EEC's NW Continental Shelf? and where are we going to on it over the next few years? In dealing with the first question, I will attempt to give a brief history of activities to date, with their results, and a short account of the hydrocarbon prospects of the area as they appear to me at the present time. In dealing with the second question, I will discuss the factors that will have a significant influence on future activities, and the likely nature and level of these activities.

Where Have We Got To?

Activities and Results to Date. Whilst geophysical surveys over the southern North Sea and some exploration drilling just off the coast of the Netherlands commenced very early in the 1960s, activities offshore of NW Europe started in earnest in 1964 – only some 15 years ago – immediately after final ratification of the 1958 Continental Shelf Convention and agreement between Governments of at least some of the international boundaries. Activities initially were mainly in the southern North Sea area, where the water depths allowed drilling with jack-up mobile units. To start with, existing jack-ups in use in other parts of the world were brought into the North Sea, but a number proved unsuitable, so that subsequently more and more jack-ups designed specifically for North Sea conditions were employed.

With the advancement of semi-submersible drilling units, activities moved gradually northwards into the deeper and more hostile waters of the northern North Sea. Without a doubt, the opportunities there encouraged the great advances made in semi-submersible design around the end of the 1960s and the building of, in fact, too many such units in the early 1970s. This led to a surplus and to quite a number being used as floating hotels (flotels) during the construction periods of the development of some of the North Sea fields.

Exploration activity reached a peak in 1975, when including Norwegian offshore waters around 50 mobile units were engaged in drilling operations, with over half in UK waters. After dropping to around 30 at the beginning of this year, the figure is currently, around 35 (excluding around 20 units being used as flotels). The decline in activity is to be deplored and has been due mainly to the decline in drilling in the UK offshore waters, brought about by the policies of the last UK Government.

Exploration has been very successful in certain of the offshore basins. Very high success ratios were achieved initially, particularly in the northern North Sea basin, but they have, as usual, declined subsequently. Between 1967 and 1974 in UK waters an exploration well on average found 67 million barrels of technically recoverable oil; after 1974 this figure declined to 33 million in 1975, 16 million in 1976 and 13 million in 1977. Over the same period of time the average size of discoveries was 293 million barrels, 139, 80 and 73 million barrels respectively.

To date, on the EEC's NW Continental Shelf, approaching 50 trillion cubic feet (tcf) of gas and 12 billion barrels of commercial oil have been proved; more technically recoverable reserves than these have been found but not yet established as commercial.

Exploration successes to date have led to the development, or planned development, of some 15 gas fields (including the Frigg Field, half of which lies in Norwegian waters) and some 20 oil fields, the aggregated production capacities of which will be around 8,000 M ft^3 of gas a day and around 2.5 M b/d of oil. However, their peak productions will not, of course, coincide. Establishing these production capacities has been a tremendous technical achievement because of both the hostile waters of the North Sea and the water depths involved – credit for this must go to the oil companies, and the contractors and suppliers supporting them. Of course, set-backs, harsh lessons and technical delays have been experienced; nevertheless, both steel and concrete marine installations, which in 1964 would have been regarded as unbelievably large, have since been designed, built and installed in water depths of up to 600 feet, and we have now laid to them much larger diameter pipelines than considered possible at that time.

Exploration successes and production achievements have resulted not by chance but because of the hard and dedicated work of highly specialised international teams utilising the expertise they had gained in many different and varying parts of the world. All the major oil companies have been involved and a large number of small oil companies. On the whole NW European Continental Shelf nearly 70 different oil companies have acted as operators, and they have been joined as co-licensees in their consortia by several hundred different oil and non-oil companies.

Whilst much has been accomplished to date, it has been done at a high financial cost, far greater than envisaged at the outset. Very roughly, offshore oil/gas expenditures on the EEC's NW Continental Shelf are now probably around $30 billion. A very large proportion of the money has come from the private sector, as distinct from the Government or public sector. The financing of these major investments provided a challenge to European financial institutions, and the cities of London and Edinburgh have developed into the main petroleum financing centres of Western Europe. Funds raised internally by oil companies have been the biggest single capital source for financing both exploration and development, with bank finance tapping the international capital markets in second place. Exploration financing, as traditionally is the case, came mainly from the internal resources of the oil companies. However, in addition, during the early stages of exploration, institutional investors (insurance companies, pension funds and merchant banks) invested in some of the smaller non-oil company licensees.

The mobilisation of the very much larger funds required for development led, in some cases, to syndicated project financing, carried out by groups of banks with a variety of recourse covenants, including the payment of overriding royalties. In 1976, during a period of financial constraint, when only the highest class borrowers could get bank finance, two smaller companies raised development capital by public issue on the Stock Exchange, the issues being underwritten by a merchant bank in both cases. However, this was an exceptional form of financing and was relatively expensive. It is generally expected, during the second phase of North Sea development, that medium-term syndicated bank financing will continue to supplement the oil companies' internal cash flow and loans raised on their corporate strength.

The Hydrocarbon Prospects as Currently Seen. At the outset, let me stress that our knowledge of the geology of the NW European Continental Shelf and its hydrocarbon prospects is still very far from complete; however, after 15 years, it is beginning to take shape. I will now discuss the prospects of some of the major sedimentary basins on the NW European Continental Shelf and its Continental Slope. (I should point out that not all the international boundaries have as yet been agreed.)

The southern North Sea basin stretches onto land in the UK, where a small gas field had been found way back in the 1930s, and into the Netherlands, where one of the world's largest gas fields, the Slochteren Field, was found at the end of the 1950s. The latter was really responsible for sparking off the industry's interest in the southern part of the North Sea. We have been successful in finding and developing gas reserves in both the UK and Netherlands part of the basin. Water depths seldom exceed 150 feet in the basin. Without doubt, there are more gas fields to be found in the geological horizons already explored, albeit small in size, and just possibly from the deeper horizons still unexplored. Higher prices for gas in the UK part are needed from our single, monopoly buyer in order to encourage exploration there. We have advised the new UK Government of this and hope to have a favourable reaction in the not too distant future.

In the northern part of the North Sea there is a large sedimentary basin that does not really stretch onto land, either in the UK or in Norway; hence we, as geologists, knew very little about it until we started exploration drilling in it at the end of the 1960s. The basin has been sub-divided in a number of ways – one such division is into the Central graben, the Viking graben, the Moray Firth and the Norwegian/Danish sub-basins. Exploration has established that the basin is an exceptionally prolific oil and gas one, with, at times, exceptionally high well productivities, exceeding, I might add, the expectations of many of even the more optimistic geologists. We have oil reservoirs with widely varying levels of associated gas (gas : oil ratios varying between 100 and 2,000), some condensate reservoirs and some non-associated gas reservoirs. To date, all NW European offshore oil production, actual and planned, is confined to this northern North Sea basin. In the UK part of this basin we have so far probably found a little over half of the potential reserves of oil. The remainder, I must stress, will be much more difficult and more expensive to find and will call for far higher numbers of exploration and appraisal wells than we have drilled to date, because the additional reserves will be contained in smaller sized reservoirs and also because our success ratios will continue to decline.

The Norwegian part of the northern North Sea basin and other Norwegian sedimentary basins still further north off the Norwegian Coast are in the Tronde area at 65°N and in the Troms area at 71°N. Moving west to the west of the Shetland Islands, we have the Faroe-Shetland basin, the eastern edge of which has now been tested by some 28 exploration wells. Some of these wells have encountered heavy oil, but, because of a number of technical difficulties (e.g.

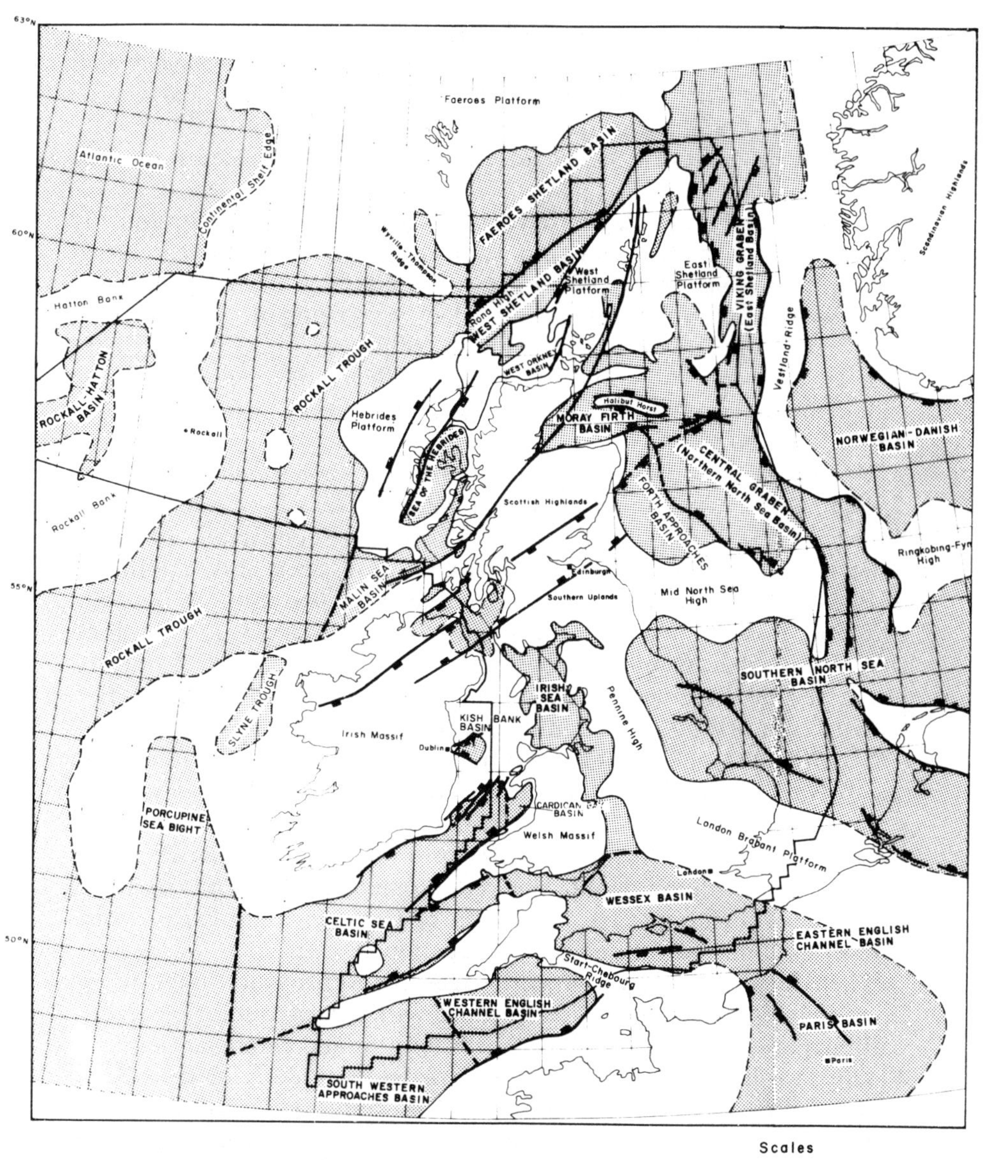

Faeroes Platform
Atlantic Ocean
Continental Shelf Edge
FAEROES SHETLAND BASIN
WEST SHETLAND BASIN
Rona High
West Shetland Platform
East Shetland Platform
VIKING GRABEN (East Shetland Basin)
Vestland Ridge
Scandinavian Highlands
Hatton Bank
ROCKALL-HATTON BASIN
ROCKALL TROUGH
Rockall
Rockall Bank
Hebrides Platform
SEA OF THE HEBRIDES
WEST ORKNEY BASIN
Halibut Horst
MORAY FIRTH BASIN
NORWEGIAN-DANISH BASIN
CENTRAL GRABEN (Northern North Sea Basin)
FORTH APPROACHES BASIN
Scottish Highlands
Ringkobing-Fyn High
MALIN SEA BASIN
Edinburgh
Southern Uplands
Mid North Sea High
SOUTHERN NORTH SEA BASIN
SLYNE TROUGH
Irish Massif
IRISH SEA BASIN
KISH BANK BASIN
Dublin
Pennine High
PORCUPINE SEA BIGHT
CARDIGAN BAY BASIN
Welsh Massif
London Brabant Platform
London
WESSEX BASIN
CELTIC SEA BASIN
EASTERN ENGLISH CHANNEL BASIN
Start-Cherbourg Ridge
WESTERN ENGLISH CHANNEL BASIN
PARIS BASIN
Paris
SOUTH WESTERN APPROACHES BASIN
63°N
60°N
55°N
50°N
Scales
50 25 50 75 100 MILES
50 25 50 75 100 KILOMETRES
Legend
WESSEX BASIN
Sedimentary Basins
Welsh Massif
Palaeozoic / Precambrian Basement Areas
Basin Boundaries
Well Defined Basement Edge (Usually Faulted)
Poorly Defined Basement Edge
Limit of UK Designated Area
Normal Fault
Transcurrent Fault

shallow depth of producing horizons below seabed, low productivity of wells and need for artificial lift) commercial reserves have still to be established. Further exploration of this basin is clearly warranted, but it will mean quite deep water exploration drilling with, if success is achieved, deep water development problems in very hostile waters.

Moving further west, we have the Rockall Trough basin and the Rockall-Hatton basin. No exploration drilling has as yet been carried out in these basins and Government policy for licensing in these deep waters is still awaited, but plans are being drawn up to at least carry out some stratigraphical drilling in one of them next year. I am afraid the hydrocarbon prospects of these basins just cannot be assessed at all at this stage and we must again await some drilling results.

Moving now south to the west of Eire, we have the Porcupine Trough basin, where some nine exploration wells have been drilled. Whilst hydrocarbon indications have been encountered, no commercial oil or gas has been found to date. Almost certainly, additional drilling will be carried out in this, another deep water basin. Between Eire and the UK, we have the small Irish Sea basin, where exploration drilling has taken place, establishing a gas field which is to be developed in the near future. A great deal of additional exploration activity cannot be anticipated in this small offshore basin.

To the south of Eire, and stretching into Cardigan Bay and into the Bristol Channel, we have the Celtic Sea basin, where some 50 exploration wells have been drilled. Unfortunately, to date results have been disappointing and, whilst both oil (in small quantities) and gas have been encountered, the only commercial find has been the Kinsale Gas Field, just to the south of Cork. I would expect some further exploration drilling, but probably not a lot.

Then further south we have the Western Approaches basin, where now some 6 wells have been drilled without encountering hydrocarbons; however, further drilling is called for and can be anticipated. Finally, we have part of the Anglo-Paris basin, underlying the central part of the English Channel. To date only one well has been attempted, but that was abandoned prior to reaching its planned depth due to technical difficulties caused by the strong currents present. A further attempt is expected shortly. To sum up on where we have got to on the EEC's NW Continental Shelf, I would say we are very much at a half-way stage; however, let me stress that we have had by far the easier half – the second half will be much more difficult and more expensive.

Before moving ahead to the second question, I would like to

discuss the water depths involved. The English Channel and nearly all the North Sea is shallower than 600 feet but, to the west of the UK, water depths for the most part exceed 600 feet – in fact over the Rockall Trough and Rockall-Hatton basins, they often exceed 3,000 feet, and at times exceed 4,000 feet. I might also comment appropriately that these waters are additionally extremely hostile, particularly during the winter months, and wave heights frequently exceed 100 feet.

Where Are We Going?

Factors which will Influence Future Activities. As already mentioned, I propose discussing first the factors that will have a significant influence on future activity.

First and foremost, there is the factor of geology, as will be revealed by both further geophysical surveys and exploration drilling. This factor is, of course, outside of our control and has already been determined over the last 300 million years or so by the Almighty. My geological training and experience has taught me to be quite humble, to avoid arrogant forecasts, and instead to be always prepared for surprises. Above all, I was taught to assume that every sedimentary basin has hydrocarbon prospects until sufficient and adequate reasons have been established to conclude that there can be none present. This means we must thoroughly explore all the sedimentary basins I have discussed, despite the water depths over them, before giving them up and discarding their opportunities. To carry out such thorough exploration will almost certainly require a drilling programme at higher levels than the current ones and one that will last into the next century.

In the United States they are drilling over 10,000 exploration wells per year, admittedly mostly on land, for additional potential reserves of around the same order of magnitude as are likely on the NW European Continental Shelf, unfortunately mostly offshore, but nevertheless again pointing towards increased levels of activity.

The next factor I wish to mention, but only briefly, is that of the price of oil and gas. We all know that, over the next few years, due to the great proportion of the world's reserves being concentrated in the OPEC countries, OPEC members, either individually or collectively, will determine the price of oil and gas. All the indications are that the price will rise in money-of-the-day and in real terms, so that again the most likely influence the factor of price will have is one of greatly increasing our activities.

The next factor we must discuss is technology. Clearly, continued development of offshore technology is called for – to combat cost escalation, to deal with deeper and deeper waters and, very importantly, to deal with the smaller and smaller reservoirs which we shall unfortunately find and have to develop in future. Specifically, I will mention a few items that, in my view, are important: (a) subsea completion and production systems; (b) less expensive fixed platforms with inexpensive means of removing them; (c) floating tethered production platforms, together with, of course, adequate riser systems. I suggest that the days of the huge fixed concrete platforms developed for the North Sea are probably over, because their costs are too high for small reservoirs. In my list, I have not included automatic drilling because I feel that, if it is to play a part, it must prove itself as a beneficial technique onshore first of all.

Another factor that will influence the future will be the social and environmental attitudes adopted by other users of the sea – notably the fishermen, the environmentalists and alas the pseudo-environmentalists. Whilst the views of the first and second must rightly be respected and considered, this should not be so in the case of the pseudo-environmentalists, who have no useful contribution whatsoever to make.

Without a doubt, if we should have too many serious blow-outs causing extensive pollution of fishing grounds or the coastlines, particularly in the North Sea, our future activities could well be adversely affected. All of us engaged in offshore activity must ensure that our operations are as safe as we can make them, and that we neither pollute the seas or leave debris on the seabed. We would thereby gain the confidence of other users and the environmentalists.

The abandonment of some of our offshore installations, such as concrete platforms, is a problem that it might be appropriate to mention here. We are being asked more and more what our solution to this problem will be. I hope that the tremendous offshore technological expertise we have built up can be used to provide positive solutions to the problem – may I suggest we should be thinking of using them to harness the inherent resources of the seas, rather than removing them, which could be many times more expensive than putting them there in the first place. In my view, we have the ability and the will to ensure that social and environmental attitudes will not adversely affect future activities, but it will call for a great deal of effort from us all.

Another factor we must not forget is the need for finance to cover the enormous investments called for. Confidence in fiscal stability is, of course, of paramount importance but, given this, the past sources of finance, i.e. oil companies' own resources and bank borrowing, should remain adequate despite the need for greater amounts.

The last factor I wish to raise – last but by no means least – is Government. Government policies will, of course, determine the opportunities available to the offshore industry. We have Governments of varying character and to complicate matters Governments that change, fortunately or unfortunately, every now and then. The industry, in taking on the high and expensive risks involved in offshore activity, needs long-term stable policies that can be depended upon. These policies must include fiscal arrangements that allow proper returns on the high risk investments inherent in offshore hydrocarbon activities.

In all EEC countries, the respective roles of the public and private sectors are constantly being debated. In the UK, BNOC was established by the last Government – the present Government is trimming its sails. In the UK and the other countries, it should be anticipated that, because of changes of Government, the roles of the public and private sectors will continue to swing backwards and forwards in the future.

Another question that must be asked is what purchasing restraints will Governments impose on oil companies. We in the UK have the Memorandum of Understanding and Code of Practice agreed between Government and industry, designed to give Full & Fair Opportunity to British industry – no more and no less. It has resulted in the build-up of the percentage of orders placed for UK goods and services to 66%; the figure may go a little higher, but probably not very much. A number of US oil industry contractors, service companies and suppliers have established bases in NW European countries, in many cases in partnership with local companies. In the UK, such partnerships are usually considered British by Government. They have been very smoothly and advantageously transferring US expertise to NW Europe. Let us hope that the expertise developed subsequently with our experience can flow back to the US in due course in a similar way.

In the UK, consideration is naturally being given as to whether, and to what extent, their offshore resources should be depleted or saved for future generations. I have been suggesting that the policies of certain extreme conservationists are abjectly defeatist and that guidance should be sought from the New Testament (Matthew, Chapter 25) where the servant who hid in the ground the one talent entrusted to him, rather than putting it to the exchangers, very rightly had it taken away from him for being slothful. Conservationism, poorly designed, can discourage exploration for more reserves, so that we would be able neither to assess nor to attain our potential reserves of indigenous wealth. It additionally reduces employment opportunities and financial opportunities.

I am afraid the factor of Government will be difficult to predict – it is perhaps even more unpredictable than my first factor of geology. In the UK, we have, I believe, convinced our new Government that the present level of exploration on the Continental Shelf is below half of what it ought to be. The offshore industry worldwide has the very necessary task of communicating with Governments to ensure that they are made aware of the true facts.

Nature and Level of Future Activities. Last year the UK Offshore Operations' Association (UKOOA) prepared and presented a paper to the Government on the necessary future activities.[1] The paper limited itself to oil only – not to detract from the importance of gas but because outside the southern North Sea gas is likely to be associated with oil very variably and, I am afraid, unpredictably. The paper assessed the required level of future exploration and development effort necessary to achieve desirable levels of future oil production as professionally as possible, based on actual UK experience gained.

Looking to the mid 1990s, it is estimated that the first 25 fields of the UK Continental Shelf will be producing less than 400 thousand b/d (20 million tons a year), as compared with the Department of Energy's anticipated demand of over 2.5 M b/d (120 million tons a year) and its forecast of possible production of approximately 2.25 M b/d (110 million tons a year) at that time – in other words, a shortfall of around 2 M b/d (100 million tons a year). (Of course, implicit in the Department of Energy's forecast of possible production is the timely use of our indigenous resources, rather than taking the incalculable gamble of conserving them for the next century.) In its paper UKOOA tackled the

[1] UK Department of Energy, 'Exploration and Development of UK Continental Shelf Oil'. *Energy Commission Paper No. 17*, 1978.

specific task of assessing the effort needed to discover enough ADDITIONAL oil to be able to produce this additional 2 M b/d (100 million tons a year). The reason for the choice of the mid 1990s was that, because of lead times, production in the 1990s will be significantly dependent on policies established at the present time, as indeed our approaching self-sufficiency now is the outcome of policies set in the 1960s. The long lead times involved in offshore oil development cannot be emphasised enough; the various stages, from licence application to peak production, can well take up to 15 years.

In order to obtain an additional 2 M b/d (100 million tons per year) in the 1990s, we should, taking into account the decline in exploration success, be drilling 60–95 exploration wells per year in UK waters. It should be noted that, whilst we drilled 75 exploration wells in 1975, the numbers have been lower since then, and in fact last year the annual figure was less than 40.

We predict that these 60–95 exploration wells per year over the period up to the end of 1987 will find between 6,000–9,000 million barrels (or 800–1,200 million tons) of technically recoverable oil, which could lead to between 18 and 32 new fields being developed in time to produce our target over a 5 year period in the middle of the 1990s.

We made two important assumptions which I must discuss. The first, of fundamental importance, was that there was the likelihood of there being enough oil in fact present on the UK Continental Shelf for our purposes. We consider that we were justified in assuming this as the total amount of technically recoverable oil we are talking about falls within estimates of potential ultimate reserves made by our Members. The second was that positive measures would be taken to lower the minimum commercial field size, or in other words the threshold field size, into the range of 150-500 million barrels. Of course, there is some over-simplification here – there will be some fields, for various geological and other reasons, that may be below or above the range.

The three main factors determining threshold size are: (1) Cost, largely dependent on technology; (2) Taxation and royalties; (3) Price of oil. The oil industry can and is doing something about the first through its extensive research programmes; the Government can do something about the second; but as already discussed the last is in the hands of OPEC.

The threshold field size has an important effect on discoveries required – the larger the threshold field size, the greater the amount of additional oil required and the greater the number of deposits we must find. Of course, when it comes to numbers of fields that must be developed, the converse is true. However, the likelihood of achieving a set production target is very markedly improved by decreasing the threshold field size, because of course the chances of finding large fields in the future are much less than the chances of finding small fields.

Having arrived at the level of exploration effort to meet our possible production target in the 1990s, UKOOA's paper went on to consider whether or not the level of effort would be realised. It concluded two types of incentives were called for and that both were needed. The first type is to ensure access to sufficient drillable exploration prospects, namely: (a) Increasing the size and timing of licensing rounds. There has also been a major decline in licensing in the UK over the last few years – before 1976 over the 12 years from 1964 the Government on average offered 221 blocks per year, since 1976 it has offered only 39 blocks per year, a decline of 87%; (b) Encouragement of farm-in opportunities. I am happy to say the present UK Government are planning to give both these incentives.

The second type of incentive, is to establish appropriate fiscal and depletion policies to encourage investment, namely by introducing: (a) Provisions to reduce the minimum commercial field size; (b) Equitable and stable terms to provide potential rewards consistent with the high risks involved, as this will ensure efforts are maintained. We are currently in discussion with the present UK Government on both of these, and are optimistic of the outcome.

Finally, where does all this take us? What will the future have in store for us? May I offer the following comments:

1. There will be an increased demand for smaller fixed production platforms and for floating platforms for both oil and gas production – due mainly to the fact that the reservoirs we shall be discovering and developing will be getting smaller and smaller.
2. There will be an ever-increasing demand for subsea completions and riser systems – due mainly to the fact that we will be moving into deeper and deeper waters.
3. There will probably be a demand for further major gas pipelines, but the demand for additional oil pipelines is uncertain – new oil discoveries are likely to be put into existing lines, and in some cases SBM's used.
4. There will be a demand for overall improvements in all offshore technology to reduce costs or at least meet inflation. This may well include new tools, both for construction and for maintenance including inspection and repair. The latter is, of course,

becoming more and more important as development takes place.

5. The prospects, in my opinion, warrant a mammoth effort that will call for the participation of a major share of the international oil industry's worldwide offshore effort, its expertise and its financial resources – until the end of the century.

In conclusion, may I say the development of the hydrocarbon reserves of the EEC's NW Continental Shelf has still a tremendous future potential ahead. We, in the oil industry, with the encouragement of Governments, look forward to realising this potential.

3 GAS: RECENT DEVELOPMENTS AND PROBLEMS OF SUPPLY

Nordine Aït-Laoussine

The theme of this paper is recent developments and problems of gas supplies, and since Algeria is basically a gas-producing country, I shall attempt to present a producer's view. But in doing so, I must make it clear that I am not speaking on behalf of OPEC. OPEC is not an organisation with a single view on all aspects of energy matters; it is a grouping of sovereign states, each with its own views on the major issues of the day and a common interest in the development of their hydrocarbon reserves for the economic wellbeing of their peoples.

I must observe at the outset that gas is very much the 'neglected child' of the hydrocarbon family, and has suffered from the popularity of the big brother called oil, to the extent of being sacrificed for his brother's growth. A perfect illustration of this state of affairs is the French expression 'gaz fatal' used for associated gas, a nuisance component from crude oil production, a sort of necessary evil that must be destroyed. Consider that we are still flaring to the sky over 200 billion m^3 of natural gas per year, two-thirds of which is in OPEC countries alone. This is the equivalent of wasting almost 4 M b/d of oil, or roughly twice the entire oil consumption in the OPEC area. Imagine how much more comfortable our present short-term energy supply situation would be if we could tap this wasted resource.

And when the 'little brother' escapes destruction and succeeds in reaching the market place it commands at the wellhead the lowest price ever paid for energy. It has for decades been sold at a fraction of the price paid on a Btu basis to the heaviest and dirtiest oil residue and has just recently approached, but never exceeded, the price paid for comparable fuels. No wonder that the industry has not, until recently, explored for gas. Most of it in fact was discovered by 'accident'. Yet, natural gas is a clean, flexible, premium fuel, with its own special role in meeting world energy needs.

In contrast to its past poor image, it is clear that gas has prospects which certainly make it the most interesting and the most promising 'child' of the hydrocarbon family, the energy resource which might save the day if given the proper care, attention and incentive. Indeed, in spite of the past attitude of the industry towards this premium fuel – or probably because of it – natural gas has quietly made tremendous gains during the last few decades.

Table 1 Share of Natural Gas in World Primary Energy Balance (%)

	1945	*1950*	*1955*	*1960*	*1965*	*1970*	*1975*	*1978*
US	15	20	26	28	32	34	29	28
Western Europe	-	-	1	2	3	8	13	17
Japan	-	-	-	1	1	1	3	4

In the United States, the share of natural gas in meeting primary energy needs has more than doubled in the 20 years from 1945 to 1965. Long-term controls on gas prices, holding them at unrealistically low levels compared to oil, have produced a distorted level of penetration, with huge quantities of gas being simply burned under industrial boilers. The progressive decline in domestic availability is tending to reverse this trend and the gas share in energy consumption has been declining since 1971, the year it reached the maximum level of 34%. In Europe, following the discovery of the Groenigen gas reserves and more recently those of the North Sea basins, and the widespread expansion of the domestic and residential gas grid in the 1960s, the increase in the gas share of primary energy has been even more striking, rising from 2% in 1960 to 17% today. Finally, with the advent of liquefied natural gas (LNG), the last remaining major consumer economy was opened to gas with the introduction of the Alaska-Japan trade in 1969. As a result, the share of natural gas in world primary energy balance has doubled in the last thirty years rising from 10% in 1950 to about 20% today. In fact, during the last five years, the gas share has continued to expand while that of coal and oil have remained nearly constant.

Turning to prospects, we shall see that the situation appears to be even more promising. To illustrate the point, I should like first to give the following facts and figures:

(a) The remaining proven reserves of natural gas are estimated at 64 trillion m^3 or 2,250 trillion ft^3 (tcf). This is equivalent to approximately 60 billion tonnes or 450 billion barrels of oil. This resource represents not only some 10% of the world's known proven reserves of fossil fuels, but also one and a half times the tar sand reserves and three times the shale oil resource base – so often quoted as the most promising alternative energy sources. Expressed in another way, the world's remaining proven reserves of gas constitute not less than 40% of total hydrocarbon known reserves (Table 2). So, we are talking here of a huge proven resource base representing a potential production of a clean flexible fuel equivalent to 60 M b/d of oil if we assume a 20-year reserve production ratio.

Table 2 Share of Natural Gas in World Fossil Fuels Proven Reserves (end 1978)

	Billion Tonnes Oil Equivalent	%
Coal	510	70
Oil	90	13
Gas	60	8
Tar sands	40	6
Shale Oil	20	3
Total	720	100

(b) But the potential is even more promising in terms of ultimate reserves. These have been estimated at four times the proven remaining reserves or about 300 trillion m^3, equal to 10,000 tcf and equivalent to approximately 250 billion tonnes, or 2,000 billion barrels of oil – in other words, approximately equal to the figure most often quoted for the ultimate crude oil resource base.

Table 3 World Hydrocarbon Resource Base: Production, Current Proven and Potential Reserves (end 1978)

	Production	*Proven Reserves*	*Potential Reserves*
Natural Gas, (trillion cubic metres)	25	64	210
Crude Oil, (billion barrels)	350	650	1,100

(c) Thus, the prospects of proving up additional reserves are brighter for gas than for oil, because the ratio of proven to potential reserves is lower for gas (31%) than for oil (60%). This is why the level of proven remaining gas reserves has doubled over the last ten years (Fig. 1), while the remaining crude oil reserves have only marginally increased during the same period. As a result, the reserve life index for gas is growing while that of oil is decreasing.

We can thus be sure that the figure of 64 trillion m^3 quoted earlier for the remaining proven reserves of natural gas seriously

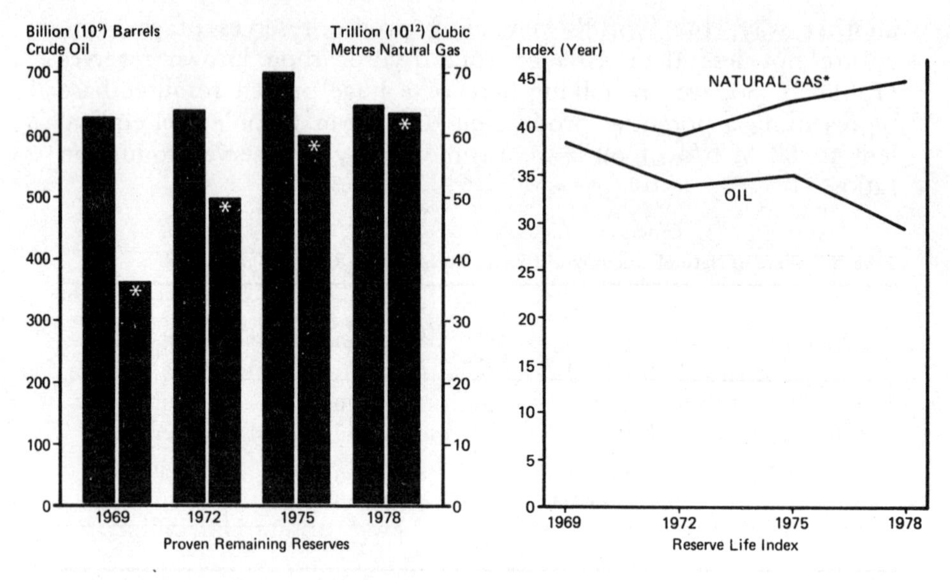

Fig. 1 Evolution of world proven hydrocarbon reserves

understates the real potential, since many gas structures have not been explored and many gas finds have not been fully delineated – as is the case in many OPEC countries. Moreover, the gas reserves proven in the OPEC area are the result of accidental discoveries because incentives to explore for gas are generally non-existent outside the consuming countries (some might say even within them). In addition, improved technology in terms of maximum drilling depths – now more than 30,000 feet – are more likely to benefit gas exploration than oil, because at these depths there is a much better chance of finding natural gas.

The 'little brother' should then command more respect and attention, certainly more than shale oil or tar sands combined. Its ability to increase its contribution to long-term energy supplies is greater than oil. Given then that we are dealing with a major energy resource, let us examine recent developments.

The geographical distribution of this resource base shows that the OPEC countries have the largest volume with 39% of total gas reserves, most of which are found in Iran and Algeria. The socialist countries of Eastern Europe and Asia come next with a share of 38% of the world total, with the remaining 23% distributed between the US, 8%, Western Europe, 6% and other areas, 9% (Table 4).

Table 4 World Natural Gas Reserves

	Trillion Cubic Metres	%
US	5	8
Western Europe	4	6
Socialist countries	24	38
OPEC	25	39
Others	6	9
Total	64	100

The data on world gas production are difficult to interpret because of the confusion which often exists between net and gross production, between the marketed production and the quantities reinjected, flared and vented. But the best estimates indicate that world gross natural gas production (Table 5) is running at about 1,800 billion m^3 (about 63 tcf), 70% of which is non-associated. Of this total 1,500 billion m^3 (52.0 tcf) or 83% is marketed, about 100 billion m^3 (3.5 tcf) or 6% is reinjected and 200 billion m^3 (7 tcf) or 11% is vented or flared. As far as the marketed production is concerned (Table 6) the United States produces the largest volume (about 38% or 20 tcf). The socialist countries come next with a share of about 35% or 18 tcf. The remaining 26% is produced in Western Europe (13% or 6.5 tcf), the OPEC area (6% or 3 tcf), Canada (5% or 2.5 tcf), and other areas (3% or 2 tcf).

Table 5 World Natural Gas Production, Gross, 1978

	Billion Cubic Metres	%
Marketed Production	1500	83
Flared or Vented	200	11
Reinjected	100	6
Total	1800	100

Table 6 World Natural Gas Production, Marketed, 1978

	Trillion Cubic feet	%
US	20.0	38
OPEC	3.0	6
Western Europe	6.5	13
Socialist countries	18.0	35
Canada	2.5	5
Others	2.0	3
Total	52	100

It is interesting to note that while OPEC's share in world proven remaining reserves is 39%, only 6% of the world marketed gas production comes from the OPEC countries. It can also been seen from Table 7 that while OPEC accounts for approximately half of world oil production, only two-fifths of the gas used worldwide for enhanced recovery is reinjected in the OPEC area. Finally, two-thirds of the gas flared or vented worldwide comes from the OPEC countries alone. OPEC countries as a whole, reinject comparatively less gas and they flare much more gas than the rest of the world combined.

Table 7 Utilisation of OPEC Gas, 1978

	Billion Cubic Metres	%
Flared	140	54
Marketed	80	31
Reinjected	40	15
Total	260	100

In international trade, we are faced with the same growing geographical imbalance between supply (either actual or potential) and demand as applies to crude oil (Table 8). This is best illustrated by the situation in the US, which consumes about 40% of world marketed gas, but which accounts for only 8% of world proven reserves. Equally, Western Europe's share in world consumption is larger than in world production or reserves, though the scale of the problem is much smaller than in the US. The only major producer which seems to have enough gas reserves to meet its expanding demand is the USSR. On the other hand, the OPEC countries which globally account for 39% of world reserves produce a relatively much smaller amount and consume even less.

Table 8 World Natural Gas Reserves, Production and Consumption, 1978 (Trillion Cubic metres)

	Reserves	*Production (marketed)*	*Consumption*
US	5	0.57	0.59
Western Europe	4	0.185	0.215
Socialist countries	24	0.515	0.465
OPEC	25	0.080	0.055
Others	6	0.150	0.130

But compared to oil, and in spite of the apparent similarity in geographical imbalance between supply and demand, the international trade in natural gas is still at a very early stage. It accounts today for about 10% of the marketed world output, whereas about 50% of crude oil is traded internationally. In 1978, international trade in natural gas amounted to about 160 billion m^3. About 85% of this was transported by pipeline, most of it for inter-regional transfers. The rest was moved in LNG form exclusively for inter-continental transfers.

The Netherlands are the main exporter with a share of 29%, or 45 billion m^3 traded in Europe. The USSR comes next with a share of 20%, or about 32 billion m^3, most of which goes to Western Europe, followed by Canada which exports about 26 billion m^3 (16% of world trade), to the US. The remaining 35% are accounted for by Norway (14 billion m^3 or 9%), Iran (8.5 billion m^3 or 5%), Brunei (7.3 billion m^3), Algeria (7.0 billion m^3), Indonesia (5.0 billion m^3) and other sources. The share of OPEC in international trade is presently about 20% or 30 billion m^3, two-thirds of which is moved in LNG form (from Algeria, Libya, Indonesia and the UAE) and the remaining one-third is transported by pipeline (from Iran).

As for the prospects, based on announced plans and probable and possible projects, there seems to be a tremendous potential for growth in international gas trade. Pipe exports are forecast to increase from $130 \times 10^9 m^3$ now to $300 \times 10^9 m^3$ in 1990, and LNG exports from $30 \times 10^9 m^3$ to $175 \times 10^9 m^3$ in 1990. It is clearly inappropriate to review each export project individually, but if they all materialise, world international trade of natural gas could increase two-fold by 1985 and three-fold by 1990. A major portion of the increase is expected to come from the OPEC countries in the form of LNG.

Despite the optimism that this potential implies, we must not ignore some of the problems that these projects are likely to face. Frankly, the significance of these obstacles is such that it would not be realistic to expect all these projects to materialise in the present economic environment. First and most important, the price paid currently for gas traded internationally offers very little attraction to the producer. The problem is felt at three levels: one level relates to the attitude of the consuming countries, so used to cheap energy and especially gas, who still believe that this premium fuel should compete with the price paid for residue or No. 6 fuel oil. At the second level the producer is faced with very high transportation costs: it has been estimated that the cost of transporting gas by pipeline over long distances is roughly twice the cost of transporting crude oil over the same distance, on a thermally equivalent basis, and that on the same basis LNG transport costs are five times higher than crude oil transportation costs. Finally, at the third level, we have the impact of the liquefaction cost in the LNG transactions. For a 10 billion m^3 project, the cost of manufacturing LNG is equivalent to about $5.0 per barrel of crude oil on a Btu basis, a level much higher than the average crude oil production cost.

As a result, the netted-back value of gas, at the wellhead, does not offer any economic incentive whatsoever to producer governments. Even after the sharp rise in oil prices of this year (1979), the economics of Gulf LNG projects simply does not yeild attractive returns. Algeria – which is fortunate in that it is much closer to the market – nets only, at today's oil prices, some $1.25 per M Btu at the wellhead, or the equivalent of $7 per barrel of oil – about one-third of the value we derive from our crude oil exports. This is, by the way, the reason why we want to maximise our gas exports via trans-Mediterranean pipelines which offer us a much more attractive economic option.

The second area of concern is the long lead time and huge investments required, especially in the case of LNG, to bring these projects into fruition. For example, a decade has elapsed between the conception and the implementation of the El Paso LNG project in Algeria. A 10 billion m^3 LNG chain costs today, in terms of the producer's facilities alone, some $3 billion; this is a very heavy burden on the national resources of the exporting country.

The attitude of governments to imports of gas is also important. Several schemes have, for example, been delayed because of the hesitations of the US Administration and the concern expressed in the consuming countries over such issues as the cost of imports, over-dependence on foreign energy sources and other considerations, such as safety, security and environmental impact. Another obstacle to overcome in order to expand international gas trade is the increased awareness of the producing countries of the desirability, under today's economics, of reinjecting gas, using it locally or conserving for future use. This issue is, of course, inextricably related to the question of price, and can only be solved in this context.

The long-term development of international trade in natural gas will therefore be determined by the mix of opportunities and difficulties which the producing countries are having to deal with in debating whether or not to proceed with their export projects. There are, on the one hand, huge reserves of gas waiting to be developed; but on the other hand, the present economic environment is not sufficiently attractive and the cost involved in developing these reserves is immense in terms of national resources. Yet within the perspective of the approaching energy crisis, it is difficult to see how natural gas can continue to be confined to the relatively minor role it has played so far. Indeed, when one looks at the natural gas supply/demand outlook of the major consuming and importing countries, we see that gas is bound to play a more important role.

In Western Europe domestic production is probably already peaking out and will start a slow decline towards the last decade of this century. Despite some useful gains in production in the United Kingdom and Norway, output from the major traditional supply source – the Dutch gas fields around Groningen – is being cut back in the interests of conservation, and will drop by perhaps 30% in the course of the next decade.

Whilst the level of imports is likely to rise very sharply – particularly between 1980 and 1985 – I believe that Western European gas consumption will become supply-limited over the second half of the next decade and onwards, unless significantly greater volumes of imports are committed. Assuming even a modest annual growth rate of some 3% per year through 1990 – which is well below the historic growth rate of nearly 20% over the last ten years – we can forecast that some five billion ft^3 per day of additional supplies must be contracted for between now and 1990. These forecasts are necessarily somewhat tentative, since there are certain uncertainties about plans to import gas into the area, but they are directionally correct. In other words, we believe that Europe faces a gas-supply shortfall at the very earliest date by which a major new project can be brought on stream. My confidence in this forecast has been heightened by the enthusiasm with which European gas supply companies have been coming to Algeria over recent months to firm up additional long-term supply agreements.

Turning to the United States, the issues are, alas, a good deal more complex. Not only are there enormous uncertainties over the future level of domestic production, but also over the willingness of the current Administration to recognise and legislate for the possible contribution that imports of LNG could make towards assuring some stability of gas supply. As far as future domestic production from the Lower 48 States is concerned, forecasts vary very widely, but a number of highly respected sources are currently postulating a very sharp decline in domestic production after 1985. And if the likely level of domestic production is shadowed in uncertainty, the contribution that can be expected from synthetic natural gas, coal gasification, and other exotic sources is equally a matter for speculation and guesswork.

I have tried in Fig. 2 to put together the various elements in the long-term US natural gas supply/demand balance to determine some pattern for the future. It can be seen that, assuming a huge level of imports from Mexico, and a contribution from synthetics which I frankly believe to be optimistic, there is absolutely no margin for growth in current levels of demand beyond 1985. Moreover, many people are taking a more pessimistic view of production from the Lower 48 States, which could only worsen the situation.

These facts make the attitude of the present US Administration in actively encouraging a return to burning gas under boilers utterly inexplicable, unless it has greater faith than the industry in the development of synthetic gas, or believes such rapid changes of direction can be made overnight. We know that – at least for the moment – LNG has been given a minor and unimportant role in the US long-term gas supply equation. The grounds given range from high prices to risks of 'political instability' in exporting countries.

But if US domestic production is really to decline by one tcf/day – or perhaps more – by 1990, it is a brave policymaker who assumes that both the technology and the hardware can be made available in time to provide more than very insignificant volumes of synthetic substitutes. The very considerable delays which seem to beset every major project in the energy field lead one to suppose that policy-

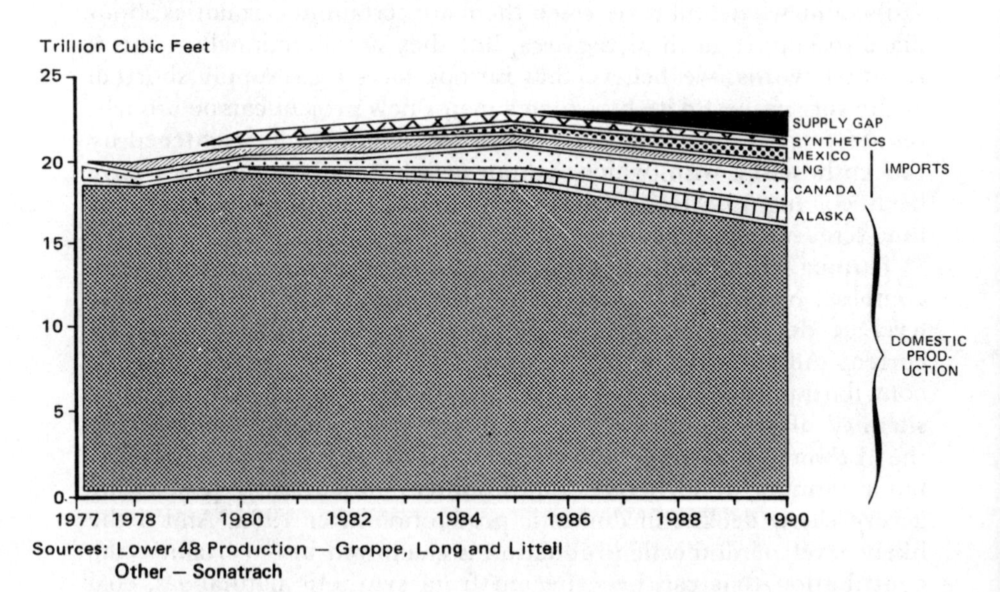

Fig. 2 US natural gas supply 1977–90

makers and planners in developed countries continue to take an over-optimistic view of their ability to substitute more exotic synthetic replacements for conventional hydrocarbon fuels. Furthermore, the recent spate of technical problems associated with nuclear power have demonstrated again the folly of placing too much hope on one single energy source achieving any considerable inroad into the dominant position held by oil and gas in world energy supply. Even if US natural gas consumption is to remain at present levels – which, of course, means a decline in its share of primary energy demand – a major role awaits LNG as the only incremental source of natural gas which has already proved its ability to provide reliable, safe and cost-competitive supplemental supply. The infrastructure exists to distribute quantities of gas which the US is unlikely to be able to produce domestically in the medium term. It would be folly to allow these facilities to fall into disuse while imports are possible.

I, therefore, see enormous opportunities for the development of international trade in natural gas: the reserves are huge and the developed countries have the market. But you will be as aware as I am that this alone is not sufficient. A way must be found of relating demand and supply with transactions made on equitable and realistic commercial terms.

At this point, I would like to place this issue within the broader context of the world energy scene. I have often in the past expressed the view that one of the major, over-riding priorities which faced the world community was the need to prepare for the coming shortage of oil. I now believe that, as a result of recent events, these few years of oil glut are already behind us. The most recent forecasts project a global oil deficit reaching 5 M b/d by 1985, and perhaps doubling by 1990. It is widely accepted that physical limitations to world oil consumption have begun to bite. In this situation, the world no longer faces the luxury of choice: we must develop world gas reserves simply because, unlike oil, the capacity to do so exists. We can no longer afford the time to pick and choose. We must develop the resources left to us today.

It is, therefore, imperative that we plan – and I use the word advisedly – to develop the world's untapped reserves of natural gas so that they become a vital element in a global strategy to avert an impending scarcity of conventional hydrocarbon production. We must not look upon oil and gas as competitors, but as complementary resources which the world community must draw upon in a proper and balanced fashion.

We must see oil and gas as complementary since that is, in reality, what they are. Algeria is fortunate in that the bulk of its gas reserves are non-associated; other countries are not so fortunate, and it is a clear and undisputed fact that, if the major industrialised economies adopt policies which effectively inhibit the development of international trade in natural gas, their access to crude oil will suffer. A recent Congressional Research Service report forecast that, if Mexico were to be inhibited from exporting its gas, not only would the US lose a supplemental gas source of up to 1.4 tcf/year, but Mexico's exportable surplus of crude oil would also fall by 0.4 M b/d. Producer governments all over the world are realising the folly of permitting vast quantities of natural gas – which, one day, must be of inestimable value to the world – to be flared to the sky each time they produce a barrel of oil. Conservation is no longer an abstract concept: it is a target which every government that seeks both to maximise the value of its own natural resources, and to join in the proper management of the world's ever-diminishing energy resource base, must strive to attain.

There are certain obvious steps we should take to create the right climate to stimulate a greater supply of natural gas. First, there must be a coherent pricing structure within the consumer economies which could, without penalising consumers of either oil or gas, provide an incentive for a more active worldwide gas exploration programme. Why should a M Btu of 'stripper' crude oil be sold in the US for about $4.29 whereas a M Btu of 'new' gas is worth only $2.24? Secondly, we must recognise the need to pay 'premium prices' for guaranteed long-term foreign availability, not merely to stimulate the search for gas, but to provide an immediate incentive for the investments which are required now in order to satisfy the energy needs of the late 1980s.

I would also like to add two further points which I think have a special relevance in considering this matter globally. First, the producing countries should not continue to bear alone the burden of the high gas transportation cost. Second, a fairer balance is needed between the producer and the consumer in terms of exposure to financial risk. The consuming countries have a special responsibility to bring about a substantial improvement in utilisation of gas reserves, and to acquire a better understanding of the ever-increasing value of this premium fuel.

I do not believe, however, that we can talk of creating the appropriate climate for a better utilisation of natural gas until exports of LNG from the Middle East to the major markets in Western Europe and the United States become demonstrably a viable proposition. The reason is not simply because the Middle East is the largest reservoir of untapped natural gas in the world, but also because we must turn to these 'frontier' sources if the shortfall in gas supply beyond 1985 is to be averted.

Price incentive is clearly the most important requirement if this is to be achieved (Fig. 3). If LNG were to be sold not on the basis of Fuel No. 6 but on terms comparable to the cost of its real global economic alternative – that is, syngas from coal – Gulf producers would be offered a real incentive to move ahead with their investments. The days when the producer governments would put gas into an LNG plant at zero wellhead value, and content themselves merely with a financial return on their investment, are over.

There are good reasons why LNG should be given this incentive. First, as mentioned before, we need to develop all our gas resources; we can no longer afford to pick and choose. Secondly, LNG is the cheapest energy source in foreign exchange terms (based on f.o.b. prices) available to consumers today and requires comparatively less investment capital to produce the same amount of energy than, for example, does syngas from coal. Thirdly, it offers considerable environmental benefits in the consuming country compared to, say, strip-mined coal. And finally, such incentives must be given now, if we are to see any improvement in the world gas supply situation before 1990.

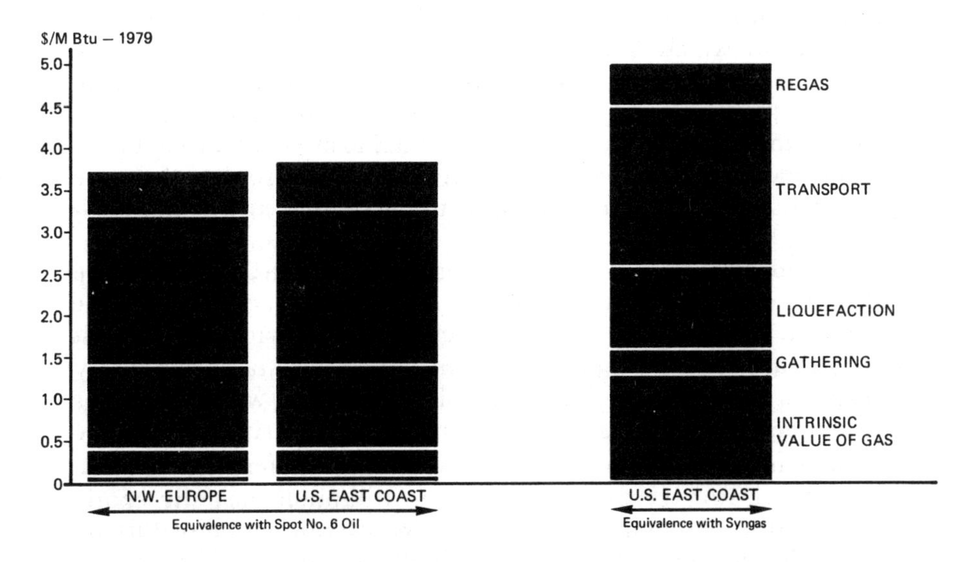

Fig. 3 Comparative economics of Gulf LNG production

Recent events in Iran, as well as the much-publicised disclosure of limitations on Saudi Arabia's production capacity, must have finally shattered the illusion that there is an infinite potential supply of crude oil. None of us here wish to see the world community facing a bleak economic future as the result of its failure to plan its energy supplies in an intelligent and rational fashion. We must 'stretch' our conventional hydrocarbon resources, and one obvious and practical way to do this is to develop our use of natural gas.

We see Europe facing a gas supply deficit of some five billion ft^3/day by 1990. Perhaps a third of this shortfall will come from the North Sea, but the whole situation is full of unknowns. In the US it remains to be seen whether synthetics, 'frontier' gas and new technology can solve the problem. Middle East LNG could make a useful contribution to supporting the market, if the terms are right. Two years ago, I suggested to the LNG Conference in Dusseldorf that we should aim at achieving gas prices in the US and Europe of around $6.50/M Btu by 1985, and may at that time have shocked some of my audience. But I still believe that this is the realistic minimum target which is necessary to encourage the development of supplementary gas supply.

My belief, that it is necessary for the major industrial economies to plan deliberately and consciously an increase in the use of natural gas within their primary energy balances, is the result of simple analysis: as the era of tight crude oil supplies is at our door, the world cannot afford to ignore the 140 billion barrels of incremental oil equivalent which is available in OPEC's gas reserves alone, quite apart from the reserves in places like Mexico, the USSR, Australia, Pakistan, Argentina etc. But it is up to the consumer economies to create the right climate, not only for a massive exploration effort to find additional reserves of oil and gas, but also for the huge investments that are required now to meet the energy needs of the 1990s.

The acid test of our ability to manage the transition from the hydrocarbon-based energy economy of the 20th century to the coal-based economy of the 21st century will, with the benefit of historical perspective, probably turn out to have been our ability to exploit in a prudent and rational manner the world's reserves of untapped natural gas. The reason for this is simply that the development of intercontinental gas projects offers fewer challenges in terms of technology, capital or utilisation than any exotic energy source of comparable size. If we cannot mobilise this resource, the longer-term energy outlook is indeed bleak. Above all, we must grasp the urgency of the situation: already Iran has cancelled the IGAT 2 project and Mexico has lost its early enthusiasm for gas sales. Unless the right climate is created to encourage international trade in natural gas, we will enter the last decade of this century having failed to mobilise one of the few resources which can offer a real contribution to the alleviation of the forthcoming energy crisis.

4 WORLD NATURAL GAS RESERVES AND THE POTENTIAL FOR GAS TRADE

James T. Jensen

The terms 'oil and gas' are often used together, no doubt reflecting their common geological origins and the similarity of the exploration, development and production processes which are utilized for both. But it is the differences between oil and gas – not the similarities – which are the keys to an understanding of the likely future commercial role which natural gas will play in the world's energy economy. Those differences suggest that natural gas – oil's 'little brother' as Nordine Aït-Laoussine calls it – will play a very different role in world energy trade than has its 'big brother' and that it will make very large demands on world capital resources in the process of doing so.

That the 'little brother' is a major potential contributor to world energy supply is indisputable. Estimated proved reserves of natural gas at the end of 1978 amounted to 2556 trillion ft^3 (tcf).[1] Natural gas thus constituted 42% of the energy content of the world's combined proved reserves of oil and gas. Furthermore, recent reserve additions for gas have exceeded those for oil. Since the oil embargo of 1973, worldwide additions to proved gas reserves reported regularly by the *Oil and Gas Journal* have constituted 55% of the combined energy content of oil and gas additions.

However, despite the magnitude of worldwide natural gas reserves and the rate at which they are increasing, the role of gas in international trade is quite small. Worldwide consumption of natural gas is less than 30% of the combined total of oil and gas. In 1978, international oil trade – primarily in tankers – was at a level of 33.8 M b/d. The total international gas trade was only 2.9 M b/d of oil equivalent; nearly all of this was carried in international pipeline systems with only about 470 thousand b/d of oil equivalent moving in tankers as liquefied natural gas (LNG). Thus, despite the major worldwide gas reserve base and the optimism about gas discoveries worldwide, LNG tanker trade is a miniscule 1.4% of oil trade.

[1] *Oil and Gas Journal*, American Gas Association, Canadian Petroleum Association, Pemex.

Transport Costs as Obstacle to Trade

The major reasons for this disparity involve the high cost of gathering and transporting natural gas when compared with oil. A quick review of candidates for the title of most expensive worldwide oil or gas projects brings to mind the Saudi Master Gas System and the Alaska Natural Gas Transportation System – both gas projects in the $15 billion class. Compare those to the Alyeska oil pipeline or the full field development of Statfjord oil field in the Norwegian North Sea (both in the $7–8 billion class) and consider whether the international banking community would agree that natural gas is oil's 'little brother'.

Low cost tanker transportation for oil has created a truly international oil market. Oil valuation throughout the world can be related to the value of the large-volume Middle East crudes, such as Arabian Light, OPEC's marker crude. The high cost of gas transportation has meant that there is no similar international gas valuation standard. Gas generally competes with other fuels, predominantly oil, and thus the process of determining the value of a natural gas supply source for any given location is very similar to the process of determining the value of an oil supply source. The determination of a market valuation for oil using quality and transportation differentials is the first step. For gas one then translates oil values into relative gas values and determines the margin required for the distribution, transportation, gathering and production of the gas to obtain a wellhead netback. If there is an insufficient netback, a non-associated gas discovery may be shut in as non-commercial and associated gas may simply be flared. In 1977, for example, the US Department of Energy estimates that over 12% of world natural gas production was flared.

Transportation economics are commonly the key to the prospects for commercialization of a gas discovery. The fact that gas may command a premium over oil and coal at market because it is clean burning and easily controlled is of little consequence if the economics of transportation to market are too high to take advantage of the premium. Most oil discoveries move to market or to export terminals by pipeline, although discoveries too small for pipelining may move by truck, particularly in North America if the distances are not too great. Small offshore oil discoveries can often be barged to a shipping terminal. But gas pipeline costs (per Btu) are generally higher than oil pipeline costs and the relative disadvantage is greater for small scale movements. For those small discoveries for which oil could be trucked or barged, there is no present practical gas transportation option at all.

Gas is at an even greater disadvantage in ocean transport. While oil tanker transportation costs are generally much lower than oil pipeline costs for comparable volumes and distances, LNG tanker transportation will commonly be more expensive than gas pipelining.

These comparisons of transport costs lead to some surprising conclusions – at least by oil standards. A large-volume, long-distance LNG movement, for example, may be less costly than a much shorter-distance, small-volume gas pipeline movement. An 8 trillion ft^3 (tcf) gas reserve in Nigeria may be 'closer' in economic terms to North European markets than an 800 billion ft^3 (bcf) gas discovery only 500 miles offshore in the North Sea.

The relative inflexibility and severe scale penalties of gas transportation systems tend to foster a conservative approach to commercial gas project development. Unless proved and long-lived reserves are available to support a pipeline or LNG project, it often cannot be financed and built. This makes many smaller-sized gas discoveries non-commercial and tends to keep gas reservoir depletion rates lower than those for comparable oil finds.

The tendency for smaller reserves to be uneconomic is particularly prevalent in long-distance international gas trade and in the tendency for many newer finds to be offshore. When the gas pipeline network first developed in the US, it benefited from some of the lowest pipeline construction costs anywhere in the world. As a result, long-distance overland gas transmission costs in the US are relatively low and even comparatively small gas discoveries can be connected to the pipeline network. When US exploration began to move offshore into the Gulf of Mexico, pipeline construction and transportation costs rose substantially. Offshore conditions in more hostile environments such as the North Sea have raised pipeline construction costs still further. Areas such as the Arctic will prove to be extremely costly as well. As a result larger and larger discoveries are potentially non-commercial. Fig. 1 illustrates typical pipeline transmission costs for US onshore and offshore, and North Sea pipelining conditions. The rapid increase in pipelining costs of less than a 200 M cf/d movement – about 1.5 tcf at a twenty-year depletion rate – is apparent. Development work has been underway for some time to find alternative ways to recover offshore associated gas. While floating LNG, methanol synthesis, or power generation facilities may make small offshore gas accumulations more economic at some future time, their commercial feasibility has not yet been demonstrated. High costs will probably apply to such areas as the US and Canadian East Coasts, suggesting that many fields which would be highly profitable onshore in North America or acceptable offshore in the US Gulf may be considered marginal and not commercialized in the Atlantic.

Gas Reserves and Trade Patterns

Gas reserves may be either associated/dissolved or non-associated. Non-associated gas, in a sense, is discretionary gas, since the discovery can be shut in and not developed until the economic climate is appropriate for commercialization. Associated/dissolved gas is more likely to be non-discretionary. If it is dissolved, it is produced along with the oil. Unless there is a market for the gas, it either must be reinjected – sometimes for enhanced oil recovery but sometimes simply for conservation until later on when there is

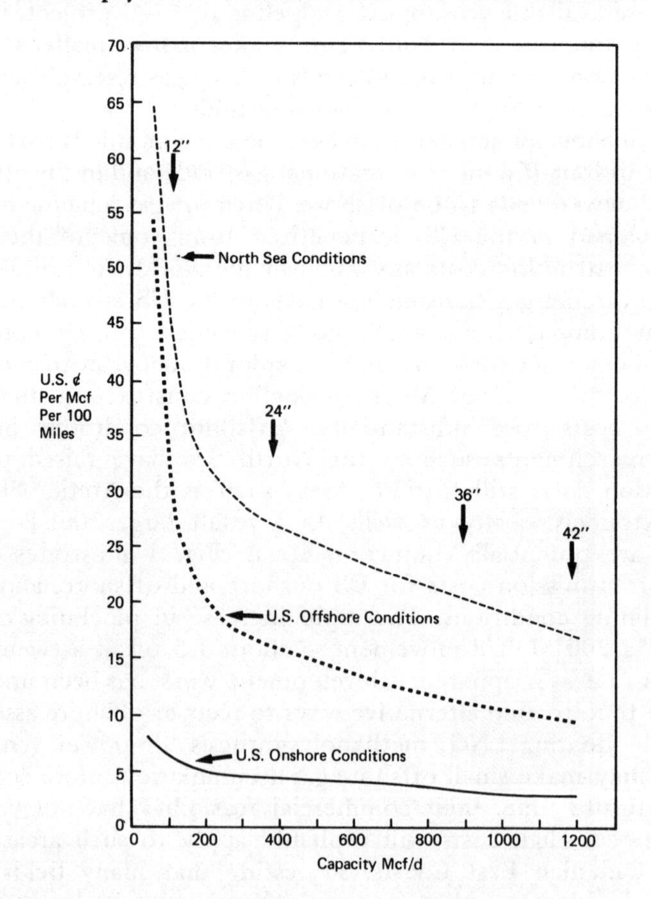

Fig. 1 Pipeline cost of service as a function of capacity (1979 dollars)

a market — or it has to be flared. Some associated gas is contained in large gas caps in oil fields where its production prematurely will deplete reservoir pressures and reduce ultimate recovery of the oil. While one usually cannot delay production of dissolved gas as a practical matter, one often cannot accelerate the production of associated gas cap gas. An estimated 28% of world gas reserves are associated/dissolved while the remainder are non-associated.

Fig. 2 shows the estimates of associated/dissolved and non-associated gas both by geographic region and political grouping. Not surprisingly, the associated/dissolved gas distribution pattern tends to follow oil's distribution pattern with OPEC showing the greatest concentration of these reserves. But very large non-associated gas reserves are found in the Soviet Union, North America and Europe as well as in Iran and Algeria. As a result, while OPEC accounts for 77% of total world and 90% of the non-Socialist world proved oil reserves, it represents only 38% of total world and 60% of the non-socialist world gas reserves.

Because of the high costs of gas transportation, it is often easier to bring consumption, in the form of gas-intensive industrial development, to the gas than it is to move the gas to market. Nearly every gas-producing country has some local market, even where the nature of its economy does not justify a reticulated gas system for household use. Most of these markets are local and highly individual. While

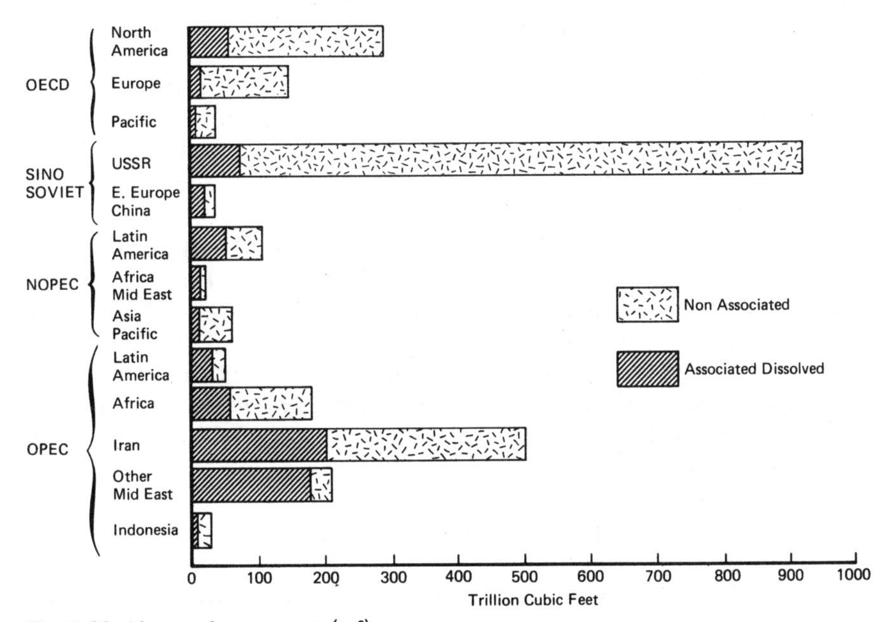

Fig. 2 World natural gas reserves (tcf)

they may have some similar characteristics, they are really not a part of an international market in the sense that refineries virtually anywhere in the world are both a part of and influenced by world oil markets.

There is a small but growing international trade in natural gas. The portion of world gas reserves which now supports this trade or, as a practical matter, can be available to expand it, is much smaller than worldwide reserve totals might suggest. Much of the gas is reserved for local markets either through direct contractual commitments or as a result of national export policies which set aside some portion of gas reserves for local needs before granting export permits. Other gas reserves are not immediately available because they are either being utilized in the production of oil or are contained in large gas caps in oil reservoirs where their premature production would prejudice oil recovery. This gas — deferred reserves in a sense — may be available at some future time but is not now a potential base for world gas trade. Still other large-scale gas reserves are not presently available because their remote location and high cost of transportation raise questions about their commercial viability.

Of the remaining gas which is not committed to local markets or deferred for one reason or another, much of it is uneconomic to utilize for either domestic or export markets because it is contained in small fields too distant to justify the gathering and transmission expense of moving the gas to market. Where such gas is associated, it will often continue to be flared. Where it is non-associated, it will often be shut-in and not developed. Thus, the portion of worldwide gas reserves which can be considered as international — either committed on existing contracts or available as large blocks as a basis for new international projects — is considerably smaller than total world reserves might suggest.

To place the total worldwide gas reserves in a market perspective, it is useful to classify them country-by-country into market status categories. Table 1 and Figs. 3, 4, 5 and 6 provide one such estimate of commercial status of worldwide gas reserves. The table and figures are based on six different commercial categories. They are:

1. Inaccessible or Flared: gas reserves which are too small or remote either to justify recovery of flared gas or full field development of non-associated gas.
2. Deferred Reserves: reserves in large gas caps or undergoing gas injection for oil recovery, such that they are unlikely to be committed to market projects until some future time.
3. Committed to Domestic Markets: gas reserves which either are contracted to domestic markets or set aside to assure that domes-

tic requirements will be covered. We do not have detailed information about many such set-asides, but have generally utilized a modified Canadian formula which provides for thirty years coverage of present marketed production.

4. Remote from Existing Market Systems: larger blocks of gas reserves which are clearly destined for a major industrial market but whose remoteness from this market raises commercial feasibility questions. Examples would include North Slope and Arctic Island gas in North America and some North Sea reserves in Europe. Some of this gas will prove feasible for commercialization and thus may later belong in the 'Committed to Market' or 'Exportable Surplus' classifications.
5. Committed to Export Markets: gas reserves on firm export contracts covering the required deliveries over the life of the contract.
6. Exportable Surplus: blocks of gas reserves which are large enough and well located enough to support export projects. In a limited number of cases, current national policy suggests that this gas

Table 1 Market Status of World Proved Gas Reserves December 31, 1978 (Trillion cubic feet)

	OECD	*NOPEC*	*Sino Soviet*	*OPEC*	*Total*	*Per cent*
Proved Reserves	464	186	945	962	2557	100.0
Less Inaccessible or Flared	3	45	120	153	321	12.6
Subtotal Potentially Commercial	461	141	825	809	2236	87.4
Less Deferred Reserves	13	2	-	279	294	11.5
Subtotal Currently Potential Commercial	448	139	825	530	1942	75.9
Less Committed to Domestic Market	304	66	371	170	911	35.6
Less Remote from Existing Market Systems	71	-	-	-	71	2.8
Subtotal Available for Export	73	73	454	360	960	37.5
Less Committed to Export Markets	49	7	15	77	148	5.8
Equals Exportable Surplus	24	66	439	283	812	31.7

Source: Jensen Associates, Inc.

will not be exported and, in other cases, discussions to commit the gas have proceeded to the point where it may no longer be available.

By this analysis, an estimated 812 tcf or approximately 32% of world reserves can be considered in the exportable surplus category.

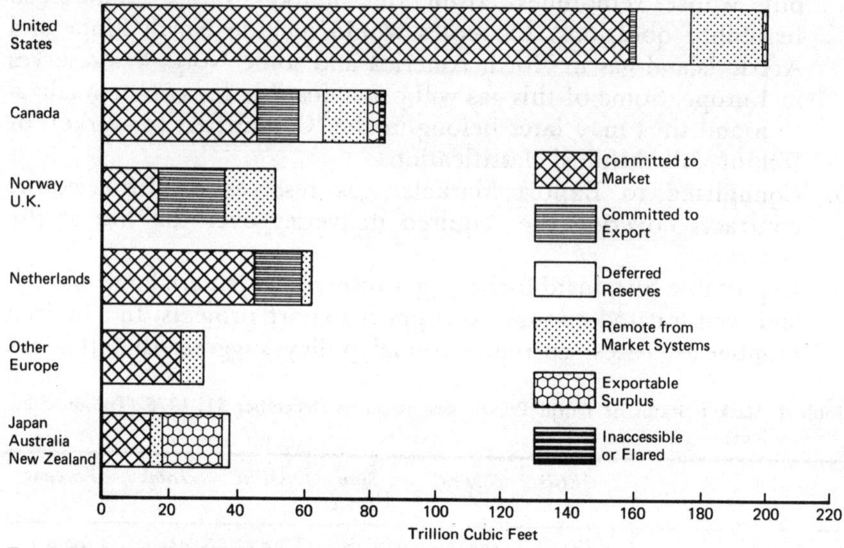

Fig. 3 Market status OECD gas reserves (tcf)

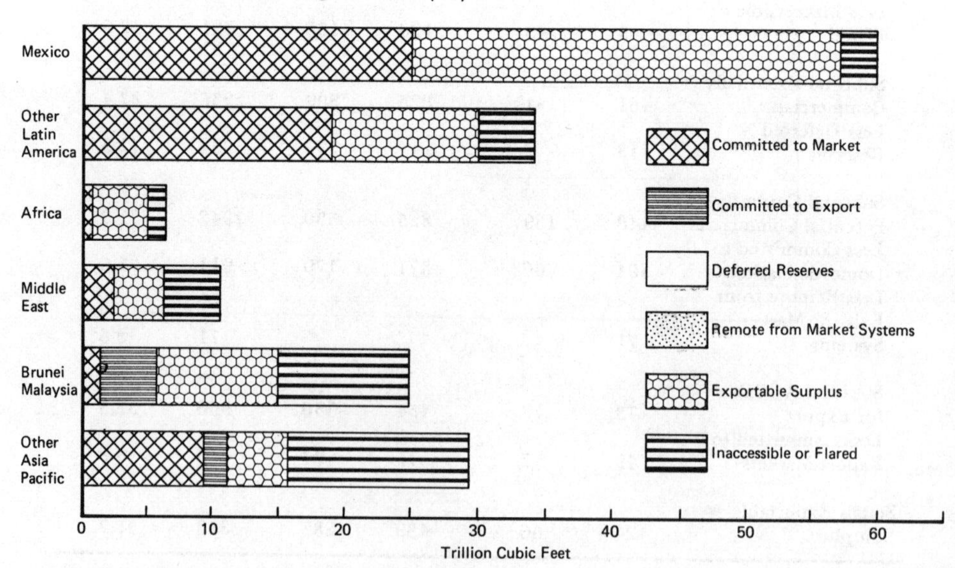

Fig. 4 Market status NOPEC gas reserves (tcf)

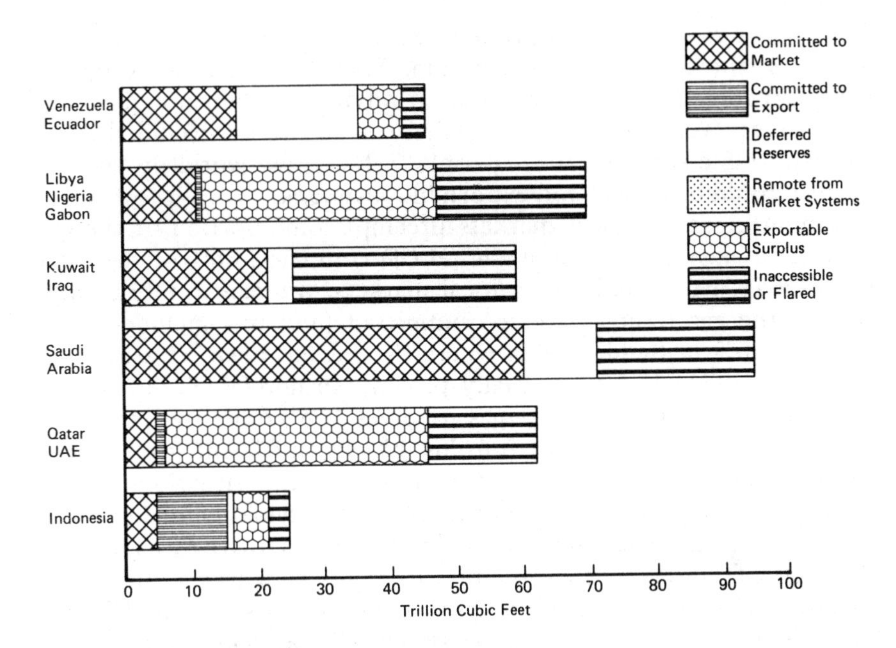

Fig. 5 Market status OPEC gas reserves (ex Algeria, Iran) (tcf)

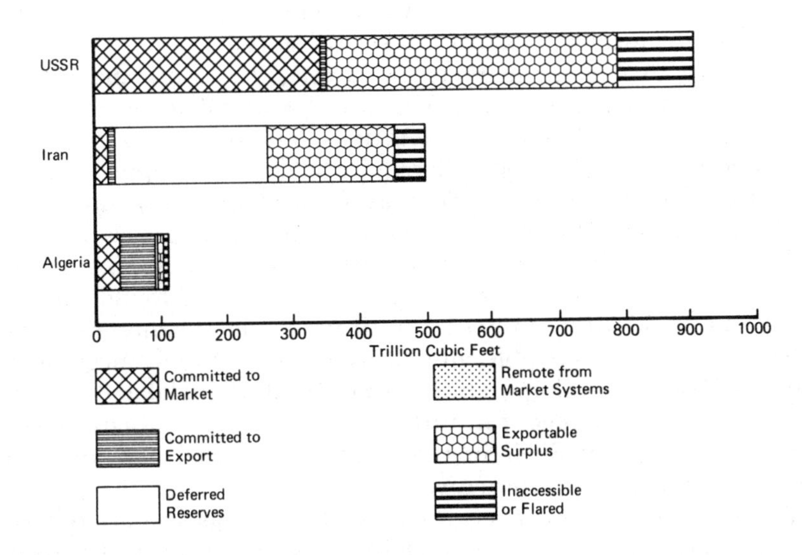

Fig. 6 Market status USSR, Iran, Algeria (tcf)

Three-quarters of the exportable surplus is concentrated in two countries, the Soviet Union and Iran. There has been some recent European re-evaluation of the risks of relying on Iranian and Soviet gas reserves. Were Iranian and Russian reserves to be excluded from the exportable surplus category, only 7.2% of the world proved gas reserves could be considered exportable surpluses.

Commitments to local markets preempt some portion of the gas reserves in each country. For the OECD countries, 66% of the total gas reserves are committed to local markets, whereas in OPEC only 18% of the gas reserves are so committed (and this includes some 60 tcf or 6% in Saudi Arabia where the gas has been set aside as a matter of policy but not actually placed). In non-OPEC developing countries (NOPEC) 36% of gas reserves are committed to local markets. In seeking export surplus blocks sufficient to support international trade, one looks for those specific situations where good blocks of gas exist beyond the likely requirements for local use.

Fig. 3 summarizes the market status for OECD gas reserves. Most OECD reserves are committed to local markets. In North America, Canada has major export commitments to the US but has classified an extra 2 tcf as potentially available for export. Fig. 3 classifies Prudhoe Bay, the Arctic Islands and Beaufort Sea gas reserves as remote from market systems. A final go-ahead on the Alaska Natural Gas Transportation System together with an agreement to export Canadian gas to the US both directly and through the pre-building of the Alaska line would move more than 25 tcf into the local and export market category.

There are two major markets in Europe, the Continent and the UK. The Continental gas market is heavily dependent on the Groningen field in the Netherlands. As it begins to go into decline in the middle 1980s, Europe will require major new supply sources. The first Norwegian North Sea gas from the Ekofisk area went to the Continent. However, the more northerly Frigg gas together with all of the UK area gas has gone to the UK. As exploration in the North Sea moves farther north, Norway has found the economic threshold of Continental projects to be steadily rising. In order to provide a credible alternative to a monopoly negotiation with the British Gas Corporation for Statfjord, Heimdal and Sleipner gas, Norway has hoped for further significant gas discoveries. A recent discovery by Shell in block 31-2 may well have provided enough gas for such an alternative.

The Continent is now supplied by Russian gas in part released from Russian needs by the Iranian IGAT-1 deliveries to the Caucasus. This was to have been institutionalized in IGAT-2 as a formal exchange agreement involving Iran, the Soviet Union and Western Europe. The upsets in Iran have not only affected the existing gas delivery levels for both the USSR and Europe but raised serious questions about the future supplies from both Iran and the Soviet Union.

The only genuine exportable surplus in the OECD is in the Pacific. Major gas reserves on the Northwest shelf of Australia are being considered as a basis for LNG projects to the US and Japan. Somewhat lesser sized reserves in the Maui field of New Zealand have been considered for LNG in the past, although the gas reserve is comparatively small for a major gas project.

Fig. 4 summarizes the market status of NOPEC gas. For this group of countries, market commitments are considerably less and the percentage of gas which is inaccessible or flared is much greater. The very large Mexican gas reserves – largely associated with the oil discoveries in the Tabasco/Chiapas region of Yucatan – are substantially in excess of foreseeable Mexican requirements. Since Fig. 4 was prepared, an initial contract with six US gas pipelines has committed a small portion of the exportable surplus to export markets. It is quite likely that these volumes will increase over time. Small exportable surpluses exist in Malaysia, Bangladesh, Bahrain, Trinidad, Argentina, Colombia, Chile, Tunisia and Thailand. The only countries in this NOPEC group which would represent a basis at present for world-scale LNG projects are Malaysia and Bangladesh. Malaysia has recently committed itself to an LNG export project to Japan.

Fig. 5 summarizes the market status for OPEC gas reserves (excluding the large Algerian and Iranian gas reserves). Qatar and Nigeria have significant non-associated gas reserves. The reserves in Qatar are located in the Northwest dome, a Permian Khuff gas discovery offshore. Nigerian non-associated and associated gas reserves are scattered. Nearly all of the rest of the OPEC gas shown in Fig. 5 is associated gas, much of which is being flared at the present time. Fig. 5 considers the reserves behind the Saudi Arabian master gas plan to be committed as a matter of policy although they have not yet been placed in specific projects.

The Permian Khuff formation, a deeper horizon than most of the oil-bearing formations in the Persian/Arabian Gulf, has been productive in Qatar and more recently in Abu Dhabi as well. Many geologists believe that the Permian Khuff will prove highly productive of gas throughout the region and lead to a sizable increase in gas reserves for ultimate export.

The three countries in the world which have the largest volumes of non-associated gas relative to their own internal needs are the Soviet Union, Iran, and Algeria. The market status breakdown for them is shown as Fig. 6. These three countries account for 635 tcf out of the total exportable surplus of 812 tcf.

Thirty-five per cent of the world's gas reserves are concentrated in the Soviet Union. The Soviet reserve estimates may be somewhat less conservatively stated than those in much of the rest of the world. They are considered as 'explored', including proved and probable as well as some possible, but they are none the less impressive in magnitude. Early Russian oil and gas exploration was concentrated in the south near the Black Sea and Caspian Sea. As exploration moved north and east, the Soviets made major gas discoveries in West Siberia, particularly on the Ob Peninsula which includes such giants as Urengoy, Yamburg and Zapolyarnoe. Approximately 75% of Russian gas reserves are located in West Siberia. Areas to the south and west, such as Turkmenistan, Uzbekistan, and the Volga-Urals region, constitute another 20%. The rest of the gas is scattered throughout the country in a number of producing basins.

The Soviet Union currently imports small quantities of gas by pipeline from Afghanistan. It had also been supplementing its more limited southern reserves by importing about one billion ft^3 a day (bcf/d) from Iran through the IGAT-1 system. It has also been delivering 1.45 bcf/d to West Germany, Italy and Austria from its northern reserves. While this is not formalized as a natural gas exchange agreement in the way that IGAT-2 was intended to be, it is similar in effect. The recent failure of IGAT-1 to deliver as planned and the cancellation of IGAT-2 which would have delivered an additional 1.65 bcf/d ultimately to Europe via the Russian exchange route has clouded the outlook for future Soviet deliveries to Europe. While most plans for utilization of Russian gas had contemplated pipeline movements, LNG projects have been discussed, both for the US East Coast out of West Siberian reserves, and for the US West Coast and Japan from the Yakutsk area of Eastern Siberia. These projects appear to be inactive at present.

Iranian gas reserves are second in magnitude to Soviet gas reserves. Approximately 210 tcf of the 500 tcf of Iranian gas reserves are associated/dissolved, with a very large portion of these contained in large gas caps of some Khuzestan oil fields. About half of the Iranian gas reserve is contained in giant non-associated gas fields, both onshore near Kangan and extending out into the central Persian Gulf. Smaller gas fields are scattered throughout the country with one concentration of reserves located near the Straits of Hormuz, around Bandar Abbas.

Oil recovery in the Khuzestan fields is particularly sensitive to bottom hole pressure decline. Before the overthrow of the Shah's government, NIOC was experimenting with a major gas injection programme which, if successful, was expected to be extended to virtually all of the Khuzestan fields. The gas injection programme would not only have postponed production from the gas caps but would have reinjected significant quantities of dissolved and nearby non-associated gas into the oil formations for later recovery. This major injection programme would have deferred, as gas caps and injected volumes, almost half of the Iranian gas reserves. Iran thus accounts for much gas which is included as deferred reserves in these estimates. Iran gas export plans had included Europe via the IGAT-2 pipeline system as well as preliminary discussions of a large LNG project from the Kangan area both to Japan and the United States. The reserves which would have been dedicated to IGAT-2 and the Kangan LNG project would have amounted to almost 21 tcf.

As it currently stands, however, the uncertainties surrounding future Iranian gas policy make it questionable that any of these projects will bear fruit within the foreseeable future. Both the IGAT-2 and Kangan projects are now cancelled, and some question exists as to whether the contract commitments under IGAT-1 will be honoured. The future of the major gas injection scheme is also in doubt. Thus, although Fig. 6 shows an estimated 188 tcf of exportable surplus for Iran, there is little likelihood that any new projects will be initiated in the foreseeable future.

Had Iran gone ahead with its plans, a number of the large gas fields which were most conveniently located to support export projects would have been committed to the gas injection programme instead. The exportable surplus reserves were, therefore, potentially somewhat more remote and therefore more costly for land-based LNG plants. These include the very large E, F, and G structures which are farther out in the Gulf, together with some of the C structure (or Pars), which extends both onshore and offshore near Kangan. Some of these reserves would be more costly to commercialize than some of the onshore gas; they might lend themselves well to the evolving technology of barge-mounted LNG liquefaction at some future time when Iran is prepared to develop an LNG export trade.

Algeria was the first country to export LNG on a commercial scale; it also has the largest and best developed programme for LNG export. The Algerian estimates in Fig. 6 are based on an Algerian proved reserve figure of 105 tcf. Approximately another

25 tcf are classed by the Algerians in the 'possible' category. The Algerian Valhyd gas development plan looks ahead to the year 2004 and is designed to handle all of the proved plus a portion of the possible. Since it is time-phased Algeria can readily scale the programme down if some of the possible reserves fail to materialize. Firm commitments for 11 LNG contracts and the pipeline across the Mediterranean to Italy account for nearly 60 tcf of export commitments. Local markets are expected to take approximately another 30 tcf. After accounting for a certain amount of oil well gas flaring, we estimate that 8.6 tcf of reserves remained in the exportable surplus category as of December 31, 1978 (and at the 105 tcf level). Most of this surplus is all but committed. The contract with Brigetta-Thyssen in Germany is treated as firm in Fig. 6. When the US Federal Energy Regulatory Commission disapproved the El Paso II and Tenneco St. John projects, a scramble developed in Europe to take over these contract commitments. The Italian pipeline together with volumes being negotiated with potential European LNG purchasers now appears to have absorbed the available volumes and Algeria can be considered essentially sold out until further future discoveries are made. The 8 tcf shown as exportable surplus in Fig. 6, though not yet firmly contracted and approved, is spoken for.

The fact that so much of the exportable surplus reserve is so remote from major markets makes an understanding of project economics especially important. One critical element in any economic evaluation is the value of the gas at market.

The Price of Gas

It is frequently argued that natural gas is a premium fuel and should command a premium price. Wether or not it does command a premium in competitive markets however, is very much a function of where it is used. While household and commercial markets – where they do exist – can often command a premium, most major international gas projects require the placing of the incremental gas in the more quickly expandable industrial and power generation markets. Some portion of the industrial market will pay substantial premiums for gaseous fuel because of its unique form value, but other users are simply buying Btus and will tend to be much more price conscious. For some portions of the market – such as metallurgical coke or liquid petrochemical feedstocks use – the form value may rest with solid or liquid fuel rather than with gas. The question of whether or not gas commands a market premium over competitive fuels is also influenced by whether it is attempting to increase its market share of the competitive fuel mix or is in the process of losing its historic share – usually through supply limitations. In the first case – characteristic of totally new market penetration, such as Europe after the Groningen discovery, or Japan in the early 1960s – gas must usually be discounted to provide an incentive for customer conversion to gas. In the latter case – more characteristic of Italy after the Po Valley gas field depletion, or the US market in the mid 1970s – gas may command a premium as users seek to avoid conversion away from gas.

The US market until recently has provided a special price premium to international trade termed 'roll-in'. The fact that domestic natural gas in the US has been price controlled at the wellhead, at prices which were below market clearing levels, has enabled importers to buy higher priced supplementary gas and 'roll-it-in' with price controlled domestic gas without loss of market share. The ability to roll-in LNG for new projects has been sharply limited by the provisions of the Natural Gas Policy Act of 1978.

Fig. 7 summarizes the percentage of the industrial and power generation market to which a given price premium (or discount) might be expected to apply. The figure is idealized and the market proportions are based on US experience. Nevertheless, since the approximate proportions of load type are surprisingly similar for most industrial countries, the curves have some universal applicability.

In attempting to develop a market for a new LNG project, the first target is often that small portion of the market which will pay high premiums for gas. Where gas has not been an important part of the energy economy this market may be very small. The early efforts to put together a small premium-priced LNG project for Japan from the Cook Inlet in Alaska encountered significant diseconomies of scale in project sizing. Only when the project was sized to go for the larger, lower-premium power generation market was the project economically viable. World scale LNG projects often must be sized to compete in the larger general-industrial fuel market rather than in selected high-premium uses alone.

The market premiums for Fig. 7 are measured against the low-sulphur fuel-oil characteristic in the market in question. In recent years, Japan has become an especially sulphur-conscious market, burning increasing quantities of LNG under boilers for environmental reasons. At the other end of the scale, Europe has tended not to pay premium prices because of its somewhat less severe sulphur restrictions and the availability of significant quantities of

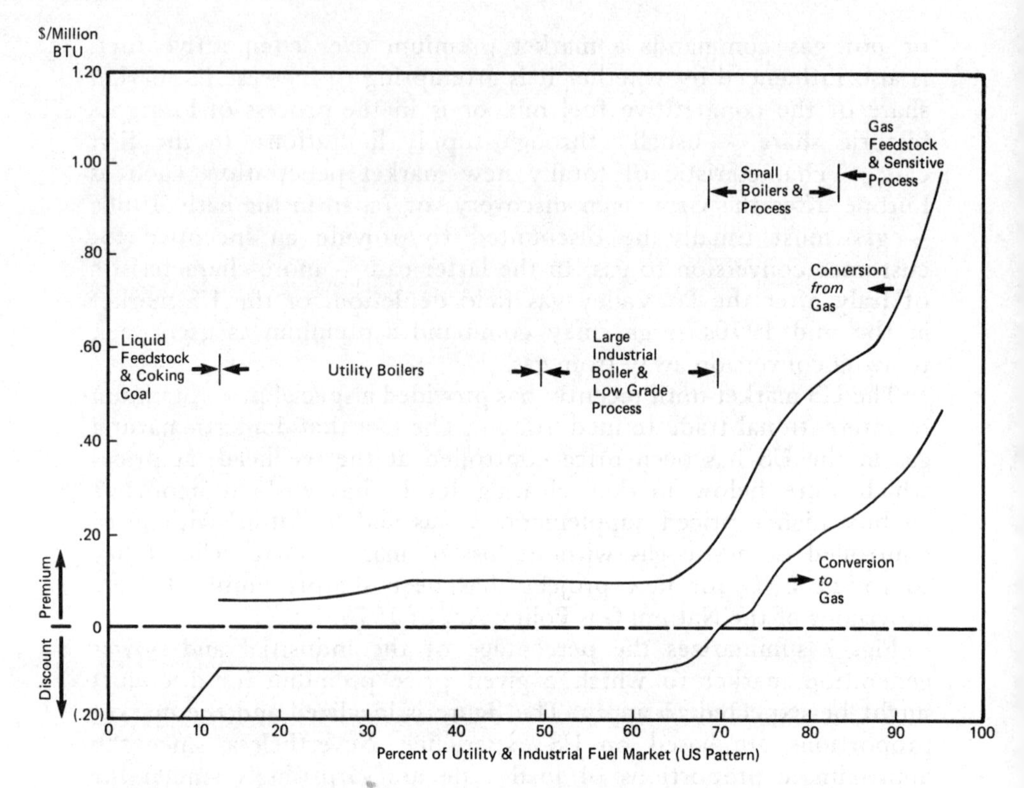

Fig. 7 Price premium (discount) relative to low sulphur fuel oil at market equilibrium

low-sulphur African and North Sea crudes in the European refinery mix. The US has been somewhat intermediate. For a world scale LNG project, where the competition is most likely to be boiler fuel, these sulphur premiums have a major influence on the final value of gas as delivered to the market.

The high costs of gas transportation over long distances has heretofore made it difficult for the major gas-consuming centres in the OECD to tap the major gas reserves of the Middle East. The proposed IGAT-2 pipeline exchange between Iran, the Soviet Union and Europe was one such venture. The Abu Dhabi/Japan LNG project was another.

Fig. 8 illustrates the likely economics and resulting wellhead netbacks for various 1983 LNG trades (at approximately 1979 price levels). Fig. 8 suggests that North Europe and the US are similar in the netbacks they provide North African supply. South Europe LNG projects would be better for North Africa. However, Algeria seems to feel that trans-Mediterranean pipeline crossings are now

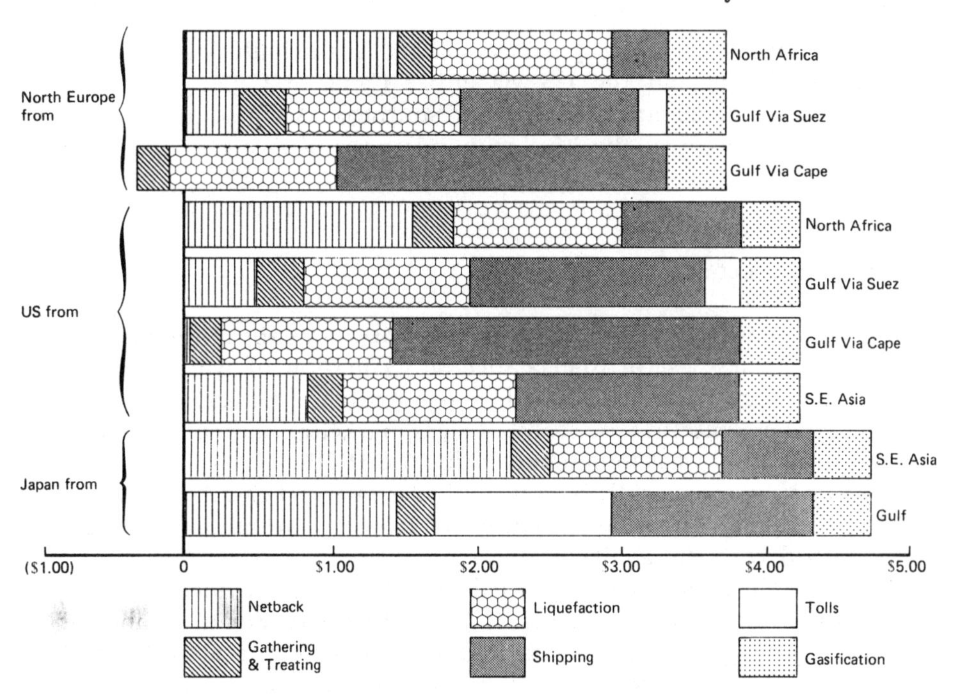

Fig. 8 **1983 Project LNG costs per MBtu (1979 dollars) including wellhead price**

feasible, and this may prove the cheapest way to move future Algerian volumes to Europe. Nigerian projects must still be LNG-based and netbacks, again, should be similar for North European and US distributors.

When looking beyond African reserves to the large reserves in the Gulf however, the problem becomes more complicated. At present oil prices, there is now a small netback from the Gulf for both Europe and the US, but it requires that the tankers take the shorter route through the Suez Canal rather than around the Cape. The political risks associated with a Suez Canal closing are far higher for LNG tankers than for oil tankers, both because of the small and inflexible nature of the LNG fleet and the fact that the long hauls are relatively much more costly for LNG.

To avoid the risks of Canal transit, by designing for an around-the-Cape haul, is to eliminate the netback in the Gulf at present oil prices. Pipeline costs from the Gulf to Europe should be lower than LNG costs, a situation the IGAT-2 Iranian/Russian exchange was designed to take advantage of. Direct pipelining is complicated by the many national boundary transit problems, all but eliminating the most economic supply system.

The netback from the Japanese market to the Gulf is better than that to Europe and the US via either Canal or Cape. Japan also provides an excellent netback from Southeast Asian sources, giving LNG trade to Japan the best economic basis of any gas-consuming country.

The US Market as a Case Study

Since gas markets are so individual, generalizations about the markets for gas are less useful than a specific evaluation of where gas fits in the energy balance for a particular market. In many ways, the US provides a useful case study of a gas market because of the major transformation it has undergone during the 1970s.

The US is the world's largest market for natural gas. At the beginning of the 1970s, a combination of growing natural gas shortages in the US with the ability of suppliers to 'roll-in' or average higher-priced LNG with price-controlled domestic gas, provided a strong incentive for the development of an LNG trade. For a number of reasons – regulatory, market, and economic – the attractiveness of the US as an LNG market has declined somewhat in the last few years. While a number of projects in operation or advanced stages of planning will clearly go forward, proposals for totally new projects are at a low ebb.

In 1970, natural gas was the fastest growing source of energy in the US. At the time of the 1954 Supreme Court decision in the Phillips case, which placed natural gas wellhead pricing under federal regulation, gas was price-controlled at the wellhead while coal and oil were not. Thus, gas took the dominant share of the US industrial market and supplied about a third of the power generation market until the first shortages began to appear about 1970. Periodic forecasts of natural gas requirements in the US have been made by a gas industry group known as the 'Gas Future Requirements Committee' (GRC). Prior to 1972, forecasts were based on the twin assumptions that there would be no shortage of supply and no change in the price relationship between gas and competitive fuels. By 1972 it was becoming evident that such forecasts of requirements were far larger in volume than most estimates of foreseeable supply. The difference between requirements and supply estimates came to be known as the 'gas gap', in effect an estimate of the excess demand created for a fuel which was price-controlled below market clearing levels.

In 1972 the GRC provided its last traditional forecast of requirements, but shifted to a consumption basis, which was defined as . . . 'usage primarily based on the availability of supply'[1] Most recent forecasts of gas demand have actually been supply forecasts in disguise, with the implicit assumption that more gas could be sold if a greater supply were available. The strongest interest in supplementary supply, such as imported LNG, dates from the shortage period of the late sixties to early seventies.

However, a major transformation of the US gas markets with substantial user conservation occurred between 1972, when widespread pipeline curtailments began, and 1977 when the first evidence of spot surpluses – often known as the 'gas bubble' – began to appear. The 1972 GRC report foresaw a 1980 requirement of 35.8 tcf, a consumption (supply) of only 27.1 tcf, and a resulting gap of 8.7 tcf. Our own present estimates of 1980 potential demand (similar to requirements) from a 1979 perspective is only 23.6 tcf – a full 12.2 tcf below GRC's earlier requirements projection. The magnitude of the reduction in demand is illustrated by the fact that our present estimate of 1980 requirements is even less than GRC's expected 1980 consumption – or supply – level. The nature of this significant transformation in US gas markets can best be understood by tracing successive GRC demand forecasts and the changing perception of demand during the period from 1972 to 1977. Fig. 9 shows the actual consumption of natural gas in the US during 1972 to 1977, compared with GRC's 1972 requirements forecast, and the 1972, 1974 and 1976 GRC consumption forecasts. The magnitude of the expected gas gap in 1972 is shown by the difference between 1972 forecasts of requirements and consumption. With a worsening gas supply situation, GRC's total consumption (or supply) forecast was reduced in 1974 and again in 1976. Actual gas consumption declined steadily between 1972 and 1977, a pattern which was consistent with GRC's increasingly pessimistic supply estimates. It was common in the industry to attribute the actual decline in gas consumption to reduced supply, particularly since there was no direct way to test the full extent of unconstrained gas demand in the face of supply shortage and curtailment. However, more detailed analysis of the underlying sectoral trends in gas consumption reveals the extent of the market transitions which were taking place.

Figs. 10, 11 and 12 compare actual consumption levels for residential, industrial and power generation uses with the GRC forecast estimates. GRC anticipated that limited supply would go preferen-

[1] Future Requirements Committee, *Future Consumption of the United States*, Vol. 5 November 1973.

tially to the residential market, so its projected shortfall between the consumption and requirements forecasts was comparatively small in 1972 (see Fig. 10) but there was a steady decline in projected residential consumption in 1974 and 1976. In part this reflected the lack of customer growth as a result of widespread moratoria on new extensions. But it also reflected user conservation, which actually exceeded the limited growth that had occurred. Thus, on a weather-adjusted basis residential demand declined from 1972

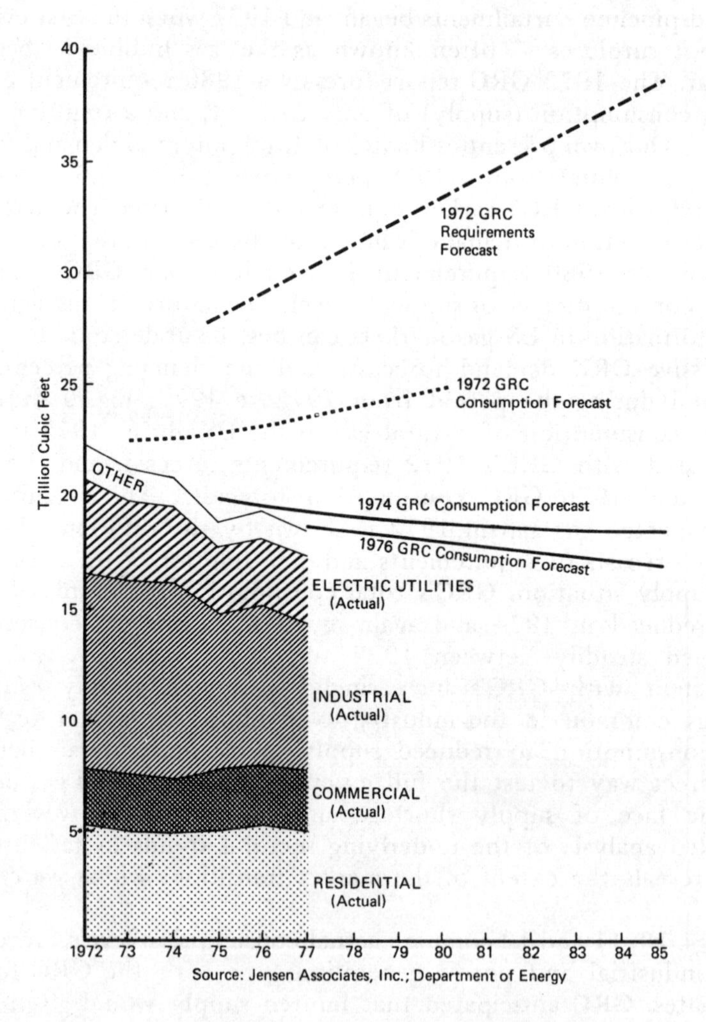

Fig. 9 US total natural gas consumption 1972–77, GRC demand forecasts 1973–85

to 1977. Since we are convinced that the trend in conservation will continue, we do not look for any significant increase in residential gas demand, even if the moratoria imposed by law were to be abolished.

The GRC expected the industrial gas load to bear a substantial share of the shortage, so its 1972 consumption estimate was considerably below its requirements estimate (see Fig. 11). Conservation in the industrial market substantially exceeded the growth

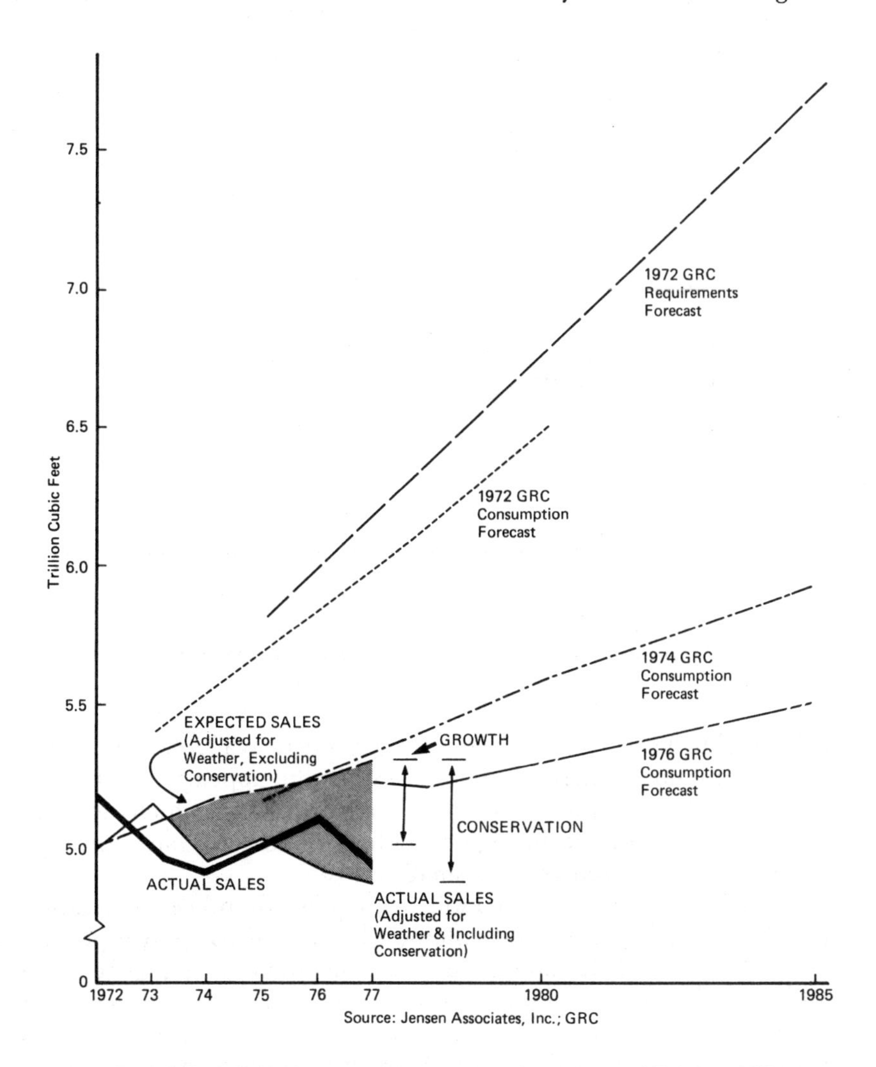

Fig. 10 **Total US residential sector natural gas market changes 1972–77, GRC demand forecasts 1973–85**

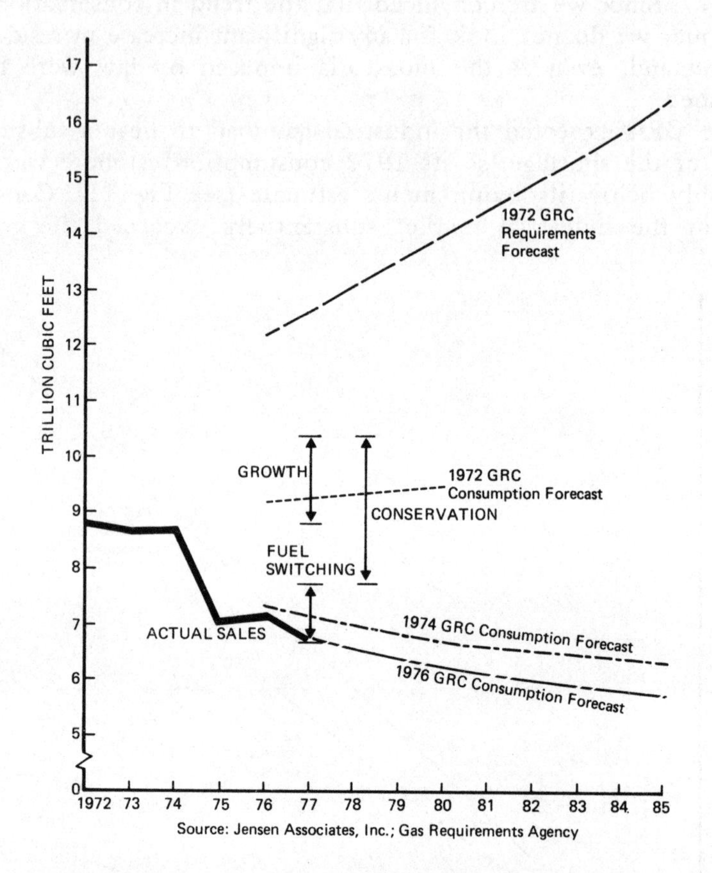

Fig. 11 US industrial natural gas consumption 1972–77, GRC demand forecasts 1976–85

which might have taken place had gas been available, and been able to maintain its market share. Despite the appearance of widespread curtailment, the amount of fuel switching which actually occurred – both forced through curtailment and voluntarily through price changes – was remarkably small.

Fig. 12 summarizes comparable estimates for power generation demand. In Fig. 12 the total volume of gas used for power generation in 1972–1977 is compared with the incremental generation attributable to oil, coal and nuclear power since 1972. Since most gas-fired power plants do not have the ability to convert to coal, one would expect declining natural gas utilization to have led to an offsetting increase in oil consumption. From 1972 to 1975, gas

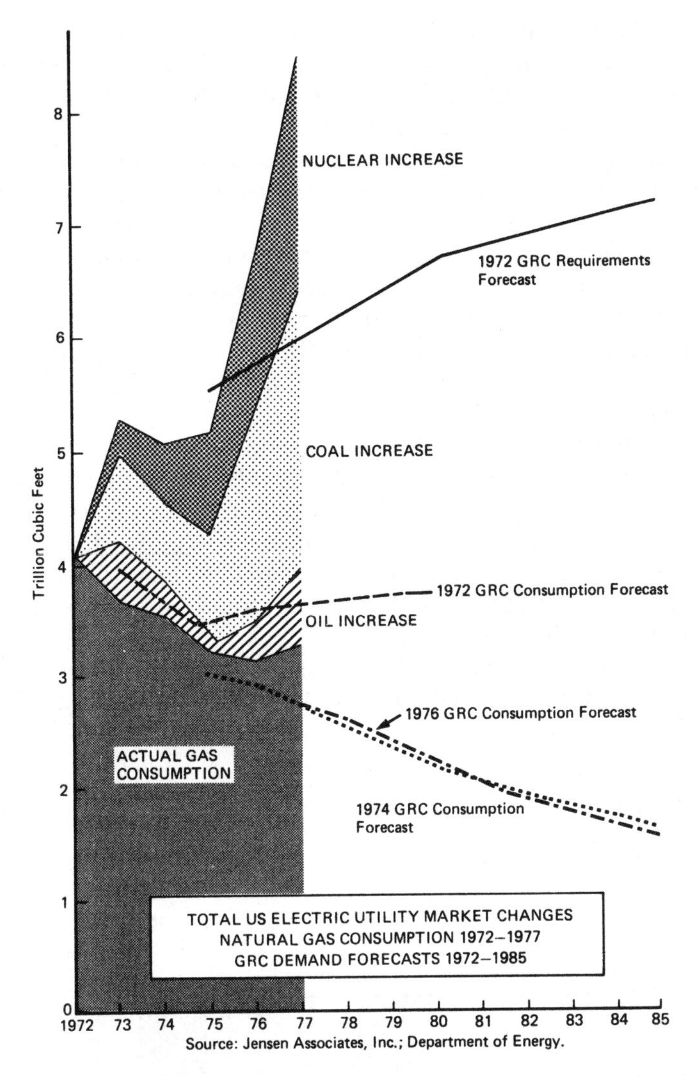

Fig. 12 Total US electric utility, market changes, natural gas consumption 1972–77, GRC demand forecasts 1972–85

and oil both declined since planned increments to nuclear and coal-fired capacity provided more generating capacity than was needed for a relatively static electric power market. Gas and oil capacity could both be used at somewhat reduced load factors. This pattern began to change in 1975 and 1976 as both gas and oil generation began to increase as the excess capacity was beginning to be absorbed.

Fig. 13 summarizes the changes which took place in the pivotal industrial market over the critical period from 1972–1977. Had gas maintained its market share, increases in business activity would have resulted in a 1.6 tcf increase in industrial gas demand. But since conservation significantly exceeded the potential growth, it required only 1 tcf of fuel switching to balance the total decline of 2.1 tcf.

Most of the fuel switching – both curtailment and price induced – was to oil, as Fig. 14 indicates. Residual fuel oil accounted for 50% and distillate accounted for another 35% of the switching away from gas. Despite the stated emphasis of the US energy policy to increase coal utilization, coal benefited hardly at all from the switching away from industrial gas use.

The switching away from gas was both curtailment and price induced. Prices rose rapidly to clearing levels in the unregulated Texas, Louisiana and Oklahoma markets, but even in the regulated interstate market gas prices advanced more rapidly than oil prices between 1975 and 1979. This is illustrated in Fig. 15 which shows competitive industrial gas and oil price trends over the period for one region – the Middle Atlantic.

While absolute levels vary from region to region, the pattern is remarkably similar throughout the US. Prior to the gas shortage

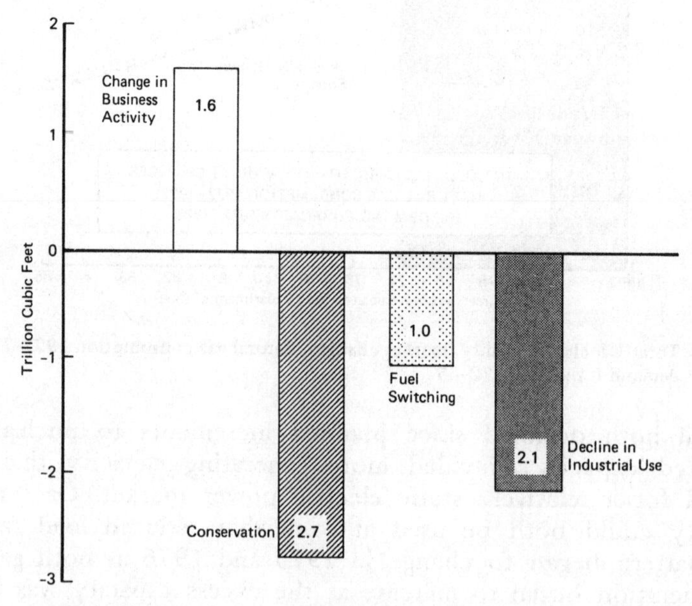

Fig. 13 Changes in industrial use of natural gas 1972–77 (tcf)

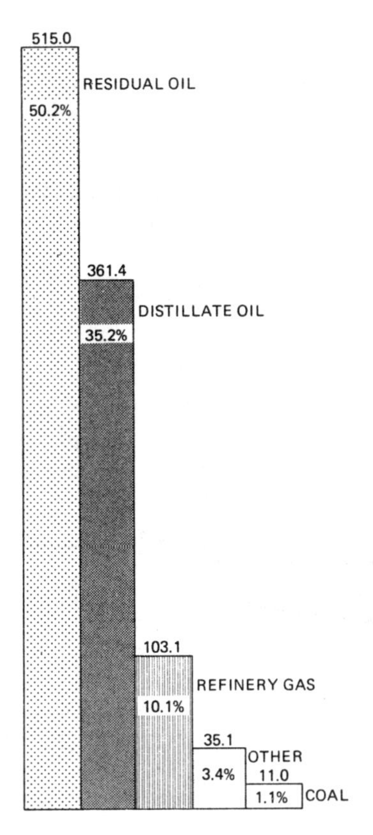

Fig 14 **Total US industrial fuel switching away from natural gas 1972–77 (Billion cubic feet equivalent)**

and the oil embargo, the price relationships between industrial oil and gas were quite stable. Under the influence of the 1973–74 OPEC price increases, industrial oil prices rose dramatically but then levelled off in real terms, as OPEC increases were moderate and the US kept price controls on domestic oil. The worst of the gas shortages occurred after the embargo at a time when price-controlled gas was much cheaper relative to oil than it had been earlier.

By 1977, the rapid increases in gas wellhead prices had narrowed the competitive pricing relationship. Thus, by 1977–78 the incentive to find supplementary gas supply such as LNG was lessened by conservation, and the risk of pricing gas out of competitive markets had increased. This pattern was compounded by the Natural Gas Policy Act of 1978 which not only provided a significant increase in gas wellhead prices, but introduced incremental pricing concepts in an attempt to clear industrial markets of excess demand. Had

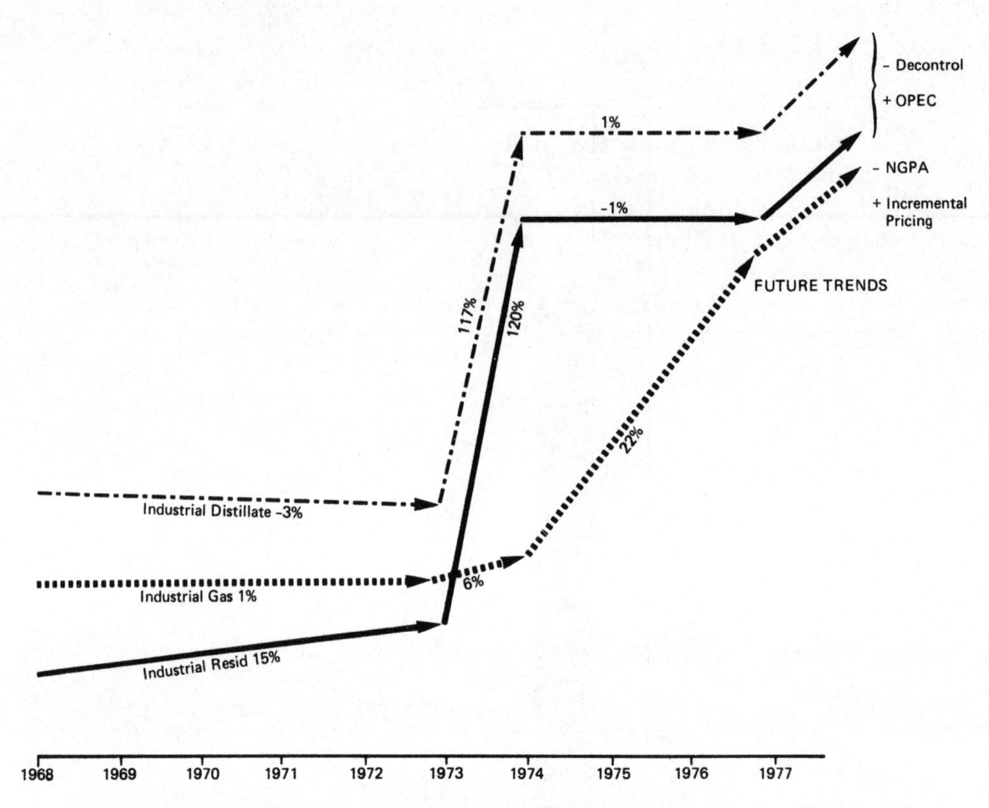

Fig. 15 Recent industrial fuel price behaviour. Price changes in real terms (% per year)

the post-Iranian oil price increases not occurred, the economic viability of new LNG projects would have been at risk.

This has now been changed by the Iranian Revolution and the sharp increases in oil prices which have accompanied it. If rapidly accelerating oil prices do not ensure that natural gas will retain US markets which oil could otherwise serve, the various actions of Administration policy to put a lid on oil imports will certainly do so. We doubt that federal administration of incremental pricing will be allowed to shed gas load in favour of oil on the basis of price alone. Much more likely is a renewed interest in gas supplements which characterized the period from 1971 to 1977.

Recent US policy has tended to discourage long-haul trade in LNG in favour of Mexican and Canadian supplies and, to a lesser extent, Western Hemisphere LNG. The oil shortage conditions created by the events of 1979 suggest that this negative policy towards gas trade may be relaxed after additional review. If so, the availability of Western Hemisphere LNG will prove quite limited

– as this paper illustrates. New long-haul LNG projects from Algeria and Indonesia – where existing US oriented projects have been concentrated – will also prove limited, at least until gas exploration is more productive. Thus, LNG interest will tend to focus on new areas such as Australia, Nigeria and ultimately on the Gulf. That these supplies will prove costly is inevitable. But that they will be more expensive than competitive oil supplies is no longer to be taken for granted in a world of rising real energy prices. We foresee a renewed interest in the United States in world LNG trade during the eighties.

5 COAL: RECENT DEVELOPMENTS AND PROBLEMS OF SUPPLIES

Michael J. Parker

Coal needs to be considered in the context of overall energy supply and demand. This paper examines several key issues on the supply of and demand for coal, and its contribution towards mitigating the world oil problem.

Coal Reserves and Resources

There is a distinction between resources (or 'coal in place') and reserves — the coal likely to be worked. All quantitative discussion of mineral reserves is fraught with difficulty, and coal is no exception. In the nature of things, there is bound to be considerable uncertainty over the global recovery factor, which will be influenced by technical and economic factors. In this respect, it is worth noting that coal exploration is more concerned with determining the economics of production than with the physical presence of coal — representing a significant difference of emphasis compared with oil exploration.

Over recent years, assessments of coal resources have shown an increasing trend, and recoverable coal reserves will clearly be enhanced by the emergence of the world oil supply problem. On a 50% recovery factor, there might be 5000 billion tonnes of recoverable coal, many times the equivalent figures for conventional oil and natural gas. In global terms, therefore, reserves are not a constraint, and there is room for very substantial long-term expansion of coal output if the economics are right and the appropriate capital is invested.

Production Prospects

There are clear indications that world coal production capacity could be broadly doubled by the end of the century. The report on coal resources produced for the 1977 World Energy Conference envisaged 5.6 billion tonnes in the year 2000 compared with 2.7 billion in 1977 – with about half of the increase coming in the World Outside the Communist Area (WOCA). However, although Australia and South Africa will show large increases, the overall WOCA increase in projected coal output is dominated by the United States (increasing from 0.6 to 1.3 billion tonnes by 2000). Indeed, the range in the various estimates which have been made of American coal production in 2000 is greater than the actual level of coal output in any other WOCA country.

Most of the increase in coal production currently being planned, at least in WOCA, is in stripped-mined coal. It is important to recognise that there are in fact two coal industries – namely deep mining and strip mining. The technology and the economics are very different (as are the environmental impacts). Furthermore, the reserve base available for strip mining (which requires shallow deposits) is very much more limited than the reserves available for deep mining. Although the data are quite inadequate, general geological considerations would lead to the view that perhaps only 10% of total coal reserves would be mineable by strip methods. Clearly, a situation in which, say, 60% of coal expansion relates to only 10% of reserves could not be stable in the long run, and would in due course raise depletion issues on strip mining. The important point is that the undeniable abundance of coal as a long-term energy source is very much a matter of deep mining. In turn this raises the question of whether enough effort is being made in developing new technologies to exploit the very large reserves of deep-mined coal.

Developing Demand for Coal

It is worth considering the incremental contribution that coal has made to world energy supply since 1973. Over the period 1973–78, primary energy consumption in WOCA increased by 370 million tonnes coal equivalent per annum, and of that increase only some 60 million tonnes were in coal. The fact that, despite the oil problem, the contribution of coal increased quite modestly during this period was due to the limitations on the rate at which coal demand can expand. To understand this, we have to look at the way the market for coal is structured. In 1977, the position was as follows (in million tonnes):

	WOCA	*CA**	*Total*
Power Stations	714	425	1130
Industrial/Domestic	222	869	1100
Total Steam Coal	936	1294	2230
Coking	296	206	502
Grand total	1232	1500	2732

* CA = Communist Areas

It can be seen that in WOCA, some three-quarters of all steam coal is used in power stations. This dominance of the power station market reflects the inherent economic 'hierarchy' of coal uses. Leaving aside coking coal, which is a specialist application, steam coal is a heat source, where the hierarchy is determined basically by the relative disadvantages against oil. These arise from the chemical and physical properties of coal as against oil. Generally, the disadvantages of coal against oil are minimised in bulk combustion, so that the order of precedence in coal markets will be: (1) Power Stations, (2) Large industrial steam-raising, (3) Small industry/ commercial/domestic heating, (4) Coal conversion to gas or liquids.

This explains the present pattern of markets – why there is a large amount of coal sold to power stations, but virtually no coal liquefied. But what of the future? Two factors might distort the hierarchy set out above:

(a) Expansion of nuclear power, which might reach a point where the net-back on power station coal was below that in alternative markets. However, given the constraints on the rate at which nuclear power can be introduced, this factor is unlikely to be of widespread importance this century.
(b) Environmental measures which distort the market to such an extent that full or partial coal conversion is preferred to direct combustion.

On this analysis (and subject to environmental regulations) there would be a major move to coal conversion when two conditions are satisfied, namely:

(a) When there is insufficient oil and natural gas to satisfy even the premium uses of oil and gas.
(b) When there is still spare coal production capacity available after optimising the non-premium substitution of coal for oil and natural gas.

In a sense, this is the mirror image of the familiar argument that oil and natural gas should be concentrated to an increasing degree on the premium uses. The development of coal will thus be complementary to the restructuring of the oil and gas markets.

Thus, while coal conversion to liquids, gas and chemical feedstocks will be very important in the long term, during the next 20 years or so, the major contribution of coal will be in bulk combustion, particularly in power stations and large industrial boilers, where it will allow the use of oil and natural gas to be minimised. The scope here is very great, and can be illustrated by the fact that the use of fuel oil in WOCA is currently over 1000 M tonnes of coal equivalent a year.

Most of the expansion of world coal output will be in steam coal, and most of this will be sold to power stations. If oil and natural gas use in power stations is to be considerably reduced, without prejudice to economic growth, and after taking realistic account of the expansion of nuclear power, the use of steam coal in power stations in WOCA could well double by the end of the century – from 700 M tonnes a year to perhaps 1500 M tonnes a year by 2000. But the dominance of power stations in the total market for coal means that total coal demand will be heavily influenced by the long lead-times required to bring new coal-fired power stations into operation – typically up to a decade. In turn, this means that most of the coal-using capability for 1990 is already determined, and that a rapid increase in the use of coal in the 1990s will require substantial programmes of power station construction or conversion from now on. The probability is, therefore, that the rate of expansion of total coal demand will be more constrained in the 1980s than will be the case in the 1990s.

Coal Trade

In respect of international trade, coal is very different from oil. Whereas some two-thirds of oil produced in WOCA is traded internationally (mainly from OPEC countries), coal trade is only some 8% of world production – and that is mainly coking coal (particularly to Japan). Steam coal is traded internationally only to the extent of 2%.

Because coal expansion is in part in replacement of oil, it is sometimes assumed that it will come largely in terms of traded coal. This is an error of perspective. So far as steam coal is concerned, there may well be a ten-fold increase in trade by 2000, but even then home use would still predominate, as can be illustrated as follows (M tonnes):

Coal demand	*1977*	*2000*	*1977–2000*
Met from home production	2180	3900	+ 1720
Traded	50	500	+ 450
Total	2230	4400	+ 2170

If world steam coal trade expanded to 500 M tonnes a year by 2000, the main importing areas would be Western Europe, Japan and other rapidly developing economies in the Far East. So far as availability for export is concerned, it would appear that world steam coal trade would not be possible without very large contributions from the United States, Australia and (to a lesser extent) South Africa. Indeed, the United States might well be the balancing factor.

Policy Implications

It could not be claimed that coal alone can solve the world oil problem. It will need to complement nuclear power and energy conservation, and the restructuring of the markets for oil and natural gas. But every analysis of the world energy situation indicates that coal, as the most abundant fossil fuel, has a vital part to play in mitigating the world oil problem, and in particular by replacing oil in bulk combustion.

However, a number of policy questions emerge from the above analysis:-

(a) Given the dominance of US coal production in WOCA, will US environmental constraints on both production and use be overcome; and will the US allow a very significant expansion of their coal resources for export even when they may still be importing large quantities of oil?

(b) Can the means of using coal in power stations and general industry be speeded up? If not there is a danger that either (i) coal will not be developed until demand is evident; or (ii) the cheapest strip-mined coal will be developed, which will tend to displace deep-mined coal rather than oil.

(c) Can coal development be left solely to market forces, in view of the long lead-times and uncertainty on the precise path of future prices?

(d) Is enough thought being given to the relative scarcity of strip-mine reserves, and the consequent relative neglect of new technologies to exploit the far more abundant deep-mine coal reserves?

These questions need weighty discussion if coal is to make its full contribution and to do so within a reasonable period of time.

6 PROSPECTS FOR NUCLEAR ENERGY SUPPLIES

Peter W. Camp

I shall attempt to depict an unbiased picture of nuclear energy today, and bring into focus forces likely to play a pivotal role in the future. To summarise, we shall find that the nuclear contribution to world energy today is small, although growing rapidly. The potential future contribution, however, is very great – nuclear has the ability to play in the 21st century a role similar to that of oil in the 20th. Realising the potential of nuclear is not dependent on technical breakthroughs or resource constraints, but on the outcome of hotly contested political issues. Finally, I will propose that success of nuclear is consistent with the interests of oil-exporting nations.

Nuclear Perspective and Supply Projections

Let me start by placing current and near-term nuclear energy supplies in perspective with world energy demand. Today nuclear energy furnishes only a minor share of world energy needs. In 1978, 220 power reactors operating worldwide supplied energy equivalent to 3 M b/d of oil. This represented perhaps 2% of world energy consumption.

Small size does not imply insignificance. In the United States, nuclear electricity generation now exceeds that from hydro power. Recent energy supply shortfalls, only a fraction of nuclear output in 1978 were memorably disruptive. Still, in the context of world energy demand, the nuclear contribution has only begun to register.

Nuclear growth is more impressive than the current level of capacity. A decade ago the nuclear stations on line were labeled 'demonstration' plants and were hardly visible in energy statistics. Today, nuclear supplies about 13% of electricity in the United States and over 10% in Switzerland, Belgium, West Germany, Japan, Sweden, and Great Britain. By 1985 nuclear capacity for the world outside centrally planned countries (WOCA) is expected to grow to about 250 GWe from today's 119 GWe. In oil equivalent, the

1985 capacity represents 7 M b/d. By 2000, the probable range of nuclear capacity is 850 to 1200 GWe – about 24 to 34 M b/d oil equivalent – and more optimistic forecasts are not uncommon. The importance of nuclear energy relates only partly to the amount of electricity generated by today's operating plants. Of greater significance is the potential growth in nuclear energy supplies available to the world to meet future demand and offset depleting fossil fuel resources.

I should warn that forecasts of nuclear growth have proved notoriously unreliable. Virtually every projection made this decade of future nuclear capacity has been reduced as time proved it to be optimistic. For example, the US Department of Energy's predecessor, the Atomic Energy Commission, predicted a 1985 worldwide installed capacity of 640 GWe as recently as 1974 – more than twice the 250 GWe we now expect just five years later. Realizing the substantial potential of nuclear energy has been an elusive task.

In an effort to assess the validity of nuclear projections, I shall attempt to identify the forces acting to advance or retard application of the technology taking a look at both the nuclear record to date and the issues challenging future expansion. First let me turn to the potential of nuclear energy – in particular, the projection of 850–1200 GWe by the year 2000. How might this be achieved? Could it be exceeded?

1000 GWe

Obtaining an installed nuclear capacity of, let us say, 1000 GWe by the beginning of the 21st century is technically feasible on grounds of resource abundance and industry capacity.

The size of world resources of uranium has not been well assessed. Very large, rich uranium deposits have been blocked out in Canada and Australia in just the past few years. Exploration has been minimal in many potentially productive areas. Any conclusion on the extent of the economically recoverable resource would be hazardous. However, for discussion purposes, one could work with an estimate of 5 to 10 M tons U_3O_8. Water reactors, the dominant type with about 90% of committed capacity, require about 5000 tons of U_3O_8 per GWe for a 30-year life on a once-through cycle. 1000 GWe of capacity commits 5 M tons in the open cycle operating mode. This capacity does not put extreme pressure on the resource base, particularly when recent design improvements with potential to reduce uranium utilization up to 40% are considered.

However, a much greater nuclear expansion – say 2000 to 3000 GWe – if achieved with water reactors could exhaust world uranium and nuclear energy alike a few decades into the next century.

Fortunately we have another type of reactor – the breeder – which is much less demanding of fissile resources. A fast breeder reactor, such as France's Phenix, can extract 50 to 100 times the energy obtained by the water reactor described above from a given quantity of uranium. In fact, it could fuel water reactors with its net gain in fissile material. One breeder might fuel two water reactors, or use the fuel gain to start-up more breeders. If thorium resources were included as fertile blankets in the breeders, the reactor-fuel system could serve for centuries on expected resource levels of uranium and thorium.

The achievement of 1000 GWe in the year 2000 would not pose a great challenge to industrial capacity. Reaching 1000 GWe requires full utilization of existing reactor supply capacity of roughly 50 GWe/year. Mining output will have to approach 150,000 tons/year, a four-fold increase from current output. Enrichment capacity must expand correspondingly. None of this is overly demanding.

The difficulties of achieving 1000 GWe worldwide capacity by the year 2000 are not technical or resource based, but political and social. Will the social fabric accept this level of expanded nuclear energy presence? This is a central problem facing the development of nuclear energy and I will return to it several times in later discussion.

The Nuclear Record

Let us now turn to the performance record of the industry, one possible indication of how the future may shape. I will relate some of the successes and point to some trouble spots. Objective assessments of market activity, economic benefits, and operating performance suggest that nuclear has achieved extraordinary acceptance and is delivering highly favourable results to electric utilities and consumers.

Positive Factors. Power generation orders over the last 10 years have favoured nuclear. Worldwide utilities have ordered some 410 GWe of nuclear units, 33% of total 10 year commitments, and the leading market share by fuel type. This broad acceptance demonstrates the confidence utilities placed in nuclear energy's favourable balance of economics, reliability, environmental impact, safety, and fuel availability over competing generation technologies.

The expectations of public utilities about the profitability of nuclear generation have been amply realized. In fact, most utilities find that their operating nuclear plants actually produce greater savings than the estimates which originally justified the investment. Nuclear units in America have generated electricity at an average cost of 1.5 cents/KWh for the past 3 years compared to 2 cents/KWh for coal and almost 4 cents/KWh for oil (base load units). The net benefit to US consumers in 1978 exceeded $3 billion, assuming nuclear was replaced by coal/oil generation at average costs – at marginal cost, the benefit would of course be greater. Looking to the future, nuclear is expected to continue to be economically more favourable than coal in the United States by a 15 to 25% edge. In cases where strip-mined coal is transported over short distances, coal generation may be marginally more economical. In regions without abundant surface coal deposits, the more common situation, the economic comparison tilts heavily towards nuclear. In France, for example, nuclear is estimated to save about 40% over a fossil alternative.

Nuclear plant technical performance parallels the economic results. Let us look at the capacity factor, the fraction of rated output actually achieved, as a measure of operating performance. Nuclear units in the United States achieved 68% capacity factor in 1978 compared to 55% for base-loaded coal units. Historical numbers also put nuclear ahead, although by a smaller margin; 1978 was an excellent year for US reactors.

Negative Factors. Although the market share, the economics, and the performance of nuclear are creditable, the industry shows signs of distress. Construction costs are increasing rapidly, suppliers are financially troubled, and the accident at the Three Mile Island plant challenges the credibility of nuclear advocates.

The cost of nuclear plants has increased much faster than general price levels. The KW of capacity brought on line in 1970 at a cost of $150/KW is expected to cost $700/KW in 1980 – an annual increase of the order of 17%. But the 1980 unit is larger than the 1970 unit, and when economies of scale are considered, the escalation rate is more like 25% per year. At least three factors are responsible for the large cost increase beyond general inflation: longer construction cycles, growing materials intensity, and an uncertain licensing process.

The regulatory process – with frequent design changes, uncertain schedules, and endless documentation – is thought to be the most important driver of increasing plant costs. Regulation and licensing directly contribute both to the lengthening of construction cycles and to growing materials requirements. The general recognition of the causal relationship of unpredictable regulations to escalation of plant cost has led to numerous proposals for standardisation of designs and stability in regulations.

New regulations have sharply increased the materials quantities needed to construct a KW of nuclear capacity. Cubic yards of concrete and tons of structural steel have almost doubled per unit of capacity from the 1960s to the 1970s. Further cost impacts follow: materials costs have increased more rapidly than inflation, and labour productivity (manhours/kw) is reduced by the increased materials quantities.

During this same period (1960s – 70s) plant size increased from 500 MW to 1200 MW, and plant costs should have benefited from economies of scale. But as you have seen – and as our more detailed analyses confirm – price escalation due to new regulations wiped away favourable scale effects.

Lengthening construction cycles, particularly in the United States, have plagued the industry. Units starting up in 1978 averaged 11 years from order to commercial operation, against a 5–6 year cycle in the early 1970s. The average construction period in America appears to be increasing about 6 months each year, with delays in operation of new units being reported almost annually for every unit. Factors at work here include longer periods to obtain construction permits, the greater materials quantities mentioned above, and the more complex engineering job driven largely by regulatory changes.

Another indicator of distress is the financial health of nuclear suppliers. The nuclear supply industry has not participated in the economic benefits provided to consumers of electricity, and is financially troubled. In the United States, 3 of the 4 active suppliers acknowledge current losses. Two of these have never recorded a single year of profit. The one supplier reporting profits also acknowledges associated uranium losses which may approach $400 M after taxes. The suppliers of the HTGR recently quit the industry in America after accumulating $950 M in losses. Around the world, the situation is little different. In Germany, AEG-Telefunken exited with losses of $1.7 billion DM, almost $1 B at present exchange value. KWU acknowledges current losses. ASEA-ATOM is unprofitable in Sweden. In Japan, neither Hitachi, Toshiba, nor Mitsubishi are operating at breakeven levels. It will be many years before the industry approaches a cumulative breakeven position.

The outlook is for further unprofitable results. Orders for new reactors have been declining since 1974 and suppliers are shipping against shrinking backlogs built in earlier years. In the United States in the past four years, cancellations have exceeded new orders. No indications of a market resurgence in the near future have yet emerged. A considerable period of capacity underutilisation faces the industry, and a thinning of supplier ranks appears inevitable.

For the advocate of a non-nuclear future, the central lesson of the accident at Three Mile Island is to confirm that nuclear energy is too hazardous to be a major energy source. The health and safety impacts of the accident do not support this conclusion. No lives were lost and the maximum radiation dose inflicted on the population near the plant was insignificant. Some encouragement can be drawn from this result – no harmful relase of radiation – even though the accident resulted from an extensive chain of human and equipment failures. Three Mile Island suggests that reactors will tolerate considerable abuse from operators, and a water reactor meltdown might just be a fable.

Three Mile Island was a financial tragedy, and the lessons for the industry are numerous. Intense efforts are being made to understand the causes of the accident and identify necessary operational changes. Remedial actions introduced by the nuclear industry include: (a) Operator training to avoid errors which could initiate a similar event. (b) Instrumentation and controls to improve operator information and response. (c) Strengthened emergency procedures and communications. (d) Funding of institutions to remain vigilant against future accidents.

Perhaps the most painful outcome of the accident for the industry is the loss of credibility nuclear advocates have suffered. Many citizens, after hearing repeated assurances that nuclear was among the safest power generation technologies available, were shocked and frightened by Three Mile Island. The industry, partly in response to the attacks of the opposition, had begun to assert the perfection of nuclear energy – a most vulnerable proposition.

The Issues

What are the issues which will determine the future of nuclear energy? They belong to technology, economics, and politics. Compared to the political challenge, the technological and economic issues are relatively minor. I will discuss the easier technical and economic areas first.

Technical Issues. No fundamental technical issues are limiting the deployment of water reactors. The basic principles in all areas of nuclear energy production are well understood. Even so, it would be surprising if a complex, highly regulated technology involving by many orders of magnitude higher levels of radioactivity than typical human activities was free of technical issues. And nuclear is not. Active work is underway in areas of safety, materials adequacy, reliability, and fuel utilisation.

Safety. Reactor safety involves the ability to control and cool the nuclear core under all conditions. Both control and cooling are under study. Reactor control systems – the sensors and mechanisms which vary power levels – are designed with safe, reliable performance as an over-riding objective. General Electric's experience in control reliability illustrates the point: in 360 reactor years of Boiling Water Reactor experience, a reactor shutdown mechanism (control rod) has never failed to perform when actuated. Still, a concern exists that the primary control mechanism could be disabled, and the adequacy of fully independent secondary shutdown mechanisms is under scrutiny. It seems clear that design modifications will result from this investigation.

Once the reactor is shut down, it still contains a potent heat source – some 10^{10} curies of radioactivity. The reactor core must therefore be cooled under both normal and emergency conditions. Extraordinary efforts are being devoted to the design and analysis of emergency cooling systems. Extensive tests of system mockups with electrical or steam heating have been conducted, but nuclear heated emergency core cooling tests have only recently begun. The LOFT (Loss of Fluid Test) facility, in Idaho, is testing a nuclear core under increasing power levels and emergency cooling conditions initiated by a pipe rupture. LOFT is expected to confirm – or challenge – methods of safety analysis. Results so far suggest that reactor design methods are conservative.

Materials Adequacy. High integrity materials are essential to isolate fission products in the core and maintain the capability of cooling systems. Water reactors have experienced a recurring, nuisance level series of materials problems, often resulting from stress corrosion mechanisms. Fuel cladding failures of this nature have spawned extensive programmes to develop designs which resist failure. PWR steam generator tubing and feed-water piping and BWR primary piping systems have shown the need for increased corrosion resistance. Design and material changes are underway to resolve these problems.

Reliability. Replacement power costs for a large reactor running to $500,000 daily have accelerated efforts to increase plant capacity factors. Analysis and elimination of the causes of unreliability, techniques to speed refuelling and maintenance outages, and designs of retro-fittable equipment to eliminate problem components, all are evident in a general up-trend in reactor capacity factors over the past few years.

Fuel Utilisation. Nuclear fuel economics generally improve as the residence time of a quantity of fuel in the reactor increases. This is particularly true in the case of an open fuel cycle. As a result, work is proceeding to develop the technical bases for higher burnups of uranium and possibly even the use of thorium in water reactors. Improved fuel materials, as mentioned above, will be required as well as the development of new fundamental data on fuel behaviour in an operating regime not previously experienced. This issue, obviously, has political overtones since improved uranium utilisation may delay the time when reprocessing and breeder introduction are essential. I will return to this topic later.

Economics. Although the objective record shows economic performance to be one of nuclear's key strengths, questions can be raised about future economic competitiveness. How can nuclear be the least costly choice in view of capital costs increasing 25% annually and fuel costs up 5-fold since 1973? And what about the costs of waste disposal?

New regulations and the licensing process, the prime drivers of nuclear capital costs, have also been affecting fossil units. Air quality requirements, which in the United States require all new coal plants to include flue gas scrubbers, have had an adverse impact on fossil fuels capital and operating costs. Just as with nuclear, the licensing process has stretched out construction cycles and allowed financial charges to increase capital costs.

Uranium prices have increased relative to coal since 1973, although recently they have actually declined in real terms. However, fuel cycle costs are only 30% of nuclear generation costs and uranium is only 45% of the fuel cycle cost at today's price level. Other fuel cycle cost elements have increased more in line with price levels in general. So the uranium price impact on nuclear competitiveness, at less than 15% of nuclear generation cost, has been cushioned.

This relatively small fuel cost component is, in an inflationary context, an advantage for nuclear. In a long term inflationary environment, the uranium fuel + operating cost, running less than half the coal fuel cost, is a much smaller base to escalate. The major cost uncertainty with nuclear (plant capital cost) is settled within 10–12 years; for coal, the exposure lasts 40 years.

The costs of nuclear core waste disposal will be uncertain until the fuel cycle is better defined and the experience base expanded. Today spent fuel is held in water pools after discharge from reactors. In the future, either the residual fertile and fissile material may be extracted and recycled, or the spent fuel disposed of directly. Direct disposal, although improbable, is the highest cost approach. Many studies, government and private, have attempted to estimate these final costs. Some surprisingly large numbers have resulted – like the Department of Energy's $260/Kg. Reduced to the cost of electricity, this is less than 1.5 mills per KWh – nuclear is currently operating at a 5 mill/KWh advantage over coal. So it seems unlikely that waste disposal economics will be decisive. It appears, then, that neither technical nor economic issues will be limiting in the future supply of nuclear energy.

Political Issues. The challenge to future nuclear growth mounted through the political process is a more central issue. Nuclear energy has become a major political lightning rod. A Swedish government fell through resisting nuclear, a US presidential candidate makes nuclear opposition a centrepiece in his platform, and hard-fought ballot referanda have been submitted to voters. The political strength and success of anti-nuclear forces are evidenced by the manipulation of the licensing process, the cancellation of some plants, declining public support, and the debates over waste disposal and weapons proliferation.

Regulatory pressures can be reflective of public concerns – a vocal, minority portion of the public questions whether nuclear plants are adequately safe even though no member of the public has been injured in the history of commercial nuclear power. Unfortunately, the licensing mechanism is also exploited in the name of safety by some who simply oppose all nuclear development, or in fact any other energy development. An effective campaign of delay, of course, directly increases project costs and erodes the economic incentive to build the plant.

The scope of mischief in the licensing process is illustrated by NRC plans to clear TMI-1 (the undamaged reactor) for operation. NRC estimates are for 335 days of hearings, although the plant is ready for operation now. The commission is also considering, rather than summarily rejecting, hearing testimony on the psychological distress caused by the accident in the nearby population.

It further plans to consider granting financial aid to parties raising questions about psychological impacts on the population. As long as there is no limit to relevant topics, licensing will continue to be lengthy and foreboding.

Many reactors have fallen prey to these pressures in the recent past. Palo Verde 4 and 5, Stanislaus, and Sundesert plants were all cancelled because of the intransigent opposition of the California State Energy Commission to nuclear construction. The New York commission's similar position also precipitated a recent cancellation (Greene County). As a result, my utility company (PG&E) intends to build a coal-fired unit at a cost of $1700/KWe – much more than the $1300/KWe capital cost of a nuclear unit even before considering that the coal plant fuel cost will be double the fuel cost of nuclear.

California's action is not only contrary to public interest, it also runs against the 1976 initiative vote where the people of California favoured nuclear by 2 to 1. The ideologically biased public appointee who gains control of the regulatory machinery seems unlimited in his ability to frustrate the public will.

But public confidence in nuclear has more recently been eroding. Since the 1976 nuclear success at the polls, where seven states voted pro-nuclear, results have been mixed. Overall, public support as indicated by opinion surveys has drifted downward from earlier 70% levels. In general, nuclear support remains above 50% at present; still, the increase in public doubt is a real concern to the industry.

Anti-nuclear effectiveness is also seen in the debates over waste disposal and weapons proliferation. The political opposition asserts that these issues are fatal flaws inherent in nuclear energy, and they have exploited both issues effectively. Yet reality seems quite different.

The basic approach to high-level waste management – concentration, processing to an insoluble form, and isolation in geologically stable media – has been known for years. Individual parts of the process, such as conversion of liquid waste into borosilicate glass, have been demonstrated. Numerous independent technical bodies who have studied the problem almost universally concur in the basic approach and its feasibility. Those who consider the waste problem manageable include the Ford Foundation, the American Physical Society, and the National Academy of Science. The missing ingredient today is action by government bodies to move ahead with development and use of disposal sites. Without action to develop these sites, the industry lacks a firm planning basis and critics can keep the waste issue alive.

Another tactic exploited by the nuclear opposition is to link power reactors with nuclear weapons. The route to a nuclear weapon through the commercial fuel cycle has not been chosen by any of today's weapon states. Water reactors produce an inferior material for weapons. Despite this fact, it has become popular among some politicians to pose the spectre of either a rogue government or a terrorist group seizing plutonium in the nuclear fuel cycle for some dark purpose. Even though the commercial fuel cycle route to a weapon is improbable, it is feasible, and therefore must be safeguarded by appropriate agreements and verification techniques. Concern about this has led the United States to attempt to re-shape the fuel cycle – an approach unlikely to be successful. It is disturbing to see a country with an abundance of both weapons and energy resources acting to deny energy-poor nations the use of a fuel resource in the name of weapons which are not wanted by those countries. One hopes that INFCE – the International Fuel Cycle Evaluation study now being completed – will propose institutional mechanisms to defuse this issue through acceptable international material controls that will allow all countries access to fissionable material for power production.

Who are the people exploiting these issues in opposition to nuclear energy? Although no single mould fits all anti-nukes, the majority have some of the following characteristics: youth, relative affluence, distrust of technology and representative government, a belief in a simple, Utopian life-style, and frequently a history of membership in activist movements such as Vietnam, the environment, or consumerism. They generally oppose existing social structures. Their appeal, however, extends beyond their political base, and until this movement crests and recedes the outlook for nuclear is unclear.

Prospects

I have attempted to sketch a picture of nuclear energy as it is today. Although small in comparison with oil, nuclear technology holds great promise. Its expansion need not be limited by constraints of resources or technology. The performance of nuclear units has been, and continues to be, extraordinarily good in terms of economic benefits, reliability, and safety of both operators and the public. At the same time, however, the industry is deeply troubled. It is financially weak, faces a long period of excess capacity, and faces a determined political opposition.

I do not propose to assert which movement will be victorious. I hope my pro-nuclear bias has not obscured the negative side of the question for you. But let me attempt to put the question in perspective before you reach personal judgements on the future of nuclear energy. Let us look at the appeal of the anti-nuclear movement and the recent dynamics of utility markets.

The anti-nuclear movement seeks to gain support primarily with a message of fear. Fear of nuclear accidents, of waste, of weapons proliferation; fear of large corporations and of technological priesthoods. Few long-range public issues are decided in the end by fear, particularly if the potential benefits are perceived by the public. Time, truth, and the growing need for alternatives to fossil energy sources should defuse the appeal of the opposition.

Power generation markets have been generally depressed since 1974. Boiler shops and turbine factories are running well below capacity along with nuclear machinery production. The cyclical peaks in the market in 1972 and 1973, which saw generation orders of 120 GWe in the United States, were followed by a sharp reduction of the anticipated requirements for energy capacity, in response to higher energy prices, lower economic growth expectations, and calls for conservation.

This change in the dynamics of electric utilities, in America decreasing from an annual demand growth of 7% to 4–5%, left most utilities with excesses of generation capacity in operation, under construction, and on order. Reserve margins – the percentage capacity installed over the annual peak – grew to 35% in America, much more than the optimum levels of 15 to 20%. Utility rate control authorities, already under criticism for passing on fuel costs, were hesitant to grant financial returns which would be adequate to raise the capital for new plants which did not appear to be needed.

Many of the plant cancellations and delays were necessary adjustments to altered economic circumstances. Capacity requirements had declined and Public Utility Companies were not encouraging expansion. Utility behaviour was appropriate. And, as is appropriate for political movements and figures, credit was taken for events that were fortuitously consistent with their programmes. The anti-nuclear movement has claimed victory, but there is also a simple rational explanation for much of the distress of the nuclear industry.

Nuclear and OPEC

Finally, I would propose that the success of nuclear energy is in the interest of the oil-exporting countries. Since nuclear is the most economic option for base load electricity generation, its preferential use in industrialized nations will strengthen those economies. The national economy which selects nuclear will, in general, experience less inflation and more favourable economic growth than one opting for higher cost electricity. The result for oil-exporting nations is stronger markets with more stable currency.

Another advantage of nuclear is to extend the life of valuable liquid fuel resources. Nuclear can substitute for liquid fuels under boilers, and save those fuels for later, higher value use. This not only extends the period for financial returns from the resource, it also extends the life of investment in distribution systems and allows time for oil exporters to transform non-renewable assets into regenerable industrial assets.

Ultimately, when the petroleum resource approaches depletion, the world will have to change to energy sources such as nuclear breeding, solar, and nuclear fusion. We will move from low capital cost, high value, depletable fuels to high capital cost, low cost, renewable fuel sources. Nuclear energy is certain to play a significant role in this future energy mix. Nations whose future role entails either acting as a source of capital or as an exporter of industrial goods have a large stake in a smooth transition.

Recognizing the common interest of nuclear development and oil producers, I would not be surprised to see more nuclear power commitments by OPEC countries in the near future. These plants would, perhaps, be smaller than the units built in industrial countries and possibly serve both to generate electricity and desalinate water. The example of the exporting nations that they consider their liquid and gaseous fuels too valuable for boiler or turbine use will not be lost on the world.

7 SOLAR ENERGY: PROBLEMS AND PROMISE

John Goodenough

As the century of inexpensive oil draws to a close, the industrialized world reluctantly looks again to coal, but this time to tide us through the transition from a dependence on fossil fuels to a utilization of the two long-term energy sources: the atomic nucleus and the sun. The frustration of our present situation was well summarized at a recent Oxford Energy Policy Club meeting by a General Electric executive: 'Nuclear-fission energy has been shown to be economically feasible, but it is politically unacceptable. Solar energy has yet to be proven economically competitive, but it is politically acceptable.' My task is to summarize some essential strategies for the utilization of solar energy in the modern world, to identify the technical obstacles to their realization, and to attempt a prediction as to when the various solar technologies may become economically viable.

Since prehistorical times, man has acknowledged his dependence on the sun. It provides warmth, the energy for photosynthesis, and rains to refresh the earth. Differential heating of the earth's surface and the earth's rotation combine to produce wind and wave power, photosynthesis provides renewable food and biomass, rains are collected in natural basins to yield hydropower. The fossil fuels themselves represent chemical energy created by photosynthesis and stored over centuries in natural collection sites. Of particular interest today are man's technologies for capturing the energy from the sun and converting it to forms that can serve his demands for domestic heating, for industrial process heat, for transport, and for electric power.

Solar Radiation

Solar radiation is nonpolluting and, for all practical purposes, inexhaustible. The sun radiates as a blackbody at $T \approx 5800$ K, which means that it has a maximum intensity at a wavelength $\lambda_{max} \approx 0.5$

μm, corresponding to a photon energy $E_{max} \approx 2.5$ eV, see Fig. 1. Absorption by molecules in the earth's atmosphere produces a spectrum like that of Fig. 2 at the earth's surface. It arrives at the outer atmosphere at a rate of 1 kW/m^2. The mean insolation, averaged over a year, decreases with distance from the equator and with the cloud cover, but it is interesting to note that it is only twice as great in

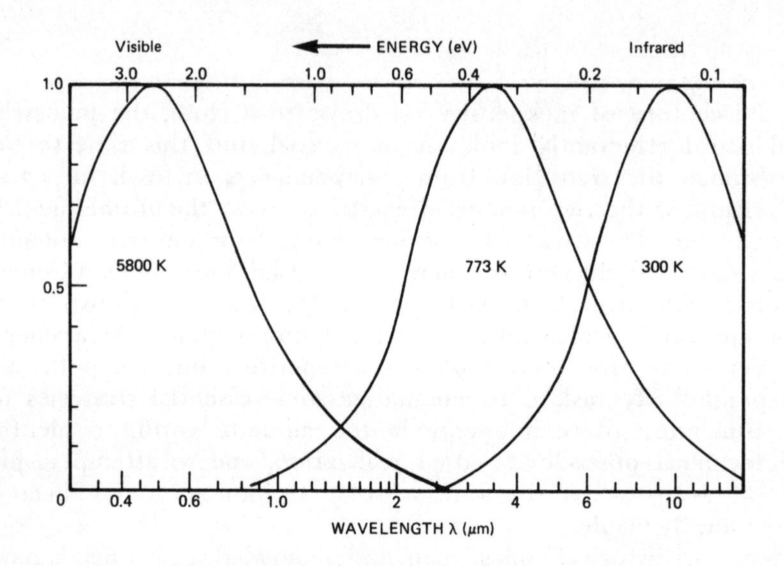

Fig. 1 **Normalized intensity of blackbody radiation from the sun (5800 K), from a hot absorber (773 K), and from a blackbody at room temperature (300 K).**

Egypt as in southern England, see Fig. 3. The land area devoted to roads and rooftops would be sufficient to supply all our energy needs were the incident solar energy converted at 20% efficiency. Therefore solar energy represents an adequate energy supply. Moreover, it is distributed, so that local collection/conversion eliminates energy-transmission costs. This feature is of special significance for an arid country with a sparse and scattered population. In addition, solar energy is directly convertible into heat, fuel, mechanical energy, and electricity. Despite these several advantages, sunlight has two properties that make economic exploitation difficult: (1) It is variable in time – diurnally, seasonally, and meteorologically, which makes energy storage an essential component of any solar-energy technology. (2) Although abundant, solar power has a low flux density even in an area of high insolation, see Fig. 3, which makes necessary large-area collectors and energy concentration.

Because direct sunlight is nearly parallel, it can be concentrated relatively inexpensively with mirrors or lenses before conversion. However, only a fraction of the sunlight reaching the earth's surface is direct, so concentration before conversion is only practical for special purposes (such as solar furnaces) or in arid areas having a high percentage of direct sunlight. In general, concentration is accomplished in the energy store after conversion.

An estimate of the economics of solar energy can be obtained by comparison with the nuclear alternative, neglecting distribution and storage costs. At 20% system conversion efficiency and a mean solar-power density of 250 W_e/m^2, a solar/electric power plant would provide, averaged over a year, 50 W_e/m^2 and require a collector area of 20 m^2/kW_e. If a nuclear reactor can be built for \$1000/kW_e, the solar/electric power plant must be built for \$50/m^2 or less to be competitive. This estimate provides a useful guide to the cost constraint even where conversion is to heat or to fuel rather than to electric power.

Solar-energy technologies would convert sunlight into four principal modes: (1) low-temperature ($T < 100°C$) heat, (2) high-temperature heat, (3) electricity, or (4) chemical energy (fuel or value-added products); they would convert wind or wave power to electricity. Each mode presents special problems.

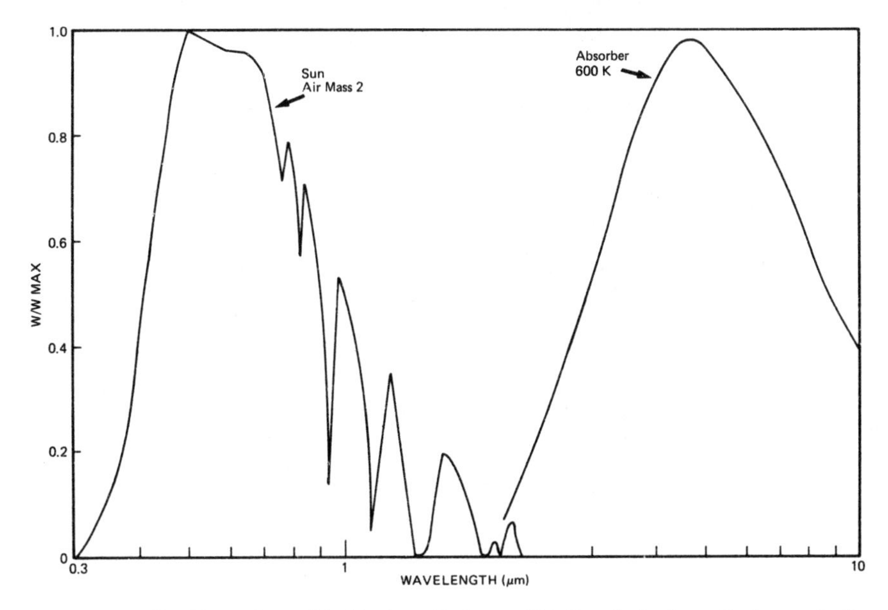

Fig. 2 Normalized intensity of blackbody radiation at 600 K and of solar radiation typically arriving directly at the earth's surface.

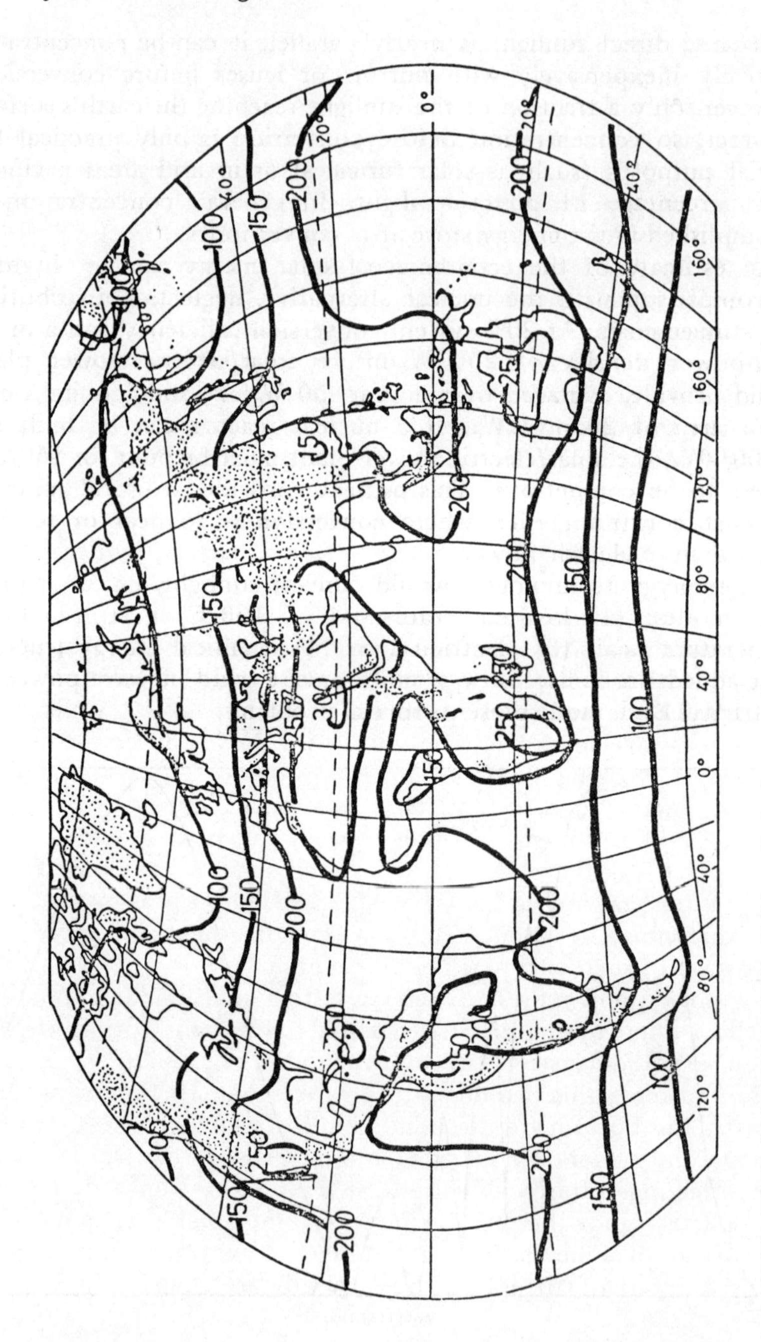

Fig. 3 Average mean global insulation on a horizontal plane at the surface of the earth (W/m^2) averaged over 24 hrs.

Low-Temperature Heat

Solar generation of low-temperature ($T \leqslant 100°C$) heat is an existing technology. Its most immediate application is domestic space and water heating. It is also used commercially for drying foodstuffs – either in the open air or more hygienically in closed solar driers – and for extending the growing season in northern latitudes. Domestic hot-water heaters have been used for many years in Cyprus, Israel, and Japan; glasshouses for growing plants have been in use for a longer time.

Passive systems can be distinguished from active systems. Each contain an object to be warmed by the absorption of sunlight and insulation to cut down heat loss. Heat loss occurs by three mechanisms: conduction, convection, and radiation. If the first two are minimized by conventional insulation, reradiation through the solar-energy inlet may be critical. This radiation loss can be sharply reduced by the use of wavelength selective surfaces either on the absorber or on the coverplate of the solar-energy inlet. Such a strategy takes advantage of the fact that blackbody radiation has an intensity spectrum with a maximum wavelength that varies inversely with the absolute temperature, see Fig. 1. The solar spectrum ($T \approx 5800$ K) is well resolved from the absorber spectra, which peak at longer wavelengths in the infrared. Active systems have, in addition, a working fluid and an energy store. The working fluid transfers energy to the store; fluid in a second circuit delivers heat from the store to the place of use.

A variety of 'flat-plate collectors' for solar generation of low-temperature heat are now marketed. These deliver heat to a working fluid that must be connected to a separate thermal-energy store. This aspect of solar energy is the most familiar.

The solar pond represents an interesting solution. It consists of a shallow pond having a black bottom for the absorption of solar energy, a liquid-density gradient to stabilize the pond against convection currents that would normally be set in motion by warmer water at the bottom of the pond, and a wavelength-selective cover that prevents evaporation and minimizes solar reflection while also reflecting infrared radiation back into the pond. Still water is a relatively poor thermal conductor, so the pond is its own energy store. Heat may be readily extracted via a working fluid in pipes at the bottom of the pond. Such ponds are under development and evaluation.

Because heat is not easily transported and can only be stored economically for up to a few days, it must be used locally and with little time delay. Industry exhausts quantities of low-temperature heat daily – a fact to which the giant cooling towers at power stations bear eloquent testimony – because of the difficulty of transporting it any significant distance. Low-temperature solar heaters are feasible only because of the distributed character of solar energy, which permits local generation of the heat where you want it. However, solar energy does not always supply the heat when you want it. For example, domestic heating is needed when the solar insolation is absent or minimal. In southern latitudes, the principal solar variation is diurnal, and inexpensive short-term stores – which use sensible heat in water or rock – can be used. In northern latitudes, on the other hand, seasonal variations introduce the need for long-term (months) thermal storage. In these locations, solar heating can only provide background heat during the winter months, and it may only be used to heat water during the summer months when it is most plentiful.

However, cooling with low-temperature heat is also possible, and this application is in phase with the maximum insolation. Coolers use the heat of vaporization of a working fluid, which is generally water. Open systems evaporate water to arid air. Closed systems use solar energy to evaporate water from a sorption medium such as a hydrated salt (zeolites may desorb over 15 weight-per cent water between room temperature and 100°C without any marked change of volume); they extract this thermal energy from the water vapour in a condenser and from the heated sorption medium by heat exchange to obtain hot water, and they extract heat from a cold store by reloading the cooled sorption medium via vapour transfer from the condensed liquid.

Solar energy can also be used to evaporate water from a dehydrating medium, thus conditioning the air in humid climates.

These technologies only await architectural innovation to provide the markets for their development. The opportunities for combined heating/cooling systems exist today in temperate, arid, and tropical zones. Many of these areas are undergoing rapid development, but unfortunately most of it with yesterday's energy-extravagant architectural designs.

High-Temperature Heat

Production of high-temperature heat for industrial processes requires concentration of the sunlight before conversion. If diffuse light is used, concentration of about 10:1 can be achieved, sufficient to provide steam efficiently. Where higher temperatures are required, mirror or lense focusing of direct sunlight must be used. An appreciable fraction of direct sunlight implies an arid climate, so this technology will probably be restricted to relatively isolated communities.

Development programmes have singled out electric-power generation as the test-bed for high-temperature technology. In this application, a heat engine is used to generate electricity and solar power is used to supply heat at the high temperature ($T > 500°C$) desired for adequate heat-engine efficiency. (The efficiency is proportional to $(T_h - T_c)/T_t$, where T_h and T_c are the hot and cold temperatures of the engine cycle.) In this application, mirror focusing is the more attractive method of concentrating the direct sunlight. Mirror-focusing schemes may be divided into two categories: single-axis concentrators that follow only the slow seasonal movement of the sun's trajectory and two-axis concentrators that also track the sun during the day. Clearly the latter must be automated, but they are capable of extraordinarily large concentration ratios.

Energy storage is critical to any high-temperature technology. This may be accomplished by thermal decomposition (pyrolysis) and/or electrical decomposition (electrolysis) of a compound into its elemental constituents. The heat of reaction $\Delta H = \Delta G + T\Delta S$ is nearly temperature independent, and the free energy ΔG supplied as electricity in electrolysis decreases with increasing temperature T. It becomes negative for $T > T_t$, where the heat $T\Delta S$ is able to decompose the compound by itself (pyrolysis only). Chemical recombination at lower temperatures ($T < T_t$) recovers the heat of reaction ΔH. The practical realization of high-temperature energy storage has not yet been achieved.

In my view, the generation of high-temperature heat from solar energy has limited application. Clearly this view is not shared by all, as can be judged from the vast sums of money being invested once again in engineering projects designed to demonstrate the economic viability of solar/thermal/electric-power generators. I believe direct solar/electric-power generation by photovoltaic cells will prove to be more competitive for most regions of the earth.

Electricity

Fig. 4 indicates the several routes by which electricity may be obtained from the sun. Wind or wavepower delivers mechanical energy that is converted directly to electricity. Since the supply is variable, energy storage is a necessary adjunct. Hydropower, already fully developed, uses natural collection systems to store mechanical energy behind a dam.

Ocean thermal gradients are created by solar melting of northern ice flows, solar heating of ocean surfaces, and ocean currents that bring the warm surface water over the colder water. This represents a constant energy source. Nevertheless, heat engines must be designed to work across a temperature difference $(T_h - T_c) \approx 20°C$, which implies large-flow heat exchangers resistant to fouling and storm damage in an open-sea environment.

Nature's other energy stores consist of chemical energy produced by photosynthesis. Replacement of the fossil fuels is the challenge we confront, and part of that replacement can be achieved by renewable biomass. Agricultural, silvicultural and urban waste represent energy stores that will be more fully utilized in tomorrow's economy. Although the development of 'energy farms' must compete with agriculture and silviculture, serious attempts have been launched to develop algae or other simple biological systems that have greater solar-energy conversion efficiencies than the *circa* 1% normally encountered in nature. Biomass is a traditional energy source that can be converted economically to high-performance fuels, and Brazil has a massive gasohol project based on sugar cane. However, land and harvesting costs will probably restrict this technology to the cycling of wastes or to regions where land and labour costs are compatible with energy farming.

The use of solar-energy absorbers to generate medium-temperature or high-temperature heat for driving a heat engine has already been discussed. Although a perennial contender, it will probably not be able to compete with the solar (photovoltaic) cell, which converts sunlight directly to electricity.

Photovoltaic cells utilize semiconductors or dyes in which absorbed sunlight excites electrons from a lower energy to a higher energy across a forbidden-energy gap E_g. The wavelength of the light required to do this is $\lambda < hc/E_g$, where h is Planck's constant and c is the velocity of light. The electric power delivered is $P = VI$, where the voltage V is less than the gap energy divided by the magnitude e_o of the electronic charge, i.e. $V < E_g/e_o$. From Figs. 1 and 2, it is clear that an $E_g < 2.0$ eV is required if a major fraction of the solar energy is to be utilized. An acceptable gap energy turns out to be $E_g = 1.5 \pm 0.3$ eV, and the voltage realizable from the cell is $V \approx 0.6E_g/e_o$. The magnitude of the current I delivered by the cell is proportional to the efficiency with which the excited electrons can be separated from the 'holes' they leave behind. This means that both the excited electrons and the holes must be mobile, each

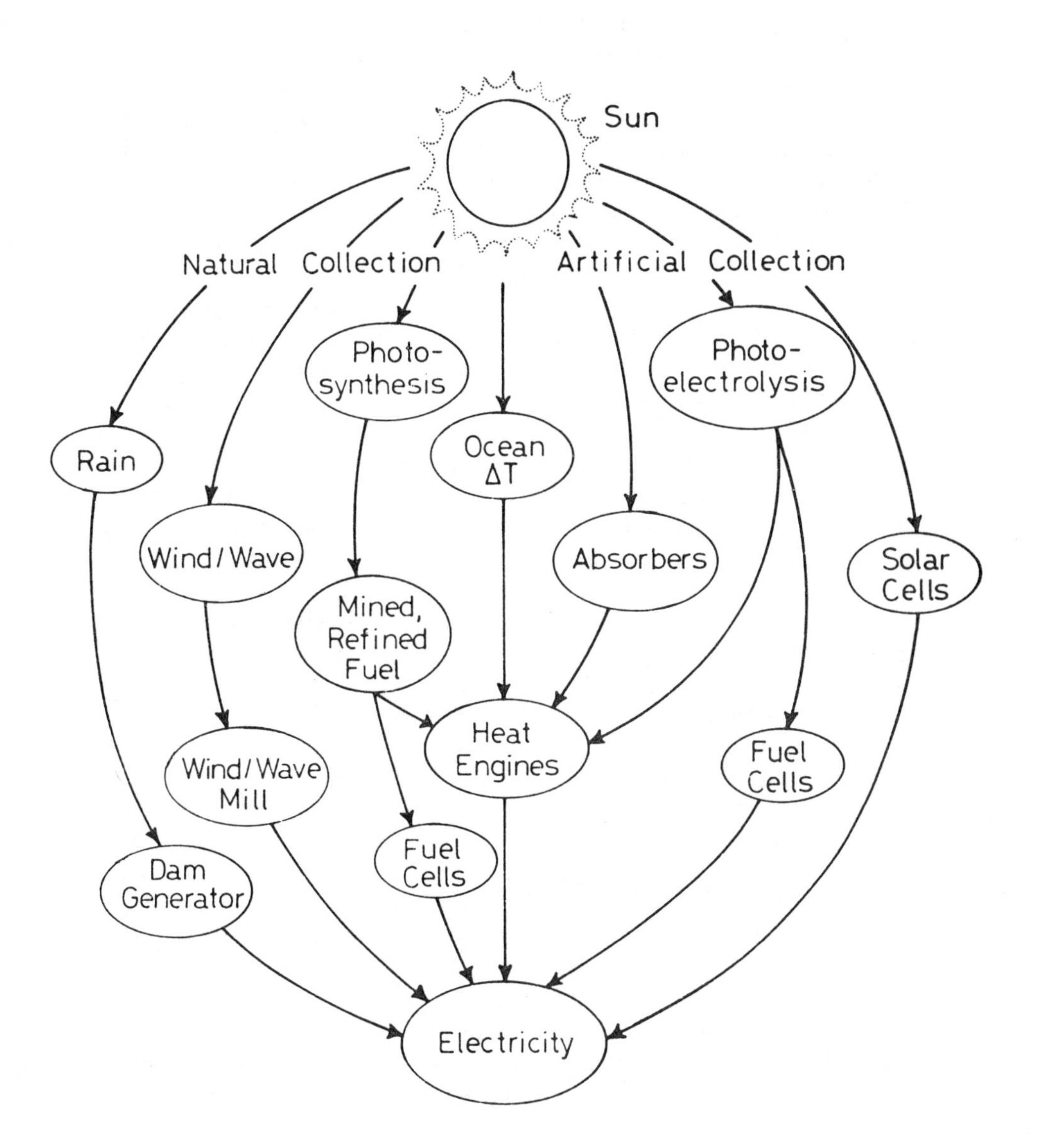

Fig. 4 Routes to the generation of electricity from solar power.

contributing to the current I. If either species becomes trapped at surfaces, individual atoms, or material imperfections, they not only fail to conduct electricity, but also scavenge the opposite species to produce electron-hole recombination. Recombination destroys the photogenerated charge carriers and therefore represents a 'loss'. Theoretical conversion efficiencies can approach 30%. Practical efficiencies of greater than 20% have been achieved with single-crystal semiconductors. Single-crystal cells are used for speciality applications such as space vehicles, but their unit-area cost has restricted terrestrial applications to isolated power stations where fuel deliveries are expensive. Nevertheless, the potential of direct solar/electric-power systems is so great that the massive world-wide effort to reduce the unit-area cost of solar cells without appreciable degradation of efficiency is well justified.

Electricity generated from a variable source requires a storage adjunct. The most versatile form of stored energy is the chemical energy of a fuel, and electricity may be converted reversibly to chemical energy with high efficiency. In its most familiar form, this is done in a secondary (rechargeable) battery. It may also be done in an electrolysis cell, the 'fuel' created by electrolysis regenerating electricity on recombination in a fuel cell. The fuel cell need not be located near the electrolysis cell since chemical energy is readily transported. Moreover, the chemical produced by electrolysis may also be utilized to produce heat, mechanical energy, or higher-value chemical products. Therefore electrolysis is an electric-power adjunct of great potential.

These considerations bring us to the direct generation of chemical energy by photoelectrolysis – as opposed to photosynthesis. This possibility is discussed in the next section. Before turning to it, I would call attention to a rather extravagant proposal to create a solar power station in space! The system would gain a factor of about 4 in mean solar insolation and simultaneously become a constant power source as it would be continually exposed to the sun. Photovoltaic cells would convert sunlight to electric power, and a microwave relay would beam the power generated to a receiving station on earth. It is difficult to imagine how such a scheme could be cost-effective, but the proposal keeps reappearing.

Chemical Energy

The chemical energy in a fuel represents the most versatile form of stored and portable energy. It is why we have become so dependent on the fossil fuels. Indeed, filling a tank with gasoline represents an extraordinarily high rate of system 'recharge'. Therefore, the conversion of solar energy to chemical energy – in the form of a fuel, a chemical feed-stock, or a value-added product – is a high-priority goal.

As mentioned above, nature accomplishes this goal through photosynthesis so one line of research is the development of simple biological systems that can perform efficiently the photosynthetic function of combining atmospheric CO_2 and water to produce oxygen (O_2) and hydrocarbons. Alternatively, electric power derived from a photovoltaic array may be used for electrolysis, and the splitting of water (H_2O) into gaseous hydrogen and oxygen is a prime electrolytic target. It requires a voltage of 1.23 V at standard pressure and temperature, so two photovoltaic cells in series (corresponding to two photons per electron transfer) may prove optimal. Hydrogen is an important chemical feedstock now derived primarily from fossil fuels. It may be combined with coal to yield the high-performance liquid or gaseous fuels now derived from oil and gas, or it may be used directly as a gaseous fuel storable in metals as a solid hydride.

Photoelectrolysis offers direct solar/chemical energy conversion. A conventional water-electrolysis cell consists of two electrodes immersed in an aqueous electrolyte. An external source of electric power, e.g. a photovoltaic array, creates the voltage between the electrodes responsible for splitting the water. Electrons pass from one electrode to the electrolyte to produce hydrogen; electrons pass from the electrolyte to the other electrode to produce oxygen. The electric current within the electrolyte is carried by mobile ions. In a photoelectrolysis cell, one (or both for two photons per electron transfer) of the metallic electrodes are replaced by semiconductors that are directly exposed to sunlight. The electrodes are connected by an electric shunt, and the voltage is supplied by photovoltaic action at the semiconductor-electrode/electrolyte interface. Photoelectrolysis of water by UV light has been demonstrated with inexpensive electrode materials, but it is proving difficult to find an equally inexpensive material responsive to visible light that is also chemically stable under the working conditions of the cell.

Homogeneous systems for the splitting of water by sunlight are also being investigated. In these, a complex in solution transfers an electron, excited by sunlight, to an aqueous solution – either directly or through another complex – to produce gaseous hydrogen, and the oxidized complex then captures an electron from the solution to produce oxygen. The two gases come off together in this process.

It is probable that inexpensive photovoltaic systems will be developed before a competitive photoelectrolysis system. A photovoltaic/electrolysis system would combine two known technologies, and concentration of electric power from a photovoltaic array should be less costly than concentration of product gases from a photoelectrolysis cell. However, the catalytic oxygen electrodes of the electrolysis cell also need marked improvement before a photovoltaic/electrolysis system can become economically attractive.

Summary

Solar-energy technology promises to play an important role in the next century, especially in countries having arid or tropical climates. Two conversion modes are singled out as particularly important: low-temperature heat and chemical energy stored in secondary batteries or a fuel.

Low-temperature technologies for heating, cooling, and drying already exist. Their development and market penetration only await imaginative architectural design and community planning. They are particularly relevant to regions of the world presently undergoing rapid building development.

The generation of chemical energy may be done indirectly via electric-power generation driving an electrolysis cell or charging a secondary battery. Alternatively, it may be done directly via either photosynthesis or photoelectrolysis. With the exception of waste biomass, which can be processed with existing technology, the most promising of these processes require scientific as well as engineering development. Therefore scientific research efforts continue to be needed to develop economically competitive solar cells, improved catalytic electrodes for electrolysis cells, longer-lived secondary batteries (particularly batteries of sufficient energy and power density for electric vehicles of acceptable performance), photoelectrolysis cells, and photosynthetic cells (either biological or synthetic utilizing biological components). By the turn of the century, one or more of these approaches should be proven in the laboratory and undergoing field testing.

Developments in solar-energy technology present excellent opportunities for indigenous industries in the developing world, but active participation in the science as well as the engineering of these technologies is needed for competitive innovation. University laboratories working in this field invite cooperative participation as well as financial support for their endeavours. The promise is for a piece of the action in a better tomorrow; the problems provide a worthy challenge to humanity.

Part II

ENERGY CONSUMPTION

8 THE DETERMINANTS OF ENERGY DEMAND

Robert G. Wilson

One could begin with the statement that the number of factors affecting energy demand verge on the infinite. A somewhat more useful observation would be that the principal determinant of energy demand in any one year, is energy demand in the preceding year. The reason, of course, is that energy demand is primarily dependent on the stock and efficiency of energy-consuming equipment in place and the habits we have acquired in using it. Capital equipment is relatively durable and tends to be used as long as available. To forecast energy demand in some future year raises the problem of how to determine what changes are likely in the stock of energy-using equipment, and the degree to which the equipment will be used.

The stock of energy-consuming equipment which exists at any moment in time is a reflection of income per capita and of the relationship of energy prices to other prices. (High income per capita and cheap energy prices entail a high energy/GNP ratio.) These two factors, relative income per capita and relative energy prices, provide a substantial part of the explanation why demand varies between areas (countries or sections within a country) at any point in time.

I thought I might begin my remarks by indicating to you some of the problems of definition and measurement in analyzing the two principal determinants of energy demand – capital stock and energy prices. I shall then discuss a range of factors – such as the composition of GNP and the climate – which have a significant impact on energy demand over and above the two principal determinants. I will note how these factors change over time, creating a need for more complex analysis as energy demand is projected into the future. I will conclude with a few notes on the price elasticities of energy.

Problems of Measurement

To begin with, we are faced with problems in measuring energy and then defining what is to be included in this concept. 'Energy' includes nuclear power, hydro power, solar power, and other sources which do not involve the actual combustion of fuel. Fuel which is burned is conventionally measured by its heat content expressed in Btus. Thus to measure nuclear, hydro, and solar on a comparable basis one must rely on some assumption about how this energy might have otherwise been obtained through conventional fuels. This enables us to convert the totals to oil or coal equivalents, the scale in which total energy demand is usually expressed. The problem of definition is usually whether to count non-commercial fuels. If, for example, cow dung and twigs are a major source of heat for an economy – then they need to be measured.

When we turn to making cross-sectional comparisons, examining how energy consumption relates to income per capita, the problems are several in defining and standardizing relevant measurements. When one country measures its national income in pfennigs and another in pounds, income must be converted to some common denominator. Exchange rates, of course, are the obvious choice, and yet these rates reflect only the relative prices of internationally traded goods. The relative prices of services, not generally traded internationally, may not be in the same relationship as traded manufactured goods, giving a distorted comparison of the true income per capita. That such exchange-rate conversions do indeed give a distorted picture has been amply demonstrated by a recent UN project. Examining what actually could be bought with per capita income, the UN found, for example, that in the UK purchasing power was actually 40% higher than a conventional conversion to US dollars would suggest. In Iran it was 143% higher. Lest you take too much comfort from the UN report, let me mention a Swiss bank study of what the wages of bank clerks would buy. They found such workers in Tehran to be only 38% better off than their incomes would suggest, while in London they were found to be 59% better off. The conclusions we draw from all this are that conventional conversion of GNP data using exchange rates can be highly distorting, but by precisely how much is difficult to say.

Another possible problem with the available measures of income is the existence of the so-called underground economy which, it is thought, has been flourishing in a number of countries where income is heavily taxed or welfare heavily tied to a means test. It has recently been suggested that as much as 7% of British GNP is not counted;

similar figures have been mentioned in the US. Likewise, the population itself is not completely measured even in countries very advanced in collecting data as the US. The US Immigration and Naturalization Service estimates, for example, that as many as 750,000 net illegal immigrants per year have swelled the country's population in recent years (1975–76) but have remained uncounted.

Even if we had a perfect series of income per capita, we would still be faced with the problem of how this income is distributed. With different distributions of income in different countries, the stock of energy-consuming equipment is also likely to differ. No one has yet come up with a really good statistical description of income distribution.

There are also problems in defining the price of energy. This is often taken to be an average, based on the price of each fuel consumed, weighted by the quantity consumed. Problems here arise particularly from the different levels of taxes that may be added to different energy forms or for different uses. With electric power, for example, we can find different prices to different users even from the same electric power company. Finally, recognizing we want to focus on the relationship between energy prices and the amount of energy consumed, we find that in residential use, the person responsible for purchasing the energy-consuming equipment, i.e. the home builder or apartment owner, may be interested only in the cost of the installation. He has only an indirect concern in the price of the fuel which will be paid by someone else – the occupant of the dwelling.

Other Determinants

In cross-section analysis (between countries or geographical regions within a country) there are a host of other factors which can have a significant impact. Climate is an obvious factor which explains why energy demand differs from one country to another. The rise of air conditioning, of course, has made it important to consider not only how cold the winters are, but how hot the summers are. (The spread of air conditioning in the US has been blamed – at least in part – for the unexpected increase in America's energy/GNP ratio in the late 1960s.)

Population density is another factor affecting energy demand. Dense population implies smaller housing per person, multi-family dwellings, public transportation, more walking, co-generation, and so on, all of which tend to bring about lower energy consumption per capita. (I do not mean to imply energy efficiency results in all cases. Regarding multi-family dwellings, for example, one should note that when heat or electricity is included in the rent, tenants have been found, not surprisingly, to use more compared to tenants who are individually metered. This simply illustrates the point that the price of energy affects consumption most directly when the consumer pays for it directly.)

The relative price of energy and energy-consuming equipment may be overshadowed in their effect on energy demand by other more fundamental costs. For example, construction and land costs influence the size of structures per capita. Insurance, maintenance costs, and excise taxes influence car ownership. (It is interesting to note that some analysts in the US auto industry anticipate that the rising cost of gasoline in the US will not put a damper in the long-term on enthusiasm for so-called recreational vehicles. This is because the cost of alternative vacations have still left 'camping' a relatively inexpensive vacation for many Americans.)

Governments have many legislative means at their disposal to influence the consumption of energy. The US auto emissions standards were tending to reduce the fuel-efficiency of US autos, while the more recent auto mileage requirements are now producing the opposite result. Interpreting these requirements has been critical in projecting the future path of US gasoline demand. Building codes, product standards, health and safety codes and speed limits are examples of other Government regulations which affect energy demand.

The composition of GNP in general is of particular significance for energy demand. The post-war era has seen international trade grow at a pace considerably above that of gross world product. Countries have increasingly concentrated on the production of those items which they do best while relying on imports produced by others for items they produce less efficiently. The industrial structure of different countries – which, of course, was never identical – has therefore diverged with some concentrating on energy-intensive products, with others producing less energy-intensive ones. Thus the export of a dollar's worth of steel or aluminium will have caused a much higher energy consumption than the export of the same value of electronics equipment.

It is clear that international comparisons of energy use in the industrial sector must be done at the industry level. And even then, *caveats* are necessary. For example, at a glance, the US paper industry, with its higher consumption of energy per unit of output, appeared in the early 1970s to be wasteful compared to that in Sweden. But, wood chips are burned in Sweden while they were processed into what was more profitable particle board in the US. The change in the cost of energy, however, has altered the economics, and today wood chips are being used as fuel in the US as well.

While the consumption of GNP is important in understanding the current demand for energy, we must also consider the age of the existing inventory of energy-consuming equipment and how rapidly – or slowly – it might be replaced. Europe and Japan lost a great deal of their national wealth during World War Two, and have grown rapidly since then. The US lost little, and has grown at a slower rate, and has an overall older inventory of capital goods. These differences make a comparison simply of general categories of assets less than fully indicative of how actual physical assets affect energy demand.

The importance of the historical growth rate is one which has not been studied in much detail and yet has intuitive logic. Let me spend a moment on an example. Consider two countries which I will call Rapid and Sluggish, and which each produced 20 houses. Assume that Rapid (29%) has been growing much faster than Sluggish (8%) such that 10 years ago Rapid built only 2 houses while Sluggish built 10. Over the period, Rapid will have accumulated only 80 houses compared with 140 in Sluggish. Furthermore, over half of Rapid's houses will have been built in the last three years, compared with only about one-third of those in Sluggish. If the incentives for conservation have been increasing over the years, Sluggish will be using much more energy than Rapid, not only because it has many more houses but also because more of them were built when the incentives for conservation were low. This also illustrates a point fairly obvious but not often made. That is, where energy costs are a factor in investment decisions, the current price of energy may be important. But what is more important is the perceived price of energy over the life of the asset at the time the asset was designed. Also, the rate of change in the stock of energy-consuming equipment can be influenced by the perceived rate of change in energy prices if an investor believes that, based on energy price, an asset should be replaced earlier than normal. It is these factors of change and perceptions of change which create part of the problem or challenge in demand forecasting.

Climate is about the only factor we have mentioned thus far which does not change perceptibly over time. However, even the way climate affects energy demand can be altered by the process of migration of people within a country, for example, the growth of the 'sun belt' in the US.

While some of the factors cited, like population density, and national wealth including the capital stock, may change only slowly, others – such as government requirements – can change very rapidly. Of particular note is that even though GNP changes slowly in total, at the margin it can change dramatically, and has in recent years.

Prior to 1973, industrial production in the OECD area tended to grow about 20% faster than GNP as a whole. In recent years, however, this relationship has generally reversed, with industrial production growing only about half as fast. While it is normal for industrial production to fall off relative to GNP during a recession, the ground lost is normally made up during the recovery. So far, however, this has not occurred.

It appears that an important factor behind the lag in industrial production has been the decline in steel production, a very energy-intensive process. In the US, Europe, and Japan steel production in 1977 was approximately 15% less than in 1973. While the rest of the world did substantially better over this period, total free world production was still down 10% between 1973 and 1977. We estimate that had steel production kept pace with GNP, energy demand in 1977 would have been some 2 M b/d oil equivalent higher than it actually was.

How long and how far this shift away from steel will go is difficult to judge, and certainly one must take into account the course of production of products such as plastics and aluminium which themselves are quite energy-intensive, and which can be substitutes for steel. The fact of the matter is that we appear to be in a period of distinct transition, in which the composition of GNP is changing significantly at the margin as the world realigns itself to the interacting forces of many changes.

The problem of projecting the growth in energy demand is, therefore, complex, and we at Exxon are not yet ready to trust the job to an econometric model. Indeed, our analyses are ranging over a wider and wider field on a more and more microscopic level.

Elasticities

This detailed approach to the determinants of energy demand is no doubt as exhausting to read about as it is to utilize. While this approach is employed for making so-called base case projections, we need other tools for developing a feel about movements in future demand. Price elasticities and econometric models are very handy tools to have around for this purpose.

Elasticities with respect to price are by definition the effect on demand of a change in price assuming no change in GNP. But in fact, the price of oil – and energy in general – does appear to affect GNP. The generally small GNP penalties that have been identified by most analysts do not measure the impact of higher oil prices on the standard of living of consumers in the oil-importing countries. As a higher oil price requires that a greater proportion of GNP be devoted to paying for oil imports, less output becomes available for home consumption.

While we have seen figures quoted for both short-term and long-term elasticities, their time frame is often unspecified. The time frame becomes relevant in dealing with the possibilities for new technology and/or new energy supply sources which can invalidate earlier elasticity calculations. Problems also arise from the difficulty in distinguishing between demand adjustments which are independent of GNP changes and those occurring as a result of GNP changes (income effect).

As I have indicated, the elasticities we have tested are only rough indicators. We use them to estimate the effects of price changes (import fees, petroleum product excise taxes) on our base case results. We have more confidence in short-term than in long-term elasticities. The latter, of course, are affected by the rate of turnover in the capital stock. As slower economic growth brought on by higher energy prices means a reduction in the turnover rate of the capital stock, or an extension of its average age, the rate of improvement in the efficiency of energy-consuming equipment is impaired. Thus, the rapidity with which an economy can adapt to higher energy prices is reduced.

Conclusion

In closing, however, I think it is interesting to note that some analysts believe that the production of oil, which has been the major swing or marginal source of energy, may be approaching a plateau. With lead times for developing alternative supplies being very long, the days of projecting energy demand *per se* may be numbered. Instead, the job may be to come to terms with a price which will bring demand into line with available supply. As you can appreciate from our discussion, short-run elasticities are much smaller than – perhaps no more than one-fifth to one-tenth of – long-run elasticities. Thus, if a surge in demand were to come up against a constrained supply, substantial price adjustments might result with only modest supply increases and substantial negative GNP effects. In these circumstances governmental intervention might substitute for part of the price response, reducing the relevance of past market behaviour employed in defining econometric models and the micro-analysis lying behind our energy projections. Should such a situation arise, government policies will be straining to try to accommodate often conflicting energy, economic, and environmental goals.

9 INDUSTRIAL ENERGY CONSUMPTION AND POTENTIAL FOR CONSERVATION

Walter Murgatroyd

This paper is intended to clarify the most important ways in which industry uses energy; indicate some of the most important energy conservation options, in the short and medium term; and suggest why industry is not responding to the need for conservation as quickly as it could.

Attention has been concentrated on major conservation options; to do otherwise would require a detailed industry-by-industry survey which is beyond the scope of this paper. Data are taken from United Kingdom sources but are broadly applicable to industrialised countries in West Europe and the US.

The Uses of Energy in Industry

Of the four major sectors – industry, commerce, domestic, and transport – into which we commonly group energy users, the industrial sector is usually the largest energy user in those countries which are to any significant degree industrialised. Fig. 1 gives an approximate analysis of the relative energy consumption of the sectors in the UK for the year 1978 – these proportions are also typical of most industrialised countries although, of course, there are small variations from country to country. Fig. 2 presents an analysis of the primary energy consumption of the UK industrial sector in the year 1976. You will see that there is no single dominant industry. The three largest consumers – Iron and Steel, Engineering and Non-ferrous Metals, and Chemicals (which includes fertilizers), each take approximately 7% of the national primary fuel consumption.

Although the ways in which industry as a whole uses energy appear at first sight to be bewilderingly large, it is in fact possible to classify them into a small number of basic categories which have many characteristics in common across the industrial groups. The following classification is useful for our discussion.

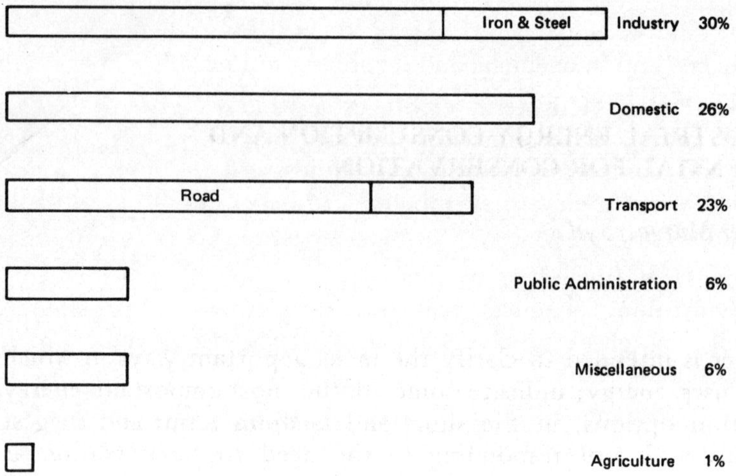

Iron & Steel Industry 30%

Domestic 26%

Road Transport 23%

Public Administration 6%

Miscellaneous 6%

Agriculture 1%

Fig. 1 Energy Consumption – UK 1978

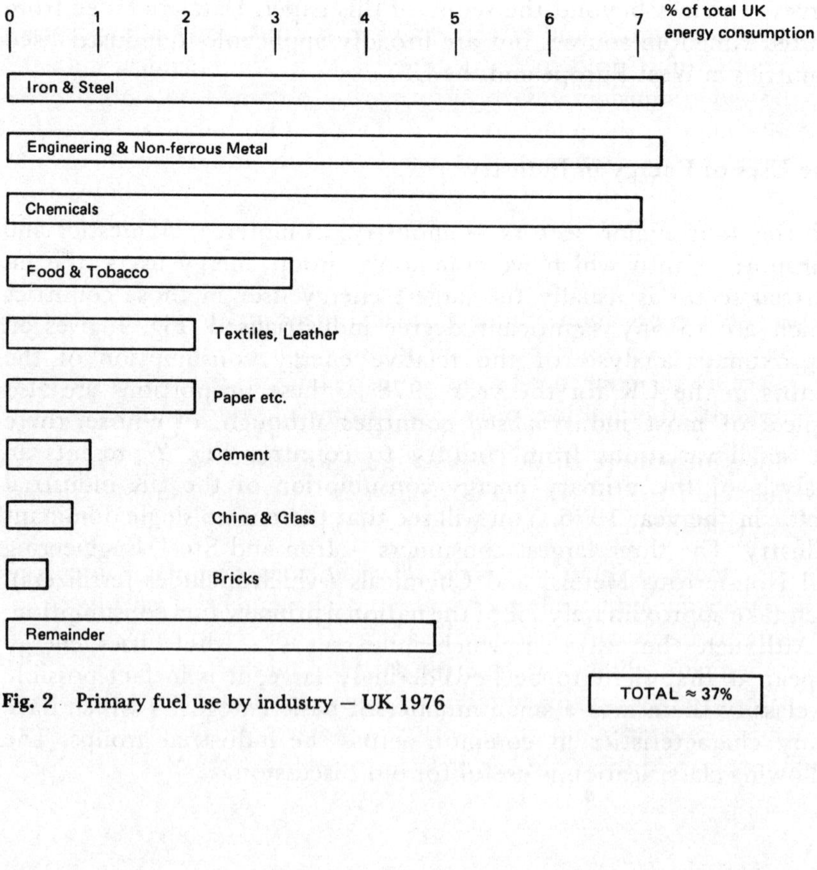

0 1 2 3 4 5 6 7 % of total UK energy consumption

Iron & Steel

Engineering & Non-ferrous Metal

Chemicals

Food & Tobacco

Textiles, Leather

Paper etc.

Cement

China & Glass

Bricks

Remainder

TOTAL ≈ 37%

Fig. 2 Primary fuel use by industry – UK 1976

1. Useful Energy Retained in Product. This is usually chemical energy and is only significant in the cases of secondary fuel producers, and in chemical and fertilizer production.
2. Motive Power. This term embraces a wide variety of devices which among other functions operate machine tools, pumps, fans, and compressors for gases, liquids and vacuum systems, conveyors and elevators, mixers, crushers and grinders, and control valves.
3. Process Heat (including cooling). Used for example in initiating and controlling chemical processes, drying textiles, paper etc., melting (e.g. glass, metals, chemicals), curing bricks, and cooking and sterilizing food.
4. Space and Water Heating (including cooling). Used to provide an acceptable working environment for personnel.

This classification is useful, because within each class the methods for producing the energy form (e.g. motive power, process and space heat) often have a common technology even though the end use varies. For example, the effectiveness with which low-pressure steam is generated and used in, say, some parts of the chemical industry can be aided considerably by information obtained on similar components in, say, the paper industry. This can be important because, compared to the domestic or transport sectors, industrial energy use proves extremely difficult to analyse in detail. We thus know considerably less about this sector than about other sectors. There are several reasons for this difficulty, including the following:

– as there is no one dominant use (cf. space heating in the residential sector or road transport in the transport sector) it is not possible to concentrate on a single analytical model such as domestic thermal loss model or automobile performance model;

– it is much more difficult to obtain statistics on the stock and characteristics of industrial equipment than for example on our stock and types of houses or automobiles;

– it is more difficult to make detailed measurements of energy flow in an industrial environment. Management is usually reluctant to allow any instrument to be fitted or any measurement to be made which will interfere with production, and in countries with strongly organized labour unions, the unions do not always understand or sympathise with the motives of anyone observing or measuring the activities of their members.

In spite of these difficulties we are gradually learning more about industrial energy flows and about the potential for improved efficiency.

Industrial Motive Power. Approximately 90% of industrial motive power is supplied by electric drives, the remainder (mostly for large compressors) is supplied directly by steam turbines. Because of the necessarily low conversion efficiency from heat to motive power (about 30%) the primary fuel requirement is significant, and exceeds 10% of the total industrial fuel consumption for all purposes. Practically all the electric drives are induction motors which, although efficient and cheap in themselves, operate at a constant speed, and one which is often higher than that required. They have to be matched to the low-speed and variable-speed demands of industrial processes. Our investigations have shown that the matching and control problems lead to serious inefficiencies in electricity use. It seems that, on average, approximately 65% of the electricity supplied to the motors is lost in the drives and control (this is approximately 6.5% of the total national energy consumption for all purposes!). There is clearly enormous scope for a better technological solution to the problem of industrial drives. We should also bear in mind the capital intensiveness of electricity supplies; the 65% losses referred to above imply that unnecessary extra generating, transmission, and distributing capacity has to be provided, the cost of which may be far larger to the nation than the cost of a more efficient electric drive.

Process Heat. It is customary to divide process heat supplies into two categories – Direct Fired and Indirect Fired. Direct firing is used principally when very high temperatures are required (e.g. in brick and pottery firing) or when the fuel is also reacting chemically to form a product (e.g. in the iron and steel and cement industries). Electrical heating is usually classed as direct firing.

Because of the difficulties of handling the hot material many high-temperature direct-fired processes are batch processes in which the material is placed in a furnace or oven which is then ignited. When the process is complete, the energy supply is discontinued and the product is allowed to cool. Usually virtually all the sensible heat within the product is wasted, losses occur from the walls of the furnace and, except in the case of electric heating, the heat in the flue gases is usually also unrecovered, although in some industries (glass melting for example) some of the heat in the flue gases is used to pre-heat the combustion air. Generally the intermittent nature of the operation makes it difficult to arrange for the rejected heat to be used for other purposes, nevertheless studies carried out by the UK Departments of Industry and Energy suggest that relatively simple modifications – modifications which would be paid back in a year or less – could lead to a 3%–5% saving in energy use. To save significantly more than this would require major changes in design or technology.

Indirect process heat is almost always supplied by steam generated in a central boilerhouse and transmitted by pipes to the processes. This permits the use of a larger and more efficient central boiler to serve a number of processes, and smoothes out the demand so that the boiler may run under steadier conditions. In principle any fuel can be used, the choice being dictated by the cost of using it and of meeting any imposed environmental constraints.

During the two decades 1950–70 there was a marked swing from coal to oil for industrial process steam production. Since 1970 coal consumption has continued to decline but has been replaced by natural gas where this was available, otherwise by oil. Changes in the UK process-heating fuel consumption are shown in Fig. 3. A broadly similar trend away from coal is found in the US. The greater convenience and flexibility of oil and gas, and the fact that the valuable space formerly occupied by coal storage and handling equipment can be put to productive use is making industry very reluctant to return to the use of coal – even in the US where coal is cheap and where there has been government pressure on industry to do so.

Some of the energy waste associated with the use of steam appears to be easily reducible. Pipe insulation in existing plant is usually not optimum, either because it was incorrectly designed or because it was optimized for an era of cheap fuel supplies. Industrial boilers tend to be badly maintained, leaks of hot water and steam are not attended to. Surveys made in the UK suggest that quite simple measures, which could usually pay-off within one year, could conserve 5%–10% of the energy supplied in many installations. Increased boiler efficiency could often be obtained through the use of a number of smaller boilers instead of one large one, because this could enable the plant to follow the steam demand with greater efficiency.

Space and Water Heating. Most of the energy here is used for space heating and in the UK and northern Europe forms a surprisingly large part (16%) of the total industrial primary fuel consumption. The proportion varies greatly between industries, being greatest in those

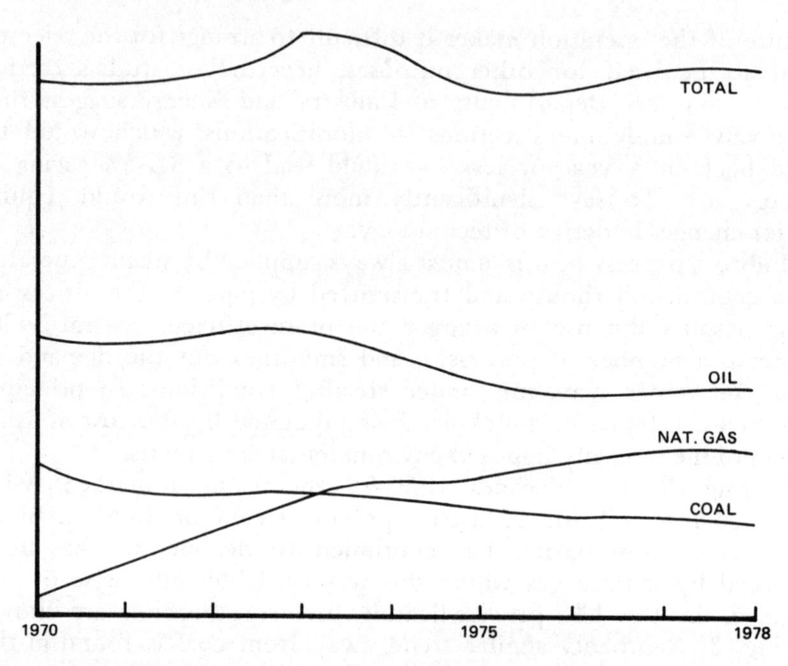

Fig. 3 UK industrial fuel consumption (excluding electricity)

– such as engineering – which have a large labour force and smallest in the large highly automated complexes such as primary chemical manufacturing plants.

Within the UK most factories are very badly designed and operated from the point of view of space heating costs. Walls and ceilings are badly insulated, large entrances are left permanently open, ceilings are high so that the heated air rises and remains near the ceiling leaving the working level cold. Often, with a properly insulated building, the waste heat from processes and machinery would be sufficient to provide heating. These facts have all been established for the UK by the Department of Energy, and are confirmed by detailed studies at individual sites carried out by Imperial College (London University).

Table 1 shows a summary of the approximate energy supplied to UK industry for the purpose discussed above. The electrical energy has been converted to a primary fuel equivalent using the average generating efficiency (29.4% including transmission losses) for the year. The process steam supply has been analyzed into two components – that supplied at a temperature below and above 180 °C – the reasons for this will be apparent later.

Table 1 UK 1976 – Fuels used for major industrial end requirements (excludes iron and steel)

	Solid	*Oil*	*Gas*	*Elec.*	*Total (PJ)**	*Total (MTOE)***
Motive Power	–	22	–	633	655	14.9
Process Heat						
1. Directly Fired	87	117	172	204	580	13.2
2. Steam						
a) <180 °C	51	177	88	–	316	7.9
b) >180 °C	14	39	17	–	70	1.6
Space and Water Heat	37	186	67	25	315	7.2
			TOTAL		1936	44.8

*Peta Joules
**Million tons oil equivalent

Fuel-conservation Opportunities

We have already noted some ways in which energy can be saved by quite simple methods in relation to process and space heating, and have mentioned a significant opportunity for motive power supplies. Within each industry and within each firm there are many specialized opportunities which are only now being authoritatively studied, and which cannot be described here. There are, in addition, some general strategies, of which I will mention two.

Process Cascading. This refers to a strategy of siting industries so that energy (especially heat) rejected by one can be used by another. Large multi-product chemical plants use this concept widely, and in fact transfer not only heat but material from one product line to another. Opportunities for cascading energy in this way are limited by the need to match the temperatures and flows. They arise more easily when steady continuous production is involved. Cascading of batch processes normally requires some form of energy storage device or very careful phasing of the processes. In short, the energy gains from cascading are offset to varying degrees by the increased need to co-ordinate processes, by the loss of some flexibility, and by the increased cost.

Industrial Co-generation. Generating electricity from fuel is an inefficient process in the sense that only 20%–40% of the useful energy in the fuel is converted into electricity, the remaining 60%–80% appearing as low temperature heat. There are sound scientific and technological reasons for this state of affairs. When there is a demand for heat and electricity they can be produced together, with an overall efficiency exceeding 80%, and it is clearly of major importance from the point of view of energy conservation that this should be done where possible. The technique is often called co-generation (sometimes, in the UK, Combined Heat and Power, and in France, Force-Chaleur).

Any of the three conventional types of power plant – steam turbines, internal-combustion turbines (gas turbines) or internal-combustion engines (e.g. diesel engines) – can be used as co-generators; it is simply necessary to ensure that the heat being rejected by the unit is at an appropriate temperature and that the unit is so situated that the cost of delivering heat and electricity is acceptable.

In the case of the steam turbine, when it is designed to produce useful industrial process heat as well as electricity, the electrical generating efficiency suffers considerably. Typical industrial back-pressure turbine co-generating units have electrical generating efficiencies in the range 12%–18% compared with 30%–40% for conventional steam turbines. This greatly reduces the effectiveness of steam-turbine co-generation as an energy efficient option. In this sense it is unfortunate that in the industrialized nations, for technical and historic reasons, electricity and heat supplies have developed separately and that electricity suppliers have adopted the steam turbine technology.

The gas turbines and diesel engines do not have this disadvantage; industrially useful heat can be produced from them without significantly impairing their efficiencies as generators of electricity. Taking account of the temperatures at which it is possible to extract heat from gas turbines and diesels, they could, in principle, supply process and space heat at temperatures not exceeding approximately 200 °C, i.e. approximately 35% of the total industrial energy demand.

From the point of view of energy conservation a useful way of comparing the co-generation options with conventional arrangements is to consider the supply of one unit of useful industrial heat from a co-generating plant, to evaluate how much electricity is generated in producing this unit of heat and how much fuel is consumed. One then evaluates how much fuel would be consumed if the same amounts of useful industrial heat and electricity had been supplied in the conventional way, i.e. by an industrial boiler and a central power station. Taking the difference one can evaluate the overall fuel savings due to using a co-generator instead of the conventional method. Table 2 shows the method, using the steam turbines, gas

Table 2 **Fuel saved by co-generation**

a) *A1 Steam-turbine co-generation*		
Industrial heat supplied	1	unit
Associated Electricity output	0.24	units
Fuel consumed	1.59	units
B2 Conventional supplies		
Industrial heat supplied	1	unit
Boiler fuel used	1.25	units
Central Power Station electricity	0.24	units
Fuel consumed	0.68	units
Fuel saved by choosing A1 = 1.25 + 0.68 − 1.59 = *0.34 units per unit of industrial heat*		
b) *A2 Gas-turbine co-generation*		
Industrial heat supplied	1	unit
Associated Electricity output	0.42	units
Fuel consumed	1.83	units
B2 Conventional Supplies		
Industrial heat supplied	1	unit
Boiler fuel used	1.25	units
Central Power Station electricity	0.42	units
Fuel consumed	1.19	units
Fuel saved by choosing A2 = 1.25 + 1.19 − 1.83 = *0.61 units per unit of industrial heat*		
c) *A3 Diesel co-generator*		
Industrial heat supplied	1	unit
Associated Electricity output	0.93	units
Fuel consumed	2.64	units
B3 Conventional supplies		
Industrial heat supplied	1	unit
Boiler fuel used	1.25	units
Central Power Station electricity	0.93	units
Fuel consumed	2.64	units
Fuel saved by choosing A3 = 1.25 + 2.64 − 2.64 = *1.25 units per unit of industrial heat*		

turbines, and diesel engines. The overall savings for each of the three options is summarized in Table 3.

Although from the fuel conservation viewpoint the steam turbine is the least favourable option, most industrial co-generating installations employ steam turbines. The reasons for this are worth looking at more closely in the context of energy policy. First, if we examine the requirements for low temperature heat and electricity in different industries we find that the electricity/heat ratio generally lies between $\frac{1}{5}$ and $\frac{1}{3}$. The steam turbine matches the lower end of this range and can burn most kinds of fossil fuel.

The gas turbine is cheaper than the steam turbine, but except where natural gas is cheap, it requires a more expensive fuel. It is

Table 3 Summary of possible co-generation options compared with conventional alternative options. (In each case, the basis is one unit of useful industrial heat.)

	Electricity Produced	*Fuel Saved*
Steam Turbine	0.24	0.34
Gas Turbine	0.42	0.61
Diesel Engine	0.93	1.25

able to produce about twice as much electricity for a given heat output, but this is often more than the factory requires. The diesel or gas engine is a more extreme case. It can produce approximately four times as much electricity as a steam turbine unit, but very few factories require so much electricity. Thus a factory investigating co-generation for its internal requirements of heat and electricity would often find that the best options were either to employ a steam turbine (if its needs were large enough) or not to co-generate at all. If, however, it had an assured market for the surplus electricity which could be produced from a gas turbine or diesel and if it could be paid at a rate approaching that offered by public electricity suppliers, the situation would be very different; co-generation would be a cost-effective option of wide application which could make a significant contribution to fuel conservation.

In most industrialised countries the public electricity supply is generated and transmitted by large organizations which enjoy monopoly or near-monopoly status. They are either state-owned or are regulated in some way by the state. The protected conditions in which they operate make it extremely difficult for a private generator to sell his electricity at what would generally be regarded as a fair price. The issues involved in determining a fair price are complex; within Europe much of the expertise and most of the required information is in the hands of the public utilities. I have been privileged to spend most of the past year looking at the institutional constraints on co-generation in several countries, but time does not permit me to enlarge on this topic. May I simply, instead, quote from a statement expressing the point of view of the Commission of the European Communities made to a Committee of the French Assemblée Nationale?

. . . it very soon becomes clear that the subject which is by far the most important, which has the largest potential for realizable economies is that of the combined production of electricity and heat

Combined production of heat and electricity by industry has had great difficulty in developing in some countries precisely because of the extremely rigid barriers which separate public producers and distributors of electricity from industrial co-generators.

The French Government has taken an important step by compelling Electricité de France to purchase co-generated electricity and by setting out the tariffs which must govern such purchases. The US Government has also passed an Act designed to encourage co-generation; the situation in that country is extremely complex because of the interaction between federal and state legislation, the different legislation between States and the existing environmental legislation. Those countries which are in the early stages of industrial expansion, and in which institutions have not yet become entrenched, have a good opportunity to arrange their affairs so as to profit greatly from co-generation.

Incentives and Obstacles for Fuel Conservation in Industry

It is very difficult to determine, from the macro-statistics available, the extent to which industry as a whole is responding to the developing fuel situation. Certainly in the US and UK, energy consumption per unit of product is beginning to fall, but the effects of new energy conservation efforts may be masked by the fact that production is not rising as fast as expected (or is even stagnant) and this has enabled (or forced) older and less efficient plants to close down.

In examining the incentives to industrialists to conserve fuel it is instructive to examine the relative importance to various branches of industry of energy costs. One way – a way which is adequate for us today – is to examine the energy cost as a percentage of the monetary output of the industry. This is shown for the UK in Table 4 which reveals that for Iron and Steel it is a major cost component, that for Chemicals and Building Materials it is large but not dominant, while for the last four branches, which between them contribute 80% of the total industrial production, the percentage is small.[1]

Although well-run organizations will always be seeking to reduce costs and although all reductions are equally acceptable, whether they stem from savings in energy, manpower, raw-materials or elsewhere, nevertheless the greatest managerial efforts, in terms of numbers of staff and of incentives, will naturally be devoted to those

[1] The energy costs shown in Table 4 are the direct costs. Energy costs are also associated with the other inputs, especially with the inputs of materials; thus the total energy cost component, which determines the long-run influence of energy prices on costs, is significantly higher than the direct costs. Nevertheless, the direct costs are the ones which are perceived as under the direct control of management and which determine the attitude to energy conservation.

Table 4 Typical industrial energy costs as a percentage of monetary output 1976 (rounded to nearest percent)

Iron and Steel	30
Chemicals	20
Building Materials	20
Textiles	10
Food	5
Engineering	5
Paper	5
Other Industries	5

Note: Food, engineering, paper and other industries account for 80% of total industrial production.

areas which are the major contributors to cost; and for 80% of industrial production, energy is not one of these areas! This fact may explain why, whereas in planning new production facilities, or in scheduling operations, quite sophisticated methods are used to optimise expenditure between raw materials, manpower requirements and capital equipment costs, energy costs are not usually included in this optimisation – they are simply accepted and added as an overhead. Energy supplies are usually handled by maintenance staff who are (or who are alleged to be) so much occupied with maintenance and refitting that they have no time to deal with many minor energy-conserving matters even though, as pointed out earlier, the pay-off is quick. Fortunately this position is beginning to change – firms are now appointing 'energy managers', though it remains to be seen how much power they have to effect changes in organisation.

Another factor which an industrialist has to consider is the high cost of losses in production due to failure of equipment or of energy supplies. The benefit of an innovation which saves, say, 20% of the firm's energy bill (a large percentage) and thus perhaps 1% of the total annual cost, may be entirely lost if it leads to a few days loss in production. In fact if significant teething troubles develop, the firm might lose its position in the market and may not survive. The effects of any innovation have thus to be estimated very conservatively. In our studies of motive power we met several cases of this. For example, the designer of the production line for a new automobile component will specify his machine-tool cutting rates very conservatively, i.e. he will over-specify the speeds really required, thinking, perhaps, that if the component is very successful he may be asked by management to increase production. The machine tools he specifies will usually be supplied by a specialist firm which in turn, in order to avoid any risk that the specification cannot be met, will adopt very conservative motor ratings. The motor manufacturer in turn will adopt a similar procedure. Thus the drives, when finally installed, will spend most of the working week operating inefficiently at perhaps $\frac{1}{3}$ or $\frac{1}{2}$ of their rated output. This problem of multiplication of design margins is a real one and a difficult one to resolve whenever a sequence of manufacturers is involved. It is one which should be remembered whenever industry is criticized for not being sufficiently responsive.

If we turn to the other extreme, to the energy-intensive industries, we find problems of another kind. These industries are usually well aware of energy costs, which form a major parameter in the design of new production facilities. They also happen to be the industries which employ large sophisticated and capital-intensive plant which may take 1–4 years to design, construct, and commission, and may have a planned life of 12–20 years. In making industrial investments of this kind, perhaps the most difficult problems to deal with are those caused by uncertainty of events over the whole lifetime of the project. During the design phase, the designer is faced with a whole range of choices which extend from major options such as which overall process to choose, to relatively minor ones such as how efficient (and therefore how expensive) should certain heat exchangers be. In most cases he has to make a choice between different combinations of capital cost, labour cost, materials cost, energy cost, reliability, and flexibility. In making this choice he has to provide the best answers he can to questions such as:

- how much will construction costs escalate during the construction period?
- at what cost will the firm be able to raise capital?
- what will be the rate of inflation during the lifetime of the plant and how will the relative prices of materials, labour, and energy change?
- will an economic stagnation or a recession, initiated, for example, by sudden changes in the price of energy, materially affect the size of the market, and if so, how will this affect the performance of the plant?
- what new plants are my competitors likely to introduce? Might they optimise their plants differently from me and if so, and if their plants fortuitously turn out to be able to respond to unexpected changes in conditions better than mine, will they corner my share of the market?

The more uncertain management is about the answers to questions such as these, the less willing it is to take risks (and this is especially true of large organisations) and so the less able it is to respond to the needs of the future – a good example of the effects of uncertainty can be seen in the oil-from-coal programme in the US.

Looking ahead in a more general way there is of course no shortage of energy supply to the earth. The solar input in its various manifestations far exceeds our current or expected future requirements. The problem is that those energy resources which in the past have been cheap and convenient – and plentiful in relation to past demands – are becoming scarce in relation to current and future demands and thus will inevitably become more valuable and expensive. The renewable energy sources, which are abundant, require considerable capital and time to exploit and if, during the transition period, economies are seriously perturbed by, for example, rising and uncertain fuel prices, there is a danger that we may not have the spare resources to accomplish the transition. The industrial sector is the one which is most sensitive to changes in the economy and especially to uncertain futures, yet it is this same sector which must in the long run provide the technology and the hardware for the transition to a wider energy base. We may liken the situation to that of a man fallen overboard in a rough sea and trying to inflate his life-jacket. If he has to stop this task too frequently in order to fight the next wave, he will eventually lack the strength to accomplish either task and will perish.

Industry is responding to the changing fuel situation. There is evidence that some of the short-term measures are being initiated. Longer-term measures – new production techniques, co-generation or a new product mix for example – require above all a period of economic stability, which in turn necessitates a long period of energy price stability, not necessarily of fixed prices, but an agreed future energy price structure in which confidence can be established.

10 ENERGY CONSERVATION: OPPORTUNITIES, LIMITATIONS AND POLICIES

Richard Eden

I shall assume that the reader is already familiar with the basic ideas of energy conservation including examples of historical or recent achievements, and technological possibilities of what might be done in the future. Such possibilities do not represent realistic opportunities for energy conservation unless they become associated with sufficient motivation for them to be developed and widely adopted.

Technical potential can provide an estimate of the resource base for energy conservation – and as with other resources some estimates are reasonable and some are ridiculous, but only those conservation options that are economically viable represent an energy 'reserve'. Improvements in technology may move conservation possibilities from the category of resources to the category of reserves, or they may make the reserves more attractive economically. However, there is uncertainty about the rate and extent of the adoption of energy conservation. Additional motivation is required to bring reserves into production – they have to be needed, and normally they would be developed in sequence beginning with those having lower cost or better pay-back.

Even when conservation opportunities are developed and adopted, they may merely represent part of the historical trend, and it may be necessary to run hard in order to keep up with past achievements. We therefore need to ask about energy conservation policies – who are the policy makers and how much might they achieve? In this paper I will discuss three aspects of energy conservation, namely, motivation, options and uncertainties, and policies.

Motivation

Fossil fuels, particularly oil and gas, are an exhaustible resource. This idea is easy for the public to grasp and is of obvious importance politically and socially: it leads to the view that energy conservation is desirable on ethical grounds in order to save fossil fuel resources for future generations. In economic terms this view would suggest that a low discount rate should be used for judging investment in energy conservation.

From the viewpoint of the government of an energy-importing country, the exhaustible resource view means that energy conservation has popular support in general terms – though the support may dwindle when particular measures such as increasing fuel prices or taxes are considered. The consumer government will be influenced by other factors, such as:

(1) Oil import costs and balance-of-payments problems. These become especially disturbing if the world price of oil is unstable and liable to sudden and unexpected increases.

(2) Avoidance of risks such as the interruption of oil imports. Even a partial interruption causing local shortages in a consumer country can have a serious political impact, and the risks of excessive dependence on oil imports may also inhibit foreign policies or have strategic implications for defence policies.

(3) Vigorous and effective energy conservation policies in oil-importing countries will not only help to stabilise longer-term expectations about world oil prices but it will provide evidence (call it a threat if you like but it is really a normal economic response) that excessive increases in the world price of oil will be counter productive, not only through causing or contributing to recessions and consequent reductions in the world oil demand but also through stimulating more rapid substitution and conservation.

Governments in oil-producing countries, like the public in consuming countries, are in favour of world energy conservation in general terms – 'oil is too valuable to burn', 'oil is our most valuable resource and depletion rates must be limited so that it is available for future generations'. Oil exporters are insistent on the need for energy conservation in consumer countries so that world demand is compatible with their own objectives to conserve oil in the ground. Yet, taxes on oil products in consumer countries (perhaps the most effective way to encourage conservation) are regarded as unfriendly if not hostile, as if the consumer government is taking from OPEC countries some of the taxes, royalties, rent or profit, that they might otherwise receive. This is a misunderstanding. Consumer government taxes, for example on gasoline, represent an internal redistribution of taxes. If automobile tyres wore out at a uniform rate the same result could be achieved by taxing tyres! The fact that the motorist is willing to pay more for gasoline through extra taxes does not mean that the additional payment could have gone to OPEC. The limits on OPEC oil prices are set also by the national ability to provide more exports to pay for oil imports and by their effect on world trade and economic growth.

Other reasons for governments of oil-exporting countries to favour world energy conservation include:

(1) Conservation will help world economic stability through reducing the chances of sudden increases in the world oil price. It is not clear that these are a disadvantage to oil exporters, but there is a risk that they would increase the chances of instability – both political and economic – in the world regions containing oil-exporting countries.

(2) The members of OPEC have an important role in aiding developing countries, many of which are dependent on increasing their imports of oil if they are to maintain a reasonable level of economic growth.

The needs of developing countries, comprising three-quarters of the world population, provide the most convincing argument for moderation by OPEC and for energy conservation, especially conservation of oil, in the developed countries which use nearly 90% of world oil supplies. Concern about exhaustible resources points to the needs of future generations, but the developing countries can point to the present needs of generations already born. Developed countries, it is said, can use their wealth and high technology to achieve a high level of energy conservation and to introduce energy technologies alternative to oil, whereas developing countries are dependent for growth on technologies of medium complexity that require the increased use of oil.

This argument does not absolve developing countries from encouraging energy conservation. This is particularly true of OPEC developing countries where oil products are often sold at prices that bear little relation either to the world need for conservation or to the long-term interests of those countries. In the poorer regions of the developing world where wood fuel is becoming an increasingly scarce resource it may become necessary to change from traditional methods of burning wood to simple but more effective alternatives. In these regions the world price of oil will be sufficient incentive for governments to seek ways of reducing the rate of growth in oil demand whilst maintaining economic growth. However, there is little doubt that if the developing world is to achieve reasonable levels of growth relative to their increasing populations they will need a greater share of the world's oil. We therefore return to the need for energy conservation in the more developed countries and to the potential for reducing oil demand in OECD countries which together use nearly 90% of the oil in the non-communist world.

This need for conservation is recognised by most governments but there are at least three major classes of obstacle to changing these good intentions into actual achievements. First, the governments themselves have other priorities besides conservation, for example the control of inflation, reductions in unemployment, other investment needs, and other balance-of-payments problems. Higher energy prices, or higher taxes on oil products, may stimulate inflationary wage demands, investment in energy alternatives to imported oil may themselves be energy intensive and lead to higher imports in the short to medium term, the encouragement of more economical cars through higher gasoline taxes could worsen a country's balance-of-payments position through increasing the imports of smaller cars or cars with diesel engines.

The second major obstacle to the implementation of energy conservation is that governments do not themselves use much energy. Energy consumption and hence decisions on energy conservation are widely dispersed through the community, so governments must rely on persuasion and possibly compulsion based on a variety of policy instruments involving fiscal measures, regulations, publicity, and support for research and development. The difficulty here is that there is considerable uncertainty about the results that can be expected from any such measures. If the pay-back is uncertain, the arguments for government intervention are weakened.

The third obstacle to conservation comes from the priorities of the many individual decision makers. The private householder considering retrofit of insulation or double glazing will note that he expects to move house within a few years and such subtleties do not observably influence the sale price of his house. The same point may be made by the housebuilder, though he might alternatively decide to use high insulation as an advertising aid. The purchaser of a new car may rate prestige, size, speed, comfort, or reliability as highly as he rates economy, particularly if the car is to be used for business purposes. The industrial manager is generally more experienced at assessing the value of other investments than those for energy conservation, and in any case he is likely to have higher priorities such as maintaining the cash flow or the working capital, urgent replacements to keep production lines running, expansion of the business, or the more efficient use of labour.

I will turn next to a discussion of some energy conservation options and the estimation of their possible consequences, and finally will return to the question of government policies.

Options and Uncertainties

The most popular option for energy conservation is the technical fix. In its more extreme form it provides a miracle solution to the energy problem in which no sacrifices are required – you can have your cake and eat it.

In assessing the potential of energy conservation through technical change it is essential to look first at past improvements in technology, partly because they give an idea about what may be possible for rates of change and better energy efficiencies, but also because past changes are built into historical trends and past relations between energy consumption and economic activity. The central theme of the technical fix philosophy is that we can do better in the future than in the past – maybe we can, prices and costs of energy will be higher, but we must be cautious about double counting. Double counting is especially likely in estimating industrial energy conservation potential, where technological change is identified and then estimated on the basis of an instant change giving (say) a potential reduction of 30% in energy demand. It is then said that such a reduction could perhaps be achieved over a 20-year period. However, many industries have historically achieved a 1.5% annual improvement in their energy-output ratio, so a 30% reduction over 20 years is no more than the trend which would be obtained by econometric analysis. For example, in the period 1950 to 1976 the energy-output ratio (measured in physical units) for iron and steel production in the UK fell by nearly 30%. This trend can be continued but it would be unrealistic to suppose it can be greatly improved on. Electricity generation in the UK has improved the average conversion efficiency from fossil fuels to electricity from about 8% in 1920 to about 22% by 1940 (unchanged to 1950), then to 31% today. This trend will not continue since further general improvements are limited by a combination of thermodynamics and requirements for reliability. In special circumstances combined heat and power is economic and gives higher efficiencies but in the UK this is unlikely to have a large effect on the 'system average'.

The second option for conservation that I wish to discuss is one that is sometimes popular with economists but not with the general public, namely to allow market forces to achieve an energy supply-demand balance. In a sense our discussion here is part of the market response – we are anticipating higher energy prices and asking (some of us are asking) whether action on conservation can help to limit any increase. However, the energy market need not be a free market and governments may intervene through fiscal measures and regulations. The key question to government, consumer, producer, and planner is the expected level of response to higher energy prices. What is the short-run energy price elasticity, and what might it become in the long run? Unfortunately there are many answers – the price elasticity depends on the fuel that is used, it depends on the sector of consumption, and it will certainly vary from one country to another and depend on the average level of income, social framework, and economic structure in each country.

Estimates of energy or fuel price elasticities vary widely, but for illustration let us consider the consequences of an increase in the average price of energy supplied to the final consumer in a developed country, on the assumption that the short-run energy price elasticity x lies between 0.2 and 0.4 and the long-run elasticity y between 0.4 and 0.8. For the simplest model, the change in energy demand Q in the short- and long-run respectively, will be given by

$$Q_{SR} = P^{-x} \quad \text{and} \quad Q_{LR} = P^{-y}$$

where P denotes the ratio of the new energy price to the old price, and Q is the ratio of new energy demand to the old demand (Q_{SR} short run, and Q_{LR} long run). With our assumptions about elasticities, if the average energy price doubles (P = 2), we obtain,

Q_{SR} lies in the range 0.87 to 0.76 (savings of 13 to 24%)
Q_{LR} lies in the range 0.76 to 0.57 (savings of 24 to 43%).

Individual companies, for example, oil companies transporting and refining oil, and chemical companies, have reported energy savings of about 20% in the six years since 1973. It is difficult to estimate the real increase in the price of energy to these companies but if they are mainly dependent on oil their costs of a unit of energy may have doubled compared with their other costs (P = 2). If the 20% saving was interpreted as a short-run price response this would imply an elasticity x = 0.32, but if we attribute 6% of the saving to a long-run trend giving 1% per annum improvement from technological change, the price-elasticity x would be only 0.2.

This illustrates one of the difficulties of estimating or using long-run price elasticities, namely the extent to which they include technical change, and it raises doubts about the relation between elasticities evaluated by cross-section analysis and those obtained using time series. Two additional problems should be mentioned. First, the question of how long is 'long-run'?; obviously it depends on the character of the energy use and whether it is for automobiles, dwellings, or industry. Secondly, the real price change P for a consumer should be measured relative to the change in the costs for energy conservation. The latter depend on the types of conservation being considered. In industry, early action on energy saving involved some engineering management time and low capital expenditure, but later action may require both more detailed engineering design work and higher capital expenditure. The costs of some capital equipment for energy conservation could move faster than energy prices if there was an excessive increase in demand perhaps as a result of an energy crisis or over-exuberant government intervention.

The third option on energy conservation that I shall discuss is government intervention. This is almost as popular as the technical fix option and is often associated with it, but it is really much wider and its popularity begins to wane when such specifics as price increases, energy taxes or emergency energy rationing are mentioned. Questions about intervention concern policy options, to which I will turn later, but here I want to note some of the essential uncertainties in government action on conservation. There are several levels of uncertainty: first, we need to know what actions government might take, then which options will be chosen and the level (for example of regulations on thermal insulation, or taxes or subsidies) and timing of decisions or legislation that may be required. Finally, and this is often ignored, we need to estimate how much would be achieved from any specific government action, and when particular levels of energy saving will be achieved. For example: housing regulations that require better thermal insulation may have little effect on the large fraction of new dwellings that has traditionally been constructed above the minimum standards. Houses designed to have low energy requirements may involve the occupiers in such low energy bills that they become increasingly casual about leaving windows open particularly if their incomes rise in real terms, and energy costs (or each individual fuel cost, since they are perceived separately) remain a relatively small fraction of total expenditure. Alternatively, energy saved through living in a low energy house may release funds for more travel, either by car or by air.

The uncertainty about the possible achievements from government intervention is compounded by the wide variations in the capital expenditure per unit of energy saved and the fact that different decision makers have different discount rates so their perceptions of the costs of saving energy vary widely. Industrial energy savings during the initial good-housekeeping stage often had a pay-back time of less than one year, but now some industrial energy managers have lists of projects awaiting financial approval and/or the availability of engineering services, each of which would give a pay-back in less than two years or better than 35% annual return. At the other end of the spectrum one has householders investing in double glazing giving a financial return from energy savings of only a few percent. The industrial manager would first observe that there are other higher priorities for capital expenditure and secondly that there is uncertainty about the pay-back period, whereas the householder would point to other incidental benefits from double glazing. Governments may use discount rates of between 5 and 10% for energy conservation investment, but there are formidable obstacles in translating this into action by industry through subsidies or regulation even if it was thought to be economically and politically desirable.

This brings me to my final two points on energy conservation options and uncertainties, namely saturation of demand and the estimation of future demand and conservation potential by activity analysis. The latter examines energy-using activities such as driving a car or heating a house and obtains future demand by estimating future levels of these activities and the efficiencies with which energy is used to perform them. Thus the total annual use of energy by a fleet of cars may be obtained from a product of three factors:

(total number of cars) × (average annual distance per car) × (average energy used per kilometre).

Estimates for future years can be made for the number of cars allowing for saturation of ownership levels, and of mileage trends and average fuel consumption efficiencies. These estimates may take account of changes in incomes and prices, and also of technical improvements in vehicles, possibly in response to government regulations rather than prices or to actual fuel shortages experienced during recent or future energy crises, and also possible changes in lifestyles that may affect the estimates.

We can try to estimate the uncertainties of results using activity analysis. The number of cars per person in the US is about 0.5, but this average conceals the variation between States (in 1970 New York averaged 0.30, Mississipi 0.33, California and Colorado 0.47, and Nevada 0.48), so there is potential for further increases. The corresponding figure in the UK is 0.26. There is uncertainty about saturation levels – times series analysis gives different results from cross-section analysis. Western Europe could saturate at 0.40 or it could reach 0.50 cars per person. In addition the rate of approach to saturation depends on economic growth and other factors, and population projections introduce further uncertainties. Typically for the year 2000, there could be an uncertainty of plus or minus 7% from a middle estimate (say 24 million) for the number of cars in the UK. This is a feature of energy conservation if the question of lifestyles is brought into the discussion. The average mileage per car involves a similar uncertainty of plus or minus 7% about a median projection (say 15,500 km) for the UK in the year 2000. Average efficiencies are harder to estimate – there was little change in the average kilometres per gallon for the UK car fleet in the period 1955–75 being about 50 km per imperial gallon, at a time when the price of gasoline was falling in real terms. It takes about 12 years to renew the entire UK car fleet and one might assume between 0.8 and 1.6% annual improvement giving a saving of between 15 and 28% by the year 2000. Thus on this factor also we have an uncertainty of plus or minus 7%.

The three factors in the energy used by cars are not entirely independent but it would be optimistic to take the low estimates for all three factors and pessimistic to take the high estimates; however if this was done the high estimate would be 1.23 times the median and the low would be 0.80 times the median, giving a ratio of 1.5 between the two estimates. Similar results can be obtained for other economic sectors. These uncertainties are comparable with those discussed earlier using price elasticities. There are two points that need to be emphasised here. First, it is possible to lean towards an optimistic or a pessimistic view by apparently reasonable choices for each of a series of factors used in activity analysis. Secondly, we simply do not know the 'best' result. Energy policy is concerned with planning under uncertainty. It is not aided by over-confident predictions about the future.

Policies

I shall conclude this discussion by examining national or government policies on energy conservation. Decisions on the use of energy are so widely dispersed through the community that one might ask whether a government can make much difference and should we not examine policies of other institutions and even individuals. The economist's answer would be that the latter are part of the market, each person optimising, or perhaps failing to optimise, in his own way. We therefore need to consider how the government can influence the market, and the criteria that affect their choice of options. I shall begin by discussing fiscal measures and conclude with regulations and standards.

The broad categories of fiscal options that are available to governments include energy pricing, taxation, and subsidies. It has become fashionable in Western economies for governments to regulate energy prices. In the US the large indigenous supplies of oil and gas permit an averaging of energy prices so that the consumer does not normally meet the full impact of an increase in the world price of oil. It is argued that the application of world oil prices to the home production of oil would bring undeserved profits to the oil producers, though there would seem to be no fundamental problem in applying a special tax to reduce windfall profits or direct them towards alternative energy supplies or conservation investment. It is interesting to speculate on the global costs of the failure in the US to allow oil prices to rise to world levels. If oil prices had been 50% higher on average over the past five years and we assume an oil price elasticity of −0.4, the oil demand in the US could have been 15% or about 3 M b/d below current demand. Under those conditions there need not have been a world shortage of oil during the revolution in Iran, and world oil prices might well have been 30% below current levels. This would have meant that the US oil import bill would have been approximately halved. Of course a lower world price for oil would not have pleased the members of OPEC, and the absence of this year's energy crisis would have weakened enthusiasm for conservation in consumer countries. The moral for consumer governments is clear, conservation is an essential part of the adjustment towards zero growth in oil demand but it needs to be maintained even if the world oil price falls in real terms. There is also a message for OPEC, namely many conservation measures have a considerable degree of permanence and their adoption by consumer countries can be greatly stimulated by sudden increases in the price of oil. Even the US is moving towards decontrol of oil prices.

There is no single or simple regime for the prices of oil products. Gasoline is taxed in most countries. An imperial gallon costs about $1 in the US, compared with between $2 and $3 in most European countries, equivalent to about $35 per barrel in the US and between $70 and $105 in Europe. In the short term this is probably the most inelastic part of the market for oil products. Motor cars cannot readily be converted to burn coal, so we must wait for technological developments for greater vehicle efficiency and for long-term changes in consumer choice towards smaller and more economical cars.

If governments were really interested in conservation and collective action is feasible under crisis conditions through the procedures agreed in the International Energy Agency, it would seem sensible to adopt collective action to impose a tax on oil products other than gasoline, since the demand for these products is more elastic to changes in price as substitutes are more readily available. However, for an individual country it is argued that taxation on oil would penalise industry compared with that in other countries. In fact it would be a transfer tax, and through reducing other taxes, on average industry need be no worse off. It is also said that collective action by consumer governments would invite retaliation from OPEC. That could only stimulate more conservation of oil use and weaken the longer-term position of OPEC.

In practice, taxation of oil products will not provide a miracle solution any more than other simple recipes. Government actions are limited by constraints which vary from one country to another and may change with time in an alarming manner. If gasoline had become expensive in the US five years ago, this would have been too early for mass production of small cars by the US automobile industry and the level of car imports could have changed so rapidly as to cause other economic problems. But taxation and decontrol of prices for oil products can make a useful contribution towards conservation of oil and the transition to other energy sources. Those countries that have moved some distance along this road can legitimately complain that, through their low prices for oil products used by their own nationals, neither the US nor most OPEC countries are making their proper contribution to this global problem.

Subsidies form the other side of fiscal measures. But as with other options for energy conservation they should be chosen so as to reinforce the market, not to oppose it. Price controls below the relevant costs of energy are a form of subsidy to one section of the community, and when coupled with exhortations to energy conservation they may be unfavourably compared to driving a car simultaneously with brakes and accelerator.

Subsidies can make their greatest impact when they are used as seed corn to stimulate the development and growth of new measures that will become fully economic as soon as they are more widely adopted. They can provide demonstrations that a particular conservation measure is economic. Equally they can show that a measure is not economic, and it is not good publicity to provide 10 cents of warm water by installing a solar system costing $20,000. One of the most valuable forms of subsidy that government can provide is support for research, development and demonstration (RD and D). These activities often involve high financial risk and, if successful, a high return, but the return from energy conservation measures will rarely come primarily to the entrepreneur who developed them, but will be shared by many consumers.

There is a difficulty about putting too much emphasis on RD

and D for energy conservation, since it may distract attention from actually getting on with the job. Many measures that can save energy are well known – better insulation, heat recovery, instrumentation, better design and better use of materials. The real problem is getting them implemented – of turning well-known ideas into actual achievements. Government subsidies need to be selective, for example, through helping publicity, education and training to show people how to act on energy conservation to meet their own best interests. These interests vary widely and a recent publication on Energy Efficiency from Shell International (1979) has graphically illustrated the wide variation amongst consumers of the economic lifetime of conservation measures, their pay-back time, and the subjective time horizon of the consumer. A modified version of this comparison is set out in Table 1.

Table 1. Payback time, subjective time horizon and equipment lifetime for various conservation measures

Conservation measure	*Pay-back time (years)*	*Subjective time horizon (years)*	*Equipment lifetime (years)*
Industry			
Good housekeeping	1–3	1–2	n.a.
Heat recovery	2–10	2–4	15
Residential			
Improved gas appliances	1–3	2–4	12
Improved central heating	1–4	2–4	12
Insulation	3–10	2–4	25
Thermostats	3–10	2–4	25
Double glazing	10–20	2–4	25
Highly insulated new house	5–15	5–10	25
Private cars	3–5	1–3	8

Source: Author's qualitative estimates based in part on a Report on Energy Efficiency by Shell International (1979).

Table 1 illustrates variations in the values of different decision makers. If governments consider these values to be wrong they may try to change them through publicity or persuasion, or by fiscal measures, regulations, or standards. But these values set by individuals are themselves related to other choices open to the decision maker. The industrialist may want to expand his business or protect his cash reserve, the householder may want to install double glazing because a new window will improve the view into the garden, or he/she may prefer to cook by electricity.

One of the situations when government intervention could be desirable, through regulations or standards, is when decision-making is divided, for example between producers of equipment and ultimate users of that equipment, or builders of houses and householders, or landlords and tenants. Landlords are rarely concerned with the energy costs that have to be met by their tenants, and they have relatively little incentive to improve the thermal insulation or heating equipment in their properties. Government regulations or a charter for tenant's rights on energy conservation could lead to a substantial saving of energy with benefits both to landlord and tenant, and to the community through the overall national benefits. Similarly regulations and standards for new buildings and for the efficiencies of motor cars, both suitably phased-in over a number of years could benefit all sections of the community. The system of penalties on manufacturers related to automobile fuel efficiencies in the US is an example of this type of intervention which appears to be a success. In contrast regulations relating to environmental protection may lead to additional requirements of energy for motor vehicles, and provide another example where values and priorities of decision makers may not favour energy conservation.

This brings me to my concluding remarks which concern complexity and uncertainty. There is no single simple miracle solution to the energy problem. Energy conservation has a role to play alongside production from all acceptable sources. The potential for energy conservation is large but, due to the complexity of values and choices by decision makers, only part of the potential will be realised. Whilst emphasising the importance of energy conservation, its realisation is not aided by exaggerated claims that underestimate the difficulties and complexities involved in converting ideas into widespread achievements. A wider range of actions is required – by institutions and individuals, and by governments; some will prove more successful than others. Each of us has a contribution to make, whether we come from an energy-exporting or an energy-importing country. Some of us can raise the price of oil, some can take to riding bicycles, others can draft regulations or impose standards for energy efficiencies.

Part III

ECONOMIC, INSTITUTIONAL AND POLITICAL STRUCTURES

11 OIL: PRICES, COSTS, TAXES

Fadhil al-Chalabi

As demand for crude oil is derived from that for refined products (and by the same token the supply of products is dependent on that of crude), the ultimate market-place for crude oil can only be traced to its end-uses, i.e. the product market. A market for crude cannot exist independently from that for refined products. Furthermore, apart from the Soviet Union, and to a much lesser extent the United States which recently became heavily dependent on oil imports, the bulk of oil consumed in the world is internationally traded, mostly imported from the major producing areas in the Middle East and Africa where low levels of economic and social development do not call for high levels of energy consumption. (Oil trade accounts for about one-fifth of total world trade.) However, the quasi-totality of world oil traded is crude rather than refined products; consumers refine imported crude oil at home, where the bulk of the world's oil-refining capacities are located.

Thus, demand for crude oil is controllable by the consumers in the same manner as supply of crude is controllable by the producers. The control of demand mainly takes the form of various fiscal and commercial policies decided by the governments of the consuming countries (taxes and levies paid directly or indirectly by end-users, quantitative controls, such as import quotas, import ceilings, speed limitations, and non-quantitative controls like those on the burning of fuels, etc.). Producers, on the other hand, control supply through pricing policies for crude oil which can be administered by them independently from demand/supply balances. Producers' policies relating to production and investment are an even more effective means of controlling crude oil supplies. They usually take the form of quantitative controls, like the various technical conservation measures, production programming, production ceilings, changes in the crude supply-mix, investment in expanded capacities, etc.

For consumers as much as for producers, the petroleum trade lies within those strategic areas of economic activity that could have a far-reaching impact on economic growth, balance of payments, inflation, employment etc. The petroleum trade is therefore closely related to world power politics. This feature has been greatly enhanced by the 'oil revolution' of 1973–74, which dramatically brought into focus the strategic importance of oil in the balance of international political power. As a result, the degree of control through policies of both consumers and producers has increased. (The change is only in degree, since the controllable nature of supply and demand has always been felt in the history of the international petroleum industry.)

Apart from some marginal markets like Rotterdam, product markets in major consuming areas are constrained, and even structured by government policies. A market, in the sense of free inter-play of supply of and demand for products, can hardly be found in most of the important consuming centres. Similarly, a market for crude oil in the same sense of free inter-play of supply and demand does not exist in reality. The narrow area of crude exchange in the 'free market' is meant only to supplement or balance the bulk of the products market. Thus, the spot market performs a balancing function in the product market but does not constitute an independent market which could be continually self-sustained.

More important is the inter-play of the policies and actions on supply and demand by governments and large economic agents, both in the consuming and producing countries. Price formation for crude oil is influenced more by the combination of consumers' policies and producers' policies, than by the market. To be sure, the impact of market forces on prices is sometimes felt, especially in times of strain on supply, as, for example, in 1979. In the long run, however, policies have a greater role in shaping the price structure than market forces, since it is consumers and producers policies which, directly or indirectly, alter both supply of and demand for crude oil.

For consumers, the price of crude oil is an input cost that determines, among other things, the price of the final products demanded in their economies. The level of crude oil prices has therefore a determining impact on the demand for products and could, for this reason, provoke policy changes in the consuming areas. However, crude oil prices are administered by producers at optimised levels, reflecting policies and shifts in controlling power rather than the direct influence of the inter-play of supply and demand, or of variations in cost structure.

When the international oil industry was completely integrated and controlled by the major oil companies (from the formation of the International Petroleum Cartel in the late 1920s until the aftermath of the Suez Crisis in the late 1950s), a market for crude oil did not exist at all. The petroleum trade was no more than inter-company exchanges of crude within the almost completely closed circuits of a world-wide integrated system controlled by the same companies. The companies' price administration (setting unilaterally the levels of posted prices) during that period served largely fiscal purposes, mainly the taxes and royalties paid to the host producing governments. History shows that the companies' pricing strategies were not linked to, nor resulted from, economic forces in the market or relative movements of supply and demand. Neither were they dictated by changes in the cost structure. They were rather motivated by the policies of the consuming nations whose aim was to acquire cheap energy and raw materials from developing countries. Hence the successive unilateral cuts in the price of Middle East oil through the adoption of fictitious pricing methods. By 1960 the price of that oil was reduced to half its equivalent in the US, whereas during the Second World War prices of the two crudes were at par. The 'tax-paid cost' per barrel incurred by the companies in lifting crude oil from oil-producing countries represented at the same time the real cost of acquisition by the consuming importing countries (especially in countries where those companies resided). Also, it reflected the producing countries' meagre share in the final value of the barrel.

The erosion of the companies oligopolistic control of the industry (since the early 1960s until the oil revolution of 1973–74) led to limited but growing crude oil exchanges outside the companies, closed channels. This structural change in the industry did not only alter the way in which the major companies administered prices themselves, but also involved the emergence of a partner in price administration, namely OPEC. Again, under the new system, the pricing of crude oil did not reflect the inter-play of economic forces relating to long-term supply and demand movements, nor variations in the cost structure. Between 1960 and 1971 posted prices were frozen in dollar terms (they actually declined in real terms) as a result of OPEC action in preventing the companies from undertaking any further price cuts. Yet during that same period, the oil reserves to output ratio in the Middle East was sharply declining (in spite of the fact that reserves multiplied many times while demand was sharply rising (see figure 1)). If purely economic forces were considered as a basis for price setting, then the oil situation during the 1960s should have warranted oil price rises to reflect the ever-narrowing gap between long-term supply and

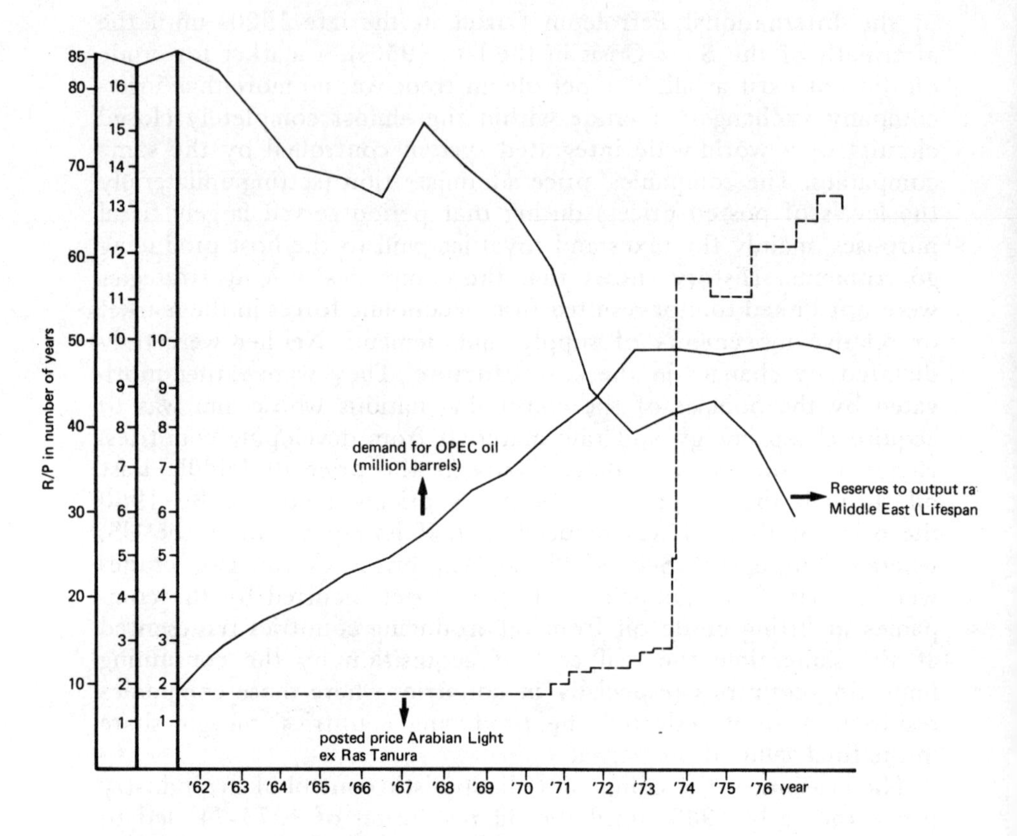

Fig. 1 Arabian Light 34°, dollar per barrel f.o.b. port of export

demand due to the accelerated shortening of the lifespan of oil reserves in the Middle East, and the growing demand for oil from those areas. Also, the evolving market price was more or less pegged to the leader's price, i.e. the companies' price set around their tax-paid cost plus a profit margin. The various phases of the history of oil prices during that period show that their levels, as ultimately set by the companies, were hardly sensitive to the variations in the market, and even to the balances in supply and demand. The actual levels were the result of company policies combined with OPEC actions. (The various policy measures taken by OPEC, including the expensing of royalties, the unification of taxation systems, etc.)

After 1973-74 the level of oil prices has been collectively determined by the governments of the oil-producing countries. Price administration by OPEC was more or less determined by short-term considerations reflecting on the one hand the balances of

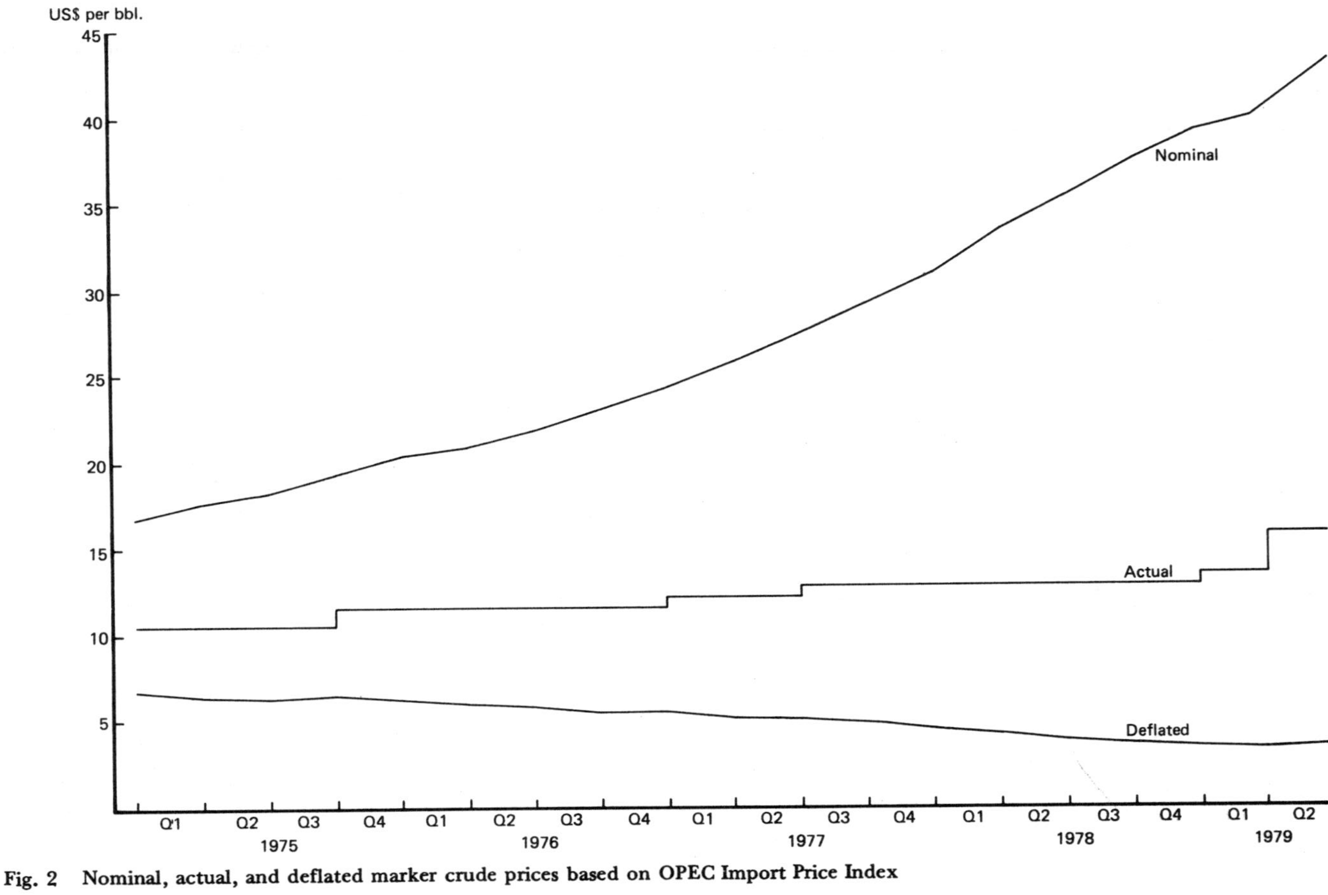

Fig. 2 Nominal, actual, and deflated marker crude prices based on OPEC Import Price Index

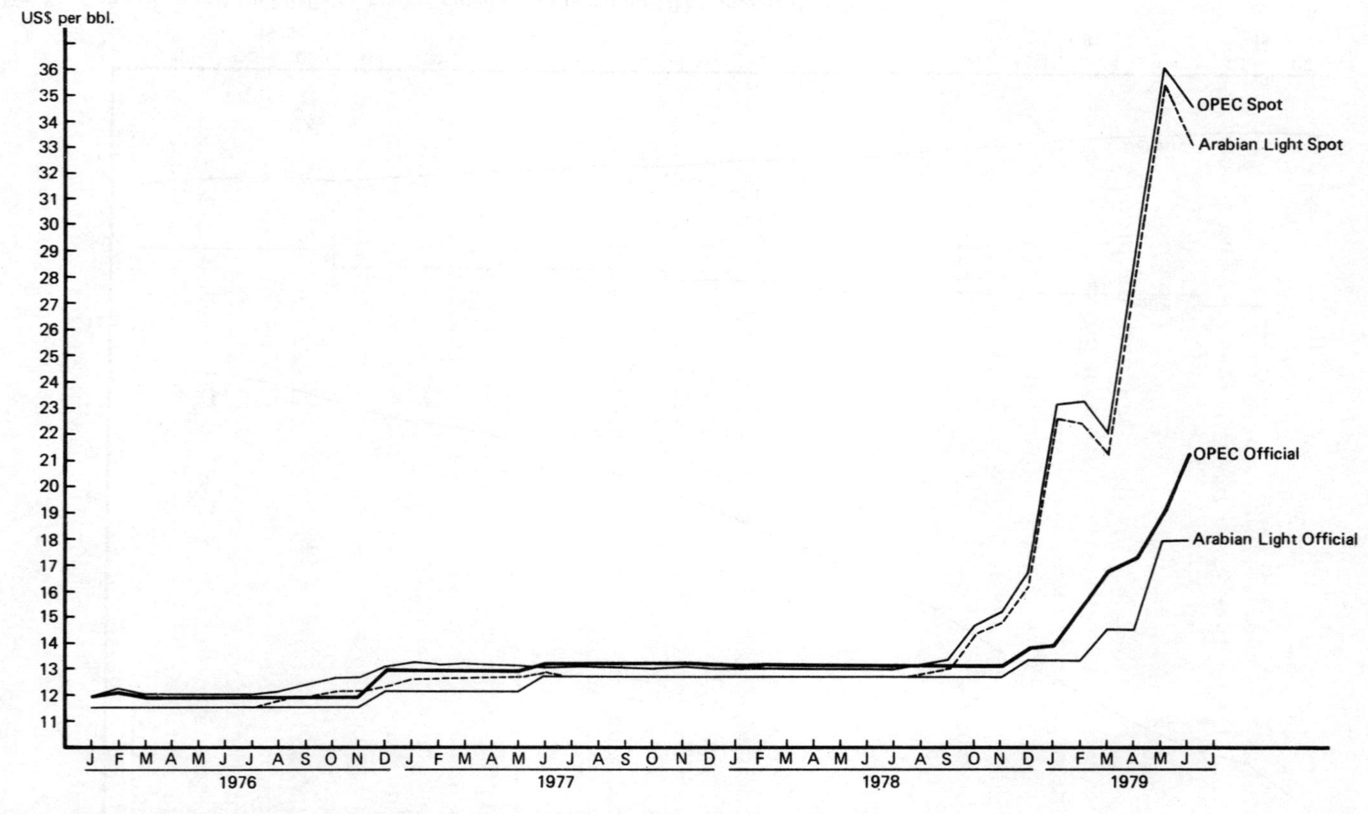

Fig. 3 Evolution of average OPEC official and spot crude oil prices vs. Arabian Light 1976–79

interests and requirements within OPEC (the price 'setters' versus the price 'takers'), and on the other hand an OPEC concern over the world economy. Certain long-term policy objectives, actually laid down in some of OPEC's political documents, like the Algiers Solemn Declaration, were rarely implemented. A good example is the continuing gap between prices in current dollars and those in real terms (1973–74 dollars) (see fig. 2). OPEC post–1973 policies on oil prices were not strictly in line with the producers' long-term objective of preserving the purchasing power of oil (to keep prices constant in real terms).

Market conditions have influenced OPEC pricing decisions only marginally. The recent history of OPEC price administration shows little sensitivity of price to the relative movements in supply and demand, or market conditions. The 10% price increase of October 1975 was decided at a time of 'market glut', whereas the price increases of 1979 were enormously below the spot market levels which resulted from market shortages (see fig. 3).

However, despite the complete take-over of pricing by OPEC and the diminished power of the oil companies, the companies have continued to play the leader's role in setting the level of prices in the spot market, especially in periods where supply and demand are fairly balanced or where quantities of crude made available are in excess of demand. By selling in the market the crude oil lifted from producing areas in excess of their downstream requirements, the companies actually set a price in that market, which other sellers have to follow. That price is greatly influenced by the tax-paid cost those companies incur in lifting the extra crude.

Prior to the complete take-over arrangements, the tax-paid cost was in fact a floor price, below which none of the companies could offer substantial quantities in the market. Obviously, the fiscal relationship that determined the level of the tax-paid cost had undergone drastic changes as a result of successive increases in the tax and royalties ratios on the equity oil (the increase of tax rates from 55% in the Tehran Agreement, to 85% in the Abu Dhabi decision of November, 1974, and also the increase of royalty rates from the traditional 12.5% to 20%). The more taxes and royalties increased the higher the cost of lifting crude, and hence the tax-paid cost, and by the same token the market price. On the other hand, the average per barrel cost incurred by the companies was influenced by the amount of oil lifted in excess of their equity entitlements. The higher the government share sold back to the companies at the market price, the lesser the average (per barrel) profit margin of the companies, hence the higher costs incurred by them. Here again, the result would be a rise in the market price level.

With the new lifting arrangements, following the complete government take-over, the companies are enjoying a flat per barrel price rebate against technological assistance and other services (like transporting lifted crude in national tankers, as in the case of Kuwait). Following these important, and subsequently, even more important changes in the structure of the industry, the companies' ability to lead the prices in the market has been enormously eroded. These changes have resulted in a radical shift in the government-company marketing relationship, whereby the status of the companies has changed from that of net sellers of crude to that of net buyers (see table 1). This situation deprives them of real power in leading price setting in the market. In fact, price leadership is gradually, but steadily, shifting to the National Oil Companies of OPEC member countries, who are emerging as the market price leaders (about 45% of the marketing of OPEC crude oil is currently controlled by the National Oil Companies of member countries).

Table 1 Estimated demand and supply of the seven major oil companies 1960–79

	Thousand Barrels per day				
	1960	*1965*	*1970*	*1975*	*1979*
Production	12,128*	16,848	26,121	24,712	18,000
Refinery Utilization	10,135*	14,764	20,660*	18,948	20,000
Oil Product Sales	10,808	15,439	22,288*	20,537	23,000
Deficit or Surplus:					
In Crude Oil	+1,993	+2,084	+5,461	+5,764	−2,000
In Products	−673	−675	−1,628	−1,589	−3,000

*Estimated by Statistics Unit, OPEC Secretariat

Producers' control of supply could also be more effectively exercised through production policies. A clear case in point was the companies' production planning under the old system of complete control by the majors. In the subsequent period when the oligopoly system was weakening, production policies affected quantities in a variety of ways, e.g. pressures exercised on the companies by some producing governments, especially Iran, to increase production during the 1960s, also pressures exercised by the companies on some governments, especially that of Iraq, which resulted in reduced production levels. Finally, after the complete take-over by producing governments, production policies took various forms that were motivated either by technical or by economic and financial considerations (production ceilings in Kuwait and Saudi Arabia, increasing financial requirements for national development reflected in higher levels of production in certain countries, changing oil supply-mix patterns, such as the 65:35 ratio in Saudi Arabia, investment policies towards expansion in production capacities, etc.).

Consumers have no less effective means of control on the market through policies to influence demand directly or indirectly. Oldest and most effective among these policy tools is the taxation system which, although varying from one country to another in the developed oil-importing countries, has resulted in a general pattern of internal pricing always isolated from the level and the structure of the OPEC price. Consumers inside the developed countries, especially those of Western Europe, pay much higher prices at the delivery end for the products they acquire than that which the producers are getting (whether in the form of world market price or in the form of government take). The huge difference in the two price structures is due mainly to the taxes and levies imposed by the governments of those countries. Another reason for that difference is the companies' profits and different additions in the product value which accumulate throughout the process of transporting and refining crude oil, distributing refined products, etc.

The gap between the price paid by consumers and that received by producers has changed over time with posted prices and tax and royalty ratios. In the mid-1960s the price that consumers were paying for the products representing a composite barrel of crude oil lifted from the OPEC area ranged between 11 and 13 US dollars per barrel, against a government take for the oil-producing countries averaging 90 US cents, so that the producers' share of the total final price paid by the end-users was no more than 8%. After the price and tax adjustments of the Tehran Agreement, this share increased marginally so that prior to the oil price revolution of 1973 it reached about 13%. Obviously, this situation has been radically changed by successive fiscal adjustments, as well as the changes in government-company relationships, including participation arrangements, and ultimately government take-over and nationalisation so that today the producers' share in the value of the final product has moved up to about one-third (see fig. 4).

In spite of those increases in the producers' share of total value of the barrel, the fiscal system in the consuming countries is still such as to create a huge gap between internal prices and international prices. The purpose of those taxes is either to encourage or discourage oil consumption (very high taxes on gasoline versus very low taxes on fuel oil) or to maximise the consumers' share in the economic

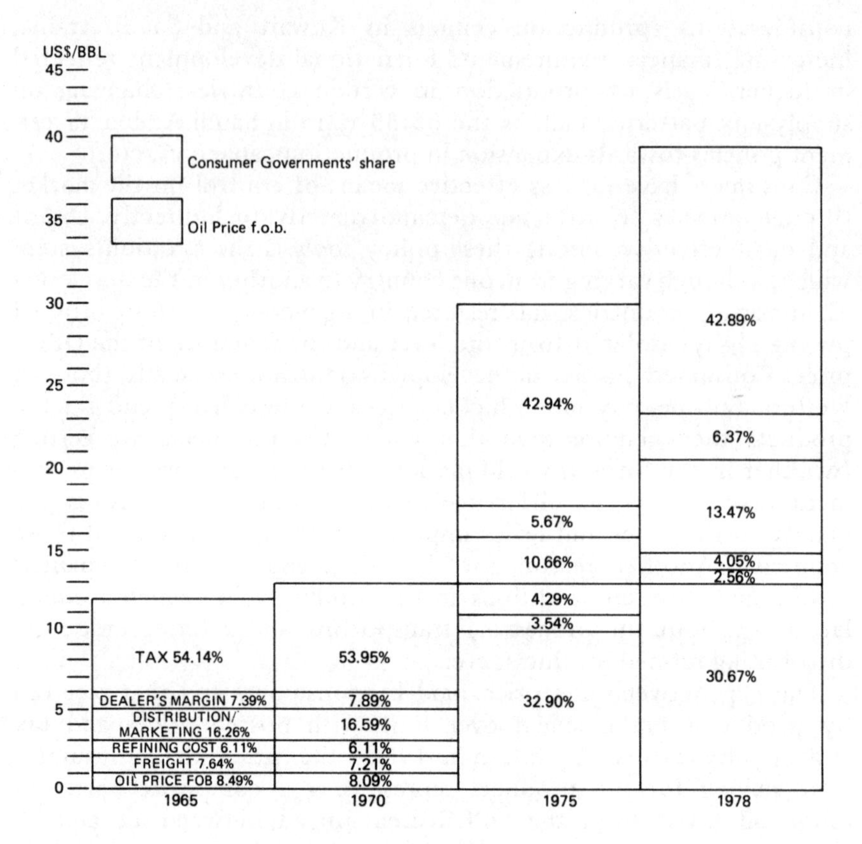

Fig. 4 Components of the price of a barrel of refined products in Western Europe

rent from the trade of oil (which actually should belong in its entirety to the oil-producing countries). This is the case of high taxes on gasoline which, because of the very low price elasticity of demand, are put at optimally high levels that do not affect the economic growth of the consuming countries. Such economic rent that the governments of the consuming countries have been reaping in the form of taxes, is usually justified on the grounds that it provides finance for the construction and maintenance of roads (from which the motorists are benefiting)!

Being a major oil producer as well as a major consumer, the US has not followed the same pattern of taxing oil products for consumption, so that the isolation of the internal price structure in the US from the international price was more important than in Western Europe. It should be stated, however, that in a country like the US, lower prices paid by end-users for all products may generate economic growth elsewhere in the national economy that could compensate for the loss of the economic rent which would have been gained through taxation. Nevertheless, the increasing dependence of that country on imported crude oil has been provoking policy reactions towards imposing higher internal taxes to be paid by the consumers.

Recently, consuming nations have been resorting to more direct and effective means of demand control through quantitative measures and mandatory policy controls. Apart from measures aimed at saving energy and putting 'the energy house into order', such as limitations on speed, regulations for heating and lighting, regulations for efficiency in energy use and other conservationist measures, consuming countries are resorting to more aggressive quantitative controls in order to regulate the volume of consumption, irrespective of the market reactions to price variations. Ceilings on imports in terms of both volume and value are already common policy practice. The IEA recent policy target of reducing current consumption by 2 M b/d is a further step towards firmly controlling demand. For the longer period, the Tokyo Summit Conference has set a new political shift in supply and demand balances, when it set the objective of reducing demand by 1985 to levels equivalent to those of 1978–79, thus barring any growth in OPEC production destined for the OECD in general.

Given that both supply and demand are increasingly controlled by producers and consumers, an approach to price planning could be adopted by the producers which would aim at achieving certain strategic objectives. Such objectives will depend on the long-term policy options adopted by producers and consumers towards either more inter-dependence or more confrontation.

One parameter for long-term price planning that could be universally accepted as a minimum, is the preservation of the purchasing power of the OPEC barrel by undertaking such price adjustments in current units of account for the pricing of oil (currently the US dollar) so as to offset the erosion in the real price, which could be due either to imported inflation from the industrialised countries, or to variations in the exchange rates of the unit of account against the other major international currencies. Solutions could be found for technical problems involved in the measurement of those variations, especially those related to world inflation. Inflation rates recorded inside the industrialised countries are much lower than inflation rates reflected in the landed cost of OPEC imports from those countries (see figs. 5 and 6 on relative price movements of the Marker Crude Unit of Account based on the OECD export price

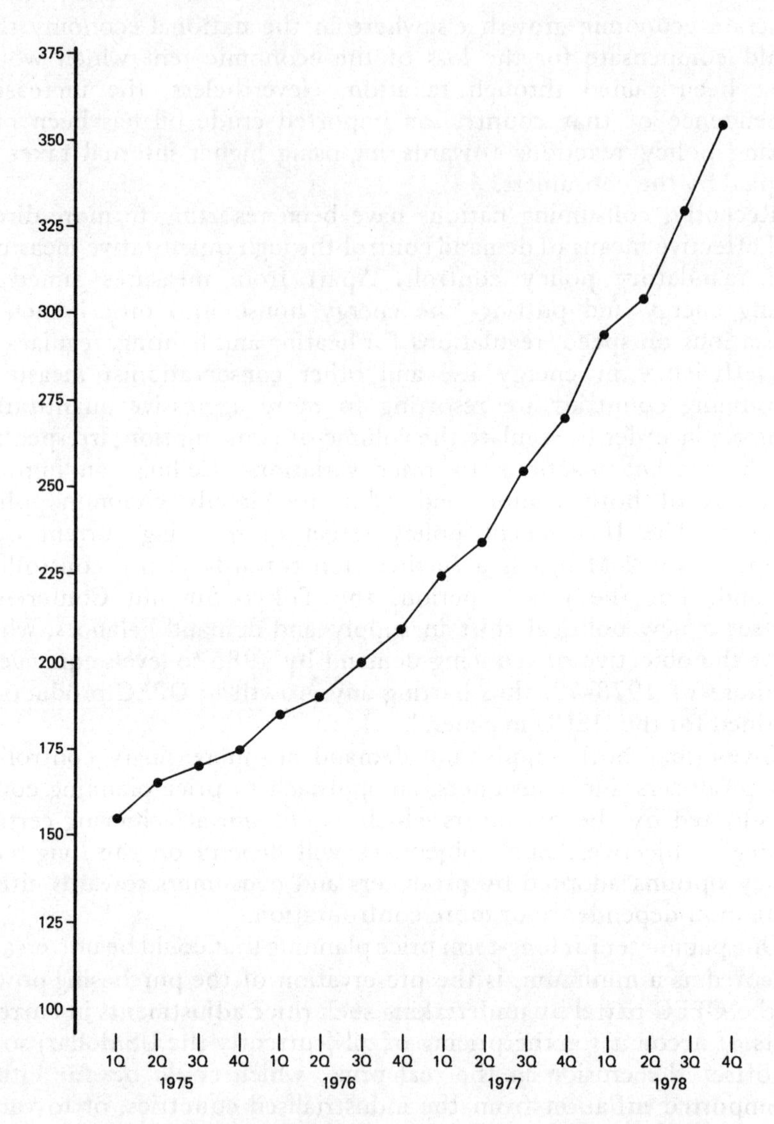

Fig. 5 Aggregated accumulated OPEC import price indices, 1975–78 (Base 1973)

indices concerned with a measure of dollar exchange variations, versus the OPEC Import Price Index). However, this concept cannot be considered strictly as a planning concept, since setting crude oil prices in real terms would not allow producers effectively to influence demand by preventing it from either over-taking supply capacities or falling behind those capacities.

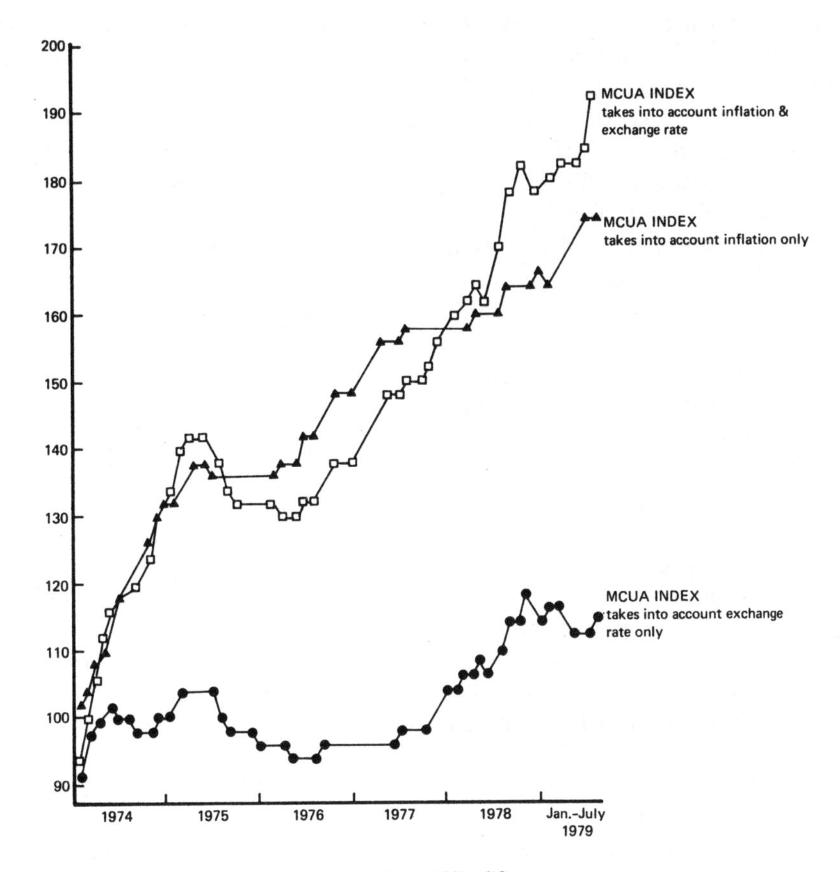

Fig. 6 Marker Crude Unit of Account indices 1974–79

A planning approach could be adopted by OPEC towards keeping world demand as long as possible in line with its capacity (the concept of capacity changes over time depending on the rate of investment). According to this approach, oil prices in real terms would be planned in such a way as to prevent world demand from growing beyond the producers' capacities (in the case of very low real prices) thus creating undue pressures on supply, but also preventing it from falling to levels that would substantially reduce the share of OPEC oil in the world energy market. Solutions could be found for technical problems that might arise in the price 'observatory' such as the measurement of price and income elasticities, price movements in real terms etc.

A closely related approach is to determine the price of oil in the light of oil depletion and its replacement by new oil. The depleted barrel of oil could possibly be replaced by a new higher cost barrel which may be obtained either through the discovery of new oilfields or through the enhanced recovery of the oil in place from the oilfields. Accordingly, the real price of depleted oil moves upward in the light of the movement of the marginal cost incurred in creating new production capacities. For example, the average cost of North Sea oil, following the latest discoveries and the development of the new oilfields, is almost double that incurred from the old oilfields. Such a cost movement should be reflected in a price movement for the barrel which is depleted from oil reservoirs. The idea is to compensate for the depletion of low-cost oil by creating incentives to invest in new oil. Furthermore, as in the previous case, this concept implies a planned approach to the required share of OPEC oil in the world energy trade. In fact, an important aspect of such an approach is that pricing policies should create at the same time both the incentive for producers to invest in new capacities, and for consumers to maintain their dependence on imported oil. As far as the producers are concerned, there is an economic interest to invest in new capacity as long as the cost of new oil is substantially below its price. In other words, oil price movement should not go below a minimum which secures an incentive for the producers to invest, nor beyond a maximum which secures incentives for consumers to reduce heavily oil consumption in favour of other sources of energy.

Lastly, oil price movements in real terms could be set in a trajectory towards parity with marginal costs of alternative sources of energy. However, this approach raises certain intellectual problems related to the definition of marginal costs of alternatives, and the nature and scale of the process of energy substitution. Costs of synthetic oil from coal (being the only complete substitute for oil) would presumably be taken as the landmark for price parity of OPEC oil. The problem, however, is that this cost is hypothetical in the sense that it is not commercially established and that it changes dramatically over time. Even with the establishment of a commercial value for the synthetic substitute, there still exists the problem of defining the scale of substitution and its timing. Furthermore, this concept could erode the long-term planning power of OPEC to control world demand for its oil in line with its future capacities.

12 THE SPECIAL CHARACTERISTICS OF OPEC AND IMPORTING COUNTRIES' NATIONAL OIL COMPANIES

Jack E. Hartshorn

Most national oil companies (NOCs) have only one shareholder, their government. All have some degree of government-conferred special advantage, preference, or monopoly. In importing countries this degree of national preference is more important than the percentage of government ownership. A bare majority shareholding, or even less, may be sufficient: BP and CFP have been more important to the oil interests of their parent governments than BNOC or Elf/ERAP may ever be. The choice of the company, at times, as a national instrument is what matters, even if it operates most of the time wholly as a private company.

This will perhaps never be true of OPEC national companies. In most OPEC countries the few vestiges of private shareholding that used to exist in some national oil operations have largely disappeared. It may be that complete government ownership is inherent in such NOCs, in that their function involves the depletion of a wasting natural resource that is the property of the nation.

The development of an NOC and its role in the world industry is thus crucially linked to its relationship with its own government. This relationship, in nature and in balance, varies enormously within both groups of NOCs. The NOC is always a chosen instrument of a government. Its management is appointed, and can always ultimately be removed, by the government. But that has not stopped certain NOCs of both kinds, from time to time, from operating far beyond any practical government control. Pertamina was one obvious example in OPEC; on the other side, there have been periods in ENI's prowess when Italian government 'control' has looked pretty tenuous. But I think there have been far more examples than ever became obvious. Indeed, it seems almost inevitable that NOC managements anywhere, as their expertise develops, will seek to consolidate and widen their own power, as distinct from that of their governmental masters.

The economic objective of NOCs will not be simply to maximise profits. What these objectives should be – i.e. what national benefits in relation to resource costs they 'should' maximise – are not easy to define, let alone apply in operational management. Accounting and efficiency auditing criteria developed for profit-maximising private enterprise may therefore not always be applicable to NOCs. But the governments of importing countries have not so far been very successful in devising other criteria to measure the performance of NOCs or other state-owned industries. This may turn out even harder for OPEC governments.

A final common characteristic of NOCs is almost tautological. They are national, not international, and in particular, not internationally integrated (though that was the original objective when some importing countries' NOCs were set up). None has any special privileges outside its own country, and only a few have much foothold abroad at all. Operationally, therefore, they tend to seek optimisation, for example of their oil logistics, at a different level from private international oil companies; also, most of their trading tends to be done at arm's length.

OPEC National Oil Companies

OPEC NOCs have certain special characteristics that those of the importing countries do not share. The industry they are involved in managing, wholly or partly, is generally the most important in their country and often the only significant one. This makes its control – by the government and/or NOC – a key lever of national economic policy; and sometimes a privileged lever of foreign policy of the country. These facts may remove the industrial and trading performance of the NOC farther from the business behaviour of profit-seeking private enterprise than the performance of NOCs in importing countries, whose industries are seldom of comparable importance within their national economies.

There is another major difference. NOCs of OPEC member countries cooperate openly, or rather their governments do, in seeking to fix and maintain prices in a way that private companies cannot. Importing countries' NOCs might be less constrained by anti-trust considerations than private oil companies have to be. But so far they have seldom openly cooperated with others of their kind to seek their common economic aims – which might of course generally be opposed to those of OPEC, i.e. to get and keep prices down.

Export Supply Responsibilities. Quantitatively, by the end of 1979 only about one-third of the crude oil trade outside North America and the centrally planned economies (CPEs) may be handled by the international major companies on 'secure' terms, i.e under contracts affected wholly or partly by preferential discounts.[1] This implies that the rest, nearly two-thirds of the crude moving in world trade, may be sold in one way or another by NOCs.

However, export supply responsibilities are wider than the NOCs' share of crude sales. These responsibilities primarily involve the control of current production allowables, of investment to develop capacity, and of exploration to increase proven reserves. It is not easy to generalise, or even to assess in individual OPEC countries, how much the NOCs as such share in, or advise on, these decisions, which are taken by their governments.

In decision-making about current and future supply, therefore, NOCs of OPEC member countries will perhaps never gain the full power and responsibility that the former private concessionaires have now surrendered to OPEC governments. For that matter, the governments will not be able to exercise these responsibilities in quite the same way that the international majors could. No OPEC government knows as much about the supply planning of any other as some of the international majors, through joint shareholdings, knew about, and indeed shared, control of supply planning in several OPEC countries. That element of the majors' supply responsibility has not passed to governments: it is simply disappearing.

Commercial Policies. These NOCs' commercial policies are as yet only developing and are already diverse between individual OPEC member countries. But the OPEC NOCs begin with some common circumstances.

Their maximum sales, along with their prices, are largely set for them. Between the national allowables and any government commitments to ex-concessionaires (and perhaps to other governments) the NOC is left with just so much oil. It can sell less, but not more (though governments, in 1979, have shown themselves ready to cut back their other contracts arbitrarily and at short notice).

As to prices, NOCs face an opposite principle: they cannot sell at less than official government selling prices, though they might be allowed to charge more whenever the market will bear it. 1979 demonstrated that OPEC collectively is not yet capable of setting ceiling prices – essentially because its member governments do not want to. Whether individual member governments want to set ceiling prices for their own NOCs' sales is not clear: perhaps a few do.

[1] This estimate was made by a major company in December, 1979 on the basis of far more comprehensive information than I could ever assemble.

These two principles set by their governments circumscribe the OPEC NOCs' commercial behaviour. But their governments have also, most of the time since 1973, contrived to give them a wonderfully easy seller's market. These governments may be able to go on doing so, even in recession, particularly if they are prepared to manipulate their allowables and keep supply below demand. (They have not yet needed to do so.) At any rate, this remains largely outside the NOCs' responsibility. Even in what might be considered matters of detail, such as price differentials, governments still seem to decide.

We have a few tentative indications, so far, of the trading practices these OPEC NOCs prefer:

Most of them appear to mix caution in general marketing with occasional opportunism, as in the spot market of 1979.

They prefer customers with refineries to traders without.

They would prefer several medium-sized customers rather than all their oil under one or two contracts.

Some have expressed a preference for dealing with NOCs of importing governments, or rather their governments have for government-to-government deals.

But only a few of the OPEC NOCs have as yet consolidated these practices into any general code. They have hardly needed to, during a period when the main commercial behaviour required from any OPEC NOC marketer has been to open his door and say to the submissive queue, 'Next, please'.

There has been little evidence yet of OPEC NOC behaviour in more difficult markets, though once or twice particular countries have revised differentials or adjusted credit terms. If the glut of crude which some OPEC spokesmen are now predicting for 1980 does occur, their marketing practices may be put to more severe tests. Otherwise, the first pilot scale experiments might come in the early eighties, when some of these NOCs will have large additional volumes of LPGs and refined products to sell. OPEC's leverage on crude will still come in useful then, but less directly.

Is There a Shift in Favour of Importing Countries' NOCs?

Certainly, government-to-government deals are now taking an increasing share of the world oil trade. It seems probable, though not certain, that this must tend to favour the importing countries'

NOCs. Also, as governments and communities' energy policies become more positive and more important in national politics, governments may assume that their NOCs will be more docile, if not necessarily more dependable, in obeying national policies than subsidiaries of foreign companies might be.

Against this trend in favour of importing countries' NOCs, for the moment, is a general political swing to the Right in some of these countries. State-ownership is a more politically polarised issue in many importing countries than in any of the diverse political entities of OPEC. Several governments in OECD countries, for example, came recently into power with commitments to cut back the privileges of their NOCs or even de-nationalise them. Some of these have had second thoughts. But even governments very eager to make direct deals with OPEC have not always chosen to channel most of the resultant imports through NOCs.

NOCs in the non-oil-producing developing countries may bring about some increase in the total share of the world oil trade moving through NOC channels. This is not because more private oil companies there may be nationalised: that process is fairly complete already. It is because these countries' consumption of oil, mostly handled by their NOCs, is one of the fastest-growing elements in world demand, though it starts from a small base. Even in recession, demand may go on growing faster than elsewhere – *if* means continue to be found for the banks and OPEC to go on financing it.

Commercial Policies of Importing Countries' NOCs. Nowadays the commercial policies of the NOCs of importing countries, at least inside OECD, do not seem distinctively different from those of privately-owned companies. Several have outstanding technical reputations in and on the fringe of the basic petroleum business. Over the years few have had outstanding luck in exploration – which has limited the repercussions of the unconventional forms of deal with host countries that they have been willing to make. But today no form of deal is too unconventional for some private companies, small or even big, to be prepared to try, provided that it gives a chance of continuing crude supply.

The accounting of these NOCs or what they publish of it is even more opaque than that of private international oil companies. But in practice, fears that such companies might be able to call upon bottomless state subsidies to compete unfairly with private oil companies hardly seem to have been fulfilled. Very often in their history, these state-favoured companies have been shorter of funds than the international majors. The majors used to have most of the control and most of the profits upstream. Now the control has gone. But 1979 does not suggest that the upstream profits yet have.

At present, the importing countries' NOCs can get less of their crude at preferential prices than the majors, but much more than the private independents do. That may enable them to increase their share of their internal markets at the expense of independents, national or foreign, who operate there. Also, NOCs have increased their share in some significant markets because the majors have chosen to pull out. (Theoretically, the upstream profits that the majors have still garnered in recent years, through getting most of their crude at official prices and/or with discounts, might have enabled them to strengthen their own market share at the expense of the independents. But with the exception of a few selected markets, they do not seem to have chosen to do so.)

There are few signs that the NOCs of importing countries are expanding much abroad – unless one counts in the exploration and production contracts that they, like many private companies, are seeking. Where they are successful, such contracts may secure crude supplies; but ownership and control of the actual foreign producing operations are likely to be firmly in the hands of the host countries' governments and NOCs from the start.

Possible Changes in the Pattern of Operations of the World Oil Industry

It is important to re-emphasise that NOCs, at either end, are simply instruments. The shift of responsibility – and primarily of control – has been from international companies to national governments. Some further alterations in the pattern of operations of the world oil industry *may* occur with the growth of government to government deals – if importing governments get more deeply involved.

How these shifts will alter the pattern of operations of world oil, nobody can yet begin to guess fully. But a few very practical elements of change are already becoming obvious.

First, crude production will probably never be developed to the annual levels that it would have been under private enterprise. The OPEC governments have a lower rate of social time preference. Crude production on the other hand might be stretched out longer: at least, in most OPEC countries that is the intention.

World trade in crude will involve many more arms' length bargains than before. This would not necessarily follow simply from the fact that NOCs, rather than international majors, serve the same third-party customers. But the majors' share of downstream as well as upstream business may go on declining. Also, in spite of NOCs' intentions the share of crude passing (perhaps more than once) through the hands of international traders may tend to rise.

International movement of crude may become, for a time at least, more expensive in terms of logistic resources such as tankers and storage facilities. OPEC governments and NOCs are now dictating less flexible lifting schedules from their export terminals. Some are imposing destination controls which limit the flexibility with which the world-wide matrix of oil trade can be optimised. All tankers are steaming more slowly because of higher bunkering costs; also, more arms' length deals involve more part-loading and discharge, and hence offset the inherent economies of scale of regular LCC operation. The recent insecurity of supply from some OPEC countries is persuading companies and countries to hold much larger stocks of crude downstream, as against upstream, in the ground and in the spare capacity of production/pipeline/terminal facilities. OPEC might fairly say that this simply represents a shift of the cost of flexibility, which its member governments had to bear before, to the importers who benefit from it. But on balance it also represents a loss of technical efficiency.

Temporarily at least, this shift also means that the world oil trade is likely to be undertaken with much less complete information than it was under the majors. One cannot pretend that complete information was very widely available. The market the majors dominated was far from fully transparent to anyone outside, on prices or anything else. But within the majors, vertical integration did make possible a constant interchange of information between the functional stages of the business; and the joint shareholdings in many concessions supplemented this with a very wide flow of information horizontally across the upstream stage. Now the majors are losing some of this information, while OPEC governments and their NOCs are as yet not getting much of either kind. It is not easy to see how to replace this interchange of information, but there is an intense need for new ways to gather and communicate it.

It is hard to argue whether this shift from international to national control will make the market more or less competitive. None of the new upstream suppliers has much influence on supplies from elsewhere, except perhaps through parallel interests in maintaining prices. The industry is becoming far less integrated, vertically or horizontally. There are, indeed, more OPEC governments than there ever were international majors, but on the other hand these governments are much more openly interested in collective price fixing than the international majors have ever dared to be during the postwar era.

Sovereignty and Bargaining Power

The ultimate bargaining strength of OPEC governments is their power to keep supply below demand; so far however they have never really used this. In their actual trading of these last few years they have benefited from their customers' recognition of this potential ultimate power: but also from something else. This has been the sheer fact of national sovereignty. The majors, still deeply beholden to OPEC upstream, have not been able to withstand this; and independent buyers have been over-awed.

It might seem that government-to-government deals should bring matching sovereign power to the buyer's side. There is little sign so far that it has. Importing governments have a special ability to throw in a lot of non-petroleum inducements, commercial and perhaps political, that oil company buyers lack. But few seem to have made much better bargains. The one potential importing NOC that might conceivably sit down across the table from OPEC with comparable trading power, the single buying agency often proposed for US crude imports, has never materialised, and seems now abandoned even as a proposal. In government-to-government bargaining on a collective scale, as in the North-South dialogue and the many proposed repeat performances, the importing side has not mustered as much common interest as OPEC can generally demonstrate.

The greatest uncertainty about this new control upstream by OPEC governments and their NOCs is in exploration, OPEC pricing has stimulated exploration elsewhere. The attitudes of OPEC governments to foreign enterprise, and their declared intentions to stretch out depletion, have however inhibited exploration by foreign companies inside OPEC countries themselves – which still probably contain the world's most prospective acreage. OPEC governments have no lack of funds to explore with, and can expect high success ratios. Some NOCs, indeed, are carrying out active programmes of exploration. Others and their governments are still seeking foreign companies to take the exploration risks, and new formulae to attract them. This may simply confirm the common view that governments are more risk-averse than private entrepreneurs – and thus unsuited to the conduct of industries involving risk. Such an aversion to risk is however one of the few luxuries that OPEC governments – whose whole wealth and position in the world derives initially from exploration – can hardly afford.

13 ENERGY AND THE FINANCIAL SYSTEM

Minos Zombanakis

I propose in this paper to touch upon a subject which is of great importance and, in the short run, possibly more dangerous than the shortage of energy itself. This is the financial system as affected by energy shortages and by recent developments in the price of energy.

We have two main problems to contend with:

(a) The increase in the price of energy and its effects on the present financial system.
(b) The additional requirements to finance development of energy substitutes and the effect of these investment requirements on the system.

The world today is served by a financial system which at best must be described as *ad hoc* arrangements between the big surplus countries and the US. Though the US is not a surplus country, it is the key to the system, since the dollar is the major reserve asset and any US policy, or any policy influencing the dollar, inevitably affects the rest of the world.

With the increase in the price of energy in 1974, the world witnessed a sudden shift of financial resources to a group of OPEC countries. Initially the shift caused deflationary pressures in the rest of the world, since oil exporters had little ability to recycle funds by buying goods from the countries which were suffering from the consequences of these vast transfers. The transfers involved recyclings through the existing financial mechanism. OPEC countries deposited their funds with international private banks or purchased instruments of indebtedness in the capital markets. The deposits were then used to finance the current deficits of various countries aggravated in many cases by the increased cost of imported oil.

From 1974 to 1978 we witnessed a process of readjustment. OPEC countries began to spend financial resources acquired through higher oil prices on imports for consumption and for domestic economic development, thus inducing a considerable increase in demand for goods from the rest of the world. The beneficiaries were the technologically advanced economies which managed not only to cover their oil deficits but to enjoy huge surpluses, to the detriment, again, of the rest of the world.

We thus have, on the one hand, a group of surplus countries consisting of a few OPEC member states and a number of strong industrialised nations like Germany, Japan, and Switzerland. On the other hand, we have a group of deficit countries consisting of weak industrialised nations and of a large number of developing countries, all tending to show a chronic structural deficit in their balance of payments. In the middle we have the US which, though a deficit country, supplies other countries with an enormous amount of liquidity by financing its deficit through the extension of credit to the rest of the world. The mismatching of maturities of deposits with maturities of loans leads to an uncontrolled increase in world liquidity through the workings of the Euromarkets.

We have now reached a position where a group of oil-exporting countries and of technologically advanced countries commands an enormous amount of excess liquidity, while another group of countries is striving to cover balance-of-payments deficits either through excess borrowing or through a continuous policy of deflation, or both. The US stays in the middle offering its private financial structure and its currency as the means of intermediation between the two groups. From time to time the US injects additional liquidity in the system by financing its deficits through the extension of credit to the rest of the world.

As we enter the 1980s, political events in Iran have demonstrated the vulnerability of the system to unstable supply of energy and unpredictable price developments. We are less confident today that we can cope with the processes of adjustment than we were a few years ago. We face higher energy prices, restricted supplies in a world which has exceeded its margins for deflation and borrowing; or a world which unfortunately does not show any willingness or political ability to reduce its level of consumption to deal with this problem.

But let us return to the financial system, the way it presently operates in dealing with country deficits. These are now financed mainly through the private financial system. Large banking institutions obtain funds from the markets, either domestic or international, and extend credit in loans of various maturity and terms. This credit does not necessarily correspond to actual demand but mainly reflects the ability of countries to borrow, and of banks to accommodate such borrowings. As a result, borrowings range in maturity from anything between a few months to 12 to 15 years. In other words, you get whatever you can wherever you can find it in order to cover structural balance-of-payments deficits.

The deficits created in the period 1974–78 were financed by this type of borrowing. As the oil price remained relatively stable between these years while inflation continued to push up the price of industrial goods, the financing of these deficits was facilitated to the point where we thought that the recycling problem was solved. New increases in the price of oil in 1979, with expectations of more to come, present us today with a new situation. At the same time the financial system, based on the dollar, demonstrates acute weaknesses, manifesting themselves in exchange-rate instability and fears that the dollar will lose its power to act as a reserve asset. Further, the current management of exchange rates attempts to deal unrealistically with the process of adjustment, thus aggravating the problem.

Many seem to think that the surplus of one country automatically covers the deficit of another and therefore, regardless of what happens, cash flows will eventually equilibrate the supply of and the demand for credit. But this is not the case. Creditworthiness remains the key to acquiring credit. And the important question raised in this context is – who undertakes the risk of credit extension? At present the international private financial institutions, motivated by profit opportunities, or by the absence of any alternative, have taken upon themselves the responsibility of covering the country deficits around the world. One wonders whether this is a healthy way to deal with the problem. We have arrived at a situation where the ability to cope with the deficit depends on the willingness of these institutions to extend credit (which is another way of saying that there is today a group of institutions which could dictate policy to countries depending on the amount of credit they wish to extend). Alternatively, one can argue that there are a number of banking institutions locked into loans to debtors of doubtful reliability around the world. If they do not extend further credit, the borrowing countries may find themselves unable to repay what they owe, and the whole game becomes so interdependent that it is impossible to predict whether the banks will eventually bring about the financial collapse of the borrowing countries, or if countries will trigger off the collapse of the banks.

I think that we made a blunder in 1974 when we treated the oil crisis at that time as a passing episode. We should have seen that it was creating structural disequilibria in the world, which should have been handled through collective schemes involving the responsibility of States rather than private Banks. I must praise the Healey Plan and the various Oil Facilities as steps in the right direction. They did not succeed because many in the US thought of them as involving distortions of the market mechanism and threats to the role of the dollar as the international instrument of transaction and asset accumulation. Those were the days when policy makers in Washington thought of world dependence on the dollar as an element of strength for the US. It has now become evident, I am afraid, that this dependence is nothing but a curse for the US, but the realization has come too late for comfort.

The effects of energy on the world financial system are thus very considerable. If countries are to continue to absorb increasing oil prices through borrowing from private institutions, the burden of servicing such debts, sooner or later, will become close to matching the borrower's current foreign exchange earnings, which, in the final analysis, is the factor which matters most. When this stage is reached, debt reschedulings become inevitable. Debt rescheduling is a problem of bank cash flows: can banks finance the rescheduling of maturities? Some institutions will be able to and some will not. But in the end this difference is irrelevant, for it is the market that will have to re-finance the debt. This calls for a lender of last resort to play the role of the ultimate supplier of credit. And today there is no lender of last resort in the international financial system.

Thus, we have worked ourselves into a situation where the financial system will have to struggle in order to survive from day to day, praying that no incident will trigger off a disruption with wide-ranging effects. What is at stake is more than the continuing abilities of deficit countries to borrow ever-increasing amounts. We are entering the 1980s burdened with this problem while facing the additional challenge of increased needs in new industries and, above all, in energy substitutes. The problem is colossal, as can be demonstrated by a simple example: in order to create the capacity to produce the equivalent of 3 M b/d of oil from synthetic fuels, an investment of about $150 billion may be required today. This shows the magnitude of the financial problem involved in reducing dependence on a diminishing commodity like oil in the years to come.

We are approaching the 1980s in an economic environment marred by inflation and to stop this plague we must consume less, save, invest and produce more. We have to invest not only to maintain the present level of consumption and prosperity but also to compensate for the decline in the conventional sources of energy, so the effort must be bigger and the savings must be greater. But to save and reinvest requires determination at the national and international level, and this determination can be created only by leadership – in the final analysis, whether one saves or spends is a political decision transmitted through national economic policy.

Can all these objectives be achieved through the existing structure without causing a major disruption? I personally think that the financial system as it is structured today cannot bear the strains that complex and massive requirements will impose upon it in the years ahead. We must try at least to relieve it of those burdens that can be handled differently and more rationally. One significant burden is the financing of balance-of-payments deficits which should be taken away from the private financial system to the maximum extent possible, and be dealt with through collective schemes involving the capital exporters of the world – both the OPEC and the technologically-advanced countries.

An important factor to be taken into consideration on the general subject of investment is the shift of financial resources in favour of the OPEC and the technologically-advanced countries. A constructive dialogue is required to help us assess the magnitude of the necessary investments, the relative advantage that various countries have today in carrying them out, and the need to locate new industries in such a way as to bring about a long-term equilibrium in world payments. Without such a dialogue it is difficult to see how the world will be able to move towards a stable and prosperous environment.

14 ENERGY AND THE BALANCE OF POLITICAL POWER

Alberto Quiros

My theme is to review energy against a power frame and a time frame. It is concerned with a situation where consumers and suppliers continue to talk past each other. They seem to be unwilling to understand each other's position, let alone to make the necessary accommodation to reduce the tension of this relationship. Perhaps one of the reasons for this is that most discussion of energy has been, until recently, very much concentrated on the economics of the situation. We have had all sorts of projections from many sources proving that oil is too highly priced, and that this is disruptive to the world economy. Other projections show that oil has not yet reached the price of the alternatives to oil. In any event, the politics of the situation have not really been considered. Oil is a political question, but the sociological and political positions of the countries involved have not been adequately discussed. There is a vacuum where we should be dealing with the inter-relationships of nations. It is this inter-relationship of nations which will create inter-dependence in a new world, and not dependence or independence. The newly coined word is precisely that, inter-dependence. Now, let us look at how this relates to power. I think that there are some four factors that I would like to cite in order that we can use them as background to our discussion.

First, if power is to be real, it must be both perceived and believed. Thus other power holders, such as nations, groups or individuals, must recognise it. Also, they must believe that it would, if necessary, be used. For example, while military might is still seen by many as supreme, the political consequences may be such as to render it unusable, as in the case of the so-called 'limited weapons' war. It is not sufficient for the power to be visible, it has to be believed and perceived in order to be effective.

Secondly, there is no monopoly of power. It is widely shared and the alignments or groupings of power-holders can change. Real power, the ability to affect others, seems in fact more widely dispersed than perhaps at any time in the world's history.

Thirdly, in any discussion of power, Machiavelli has a view to be considered, such as his understanding that to take actions which make others powerful can have serious effects. Machiavelli had no doubts – 'whoever is responsible for another's becoming powerful ruins himself'. It is very important, and I call your attention to this, because in the discussion of energy and power we will see how power has been transferred from countries to other countries, from groups to other groups, both backward and forward. In so doing, every group that transfers its own power, even if not ruining itself, places itself in a position which is much weaker than it was before.

Fourthly, while we have tended to think of national power as a balanced mixture of military, economic, and political strengths, there are many examples today of power derived largely or almost entirely from one predominant factor. It could be economic capacity and financial strength, as in the case of Japan. It should be sheer mass, as in the case of China. It can be the control of strategic resources, as in the case of Saudi Arabia and most OPEC countries.

With these four themes as background, the story of energy-related power can be seen to have developed in cycles which seem very reminiscent of former power struggles between nations. A power structure which appeared to be settled is gradually changed by underlying factors of relative power and there is a steady build-up within the structure of tensions which are not accommodated either by events or deliberate decision. Then, abruptly, realities assert themselves and there is upheaval, followed by a period of adjustment and at least an apparent accommodation. However, one or another side then fails to maintain its position so that the power relationship again fails to reflect reality. There is renewed tension, leading in turn to renewed upheaval and adjustment. The cycle goes on and, in the case of energy, you can see how this happened. There was tension before OPEC took control in 1973. There was a major adjustment and apparent accommodation during 1973 and 1974. We had renewed tension after 1975. In 1978 and 1979 there was an apparent further adjustment. This has indeed happened historically in the energy field. In spite of this fact we keep being surprised where we should no longer be surprised. We all should know by now, that oil is, perhaps, the most accident-prone industry in the world.

I would like to select from different places in my original paper* which is the source of this presentation. This will provide the basis from which to discuss future possible scenarios, looked at from political, sociological and economic viewpoints. I would like to

[1] Published in *Foreign Affairs*, Summer 1979.

recall, for instance, the decision, apparently taken in the early 1960s, to keep cheap foreign oil out of the US. This could hardly be said to have been economically determined, internal politics were very much to the fore. What was the result? The US sat, apparently safe, behind its tariff barriers with its production controlled by the Texas Railroad Commission. If there ever was a model for any future cartel, it was the Texas Railroad Commission. The decision was taken despite the knowledge that domestic reserves were not unlimited. What went unnoticed at this time, was that the US was by this means establishing the ultimately very dangerous policy of 'drain America first'. Also, as described by Machiavelli, in so doing the US unconsciously set in train the very process described of making OPEC powerful, thus leading to its own problems of the present day.

Looking back in history, you can then see that the transfer of power from consumers to producers was started precisely by moves like this. Foreign oil could not go into the US, but it had been found and had to go somewhere. So where did it go? It went to Japan, and it went to Europe. It displaced, in many places, indigenous natural resources such as coal. All these economies then became dependent on the cheap foreign oil that was kept out of the US. The US at the same time was using its limited reserves, draining its reserves, just preparing itself for the kill. Then, when it really needed energy to grow, it did not have it anymore. So now it too had become dependent on foreign oil. The price could now be raised because dependency was established and power had been transferred.

But the history is still a cyclic one. After the events of 1973, many people thought that the problem was solved. OPEC was now powerful. Some five years later, 1978, most people were talking about there being a lot of oil around. Prices were not able to match the pace of inflation. In real terms, the price of oil was lower in 1977 and 1978 than it was in 1974. So again, the cycle of power had been transferred. Why? The OPEC nations, among other things, became themselves dependent on the consumers by importing excessive technology, setting up grandiose development plans and trying to do too much too soon. Therefore they created a need for this new economic input that they were receiving, and then they could not live without it. Thus, dependency was established in another way.

It is no good just telling people that this should not have happened. People wish to progress, people wish to develop. It is not a question of whether to develop or not, that would be unrealistic, it is a question of at what pace they should develop. Walter Rostow quotes a black colleague: 'The disadvantaged of this world are about to buy tickets for the show. They are quite unmoved by the affluent emerging from the theatre and pronouncing the show bad; they are determined to find out for themselves.' It is no good telling people not to seek development. On the other hand, there are a lot of people that think we should go back to the bucolic societies. I think it is a rationalisation of failure to propose such a return. Many people forget that at the turn of the century, most of the haves of Europe were saying much the same thing. They saw little need for change or growth. The turmoil that hit Europe, the social upheaval, at the end of the last century, was a direct consequence of this attitude. At least in the US where the attitude was also prevalent, they could all go west. In Europe they had no place to go and trouble boiled over.

Furthermore, is it really true that there is so much energy around? Well, it depends who you ask. Also it depends how energy is used and for what purpose it is used and how much each individual might have a right to use. If you look at a typical projection, and many other projections are very much the same, in the year 2000 every North American is expected to use the equivalent of 62 barrels of oil per year. Every European is going to use 32, which is already only half. The OPEC countries are expected to use 10. But the other countries, which are neither developed nor OPEC, if they are lucky, are going to use 4. It is against projections of social activity that energy demand is established. Can these really be expected to hold? How long is the world going to accept a society when only very few have 62 barrels of oil a year? The unit is not really oil, it is development, it is food, it is basic growth. Can there be 62 barrels for some and 4 barrels for the others, for half the world? If you believe that this can be maintained, then perhaps there is enough energy around for a very long time. If you don't believe that this can be held, then we will have to accept sooner or later that everybody will need a lot more than 4. I think that it is important to bear in mind that the so-called balance or imbalance of supply and demand is a very subjective thing.

It is not enough to count the heads and count the barrels of oil, it is much deeper than that. I would also like to reflect for a moment on how much the Iranian crisis which has just occurred parallels the 1973–74 crisis. If you remember, the Tehran Agreement of 1971 was planned to provide for a moderate increase in real prices, only to find itself overtaken by events. In 1978 the December pricing decision by OPEC also would have resulted in a moderate increase in real prices. Even this took consumers by surprise and resulted in a flood of protests, mainly from the US. However, the revolution in Iran, and the cutting off of Iranian supplies, soon overtook the implementation of the December decision and produced a much more dramatic change. The reaction of the consumers did little to help their case. They bid the price of spot sales to record levels, again repeating the experience in 1973–74. Sales at such prices, despite their small percentage of the total market, received much media coverage in the producer nations and gave rise to political pressure on producer governments to obtain the full market effects of these benefits for their people. Surcharges were then the inevitable result. If you remember, everybody in the producer countries was reporting that oil was selling at $28 then $38 and $46 – while the marker crude price was $14 or $16. What could be happening? Why were the multi-nationals taking all the profits? Multi-nationals are always to blame, or if not them the brokers are at fault. This was the prevailing view in the producing countries. If you sat in the consuming countries, of course, it was OPEC's fault that these things were happening. Whatever the cause, the reality of the situation is that it is very difficult, almost impossible, for any producer government to hold to the marker price while oil is traded by consumers bidding against each other at $40 a barrel on the spot market. This has a major effect, a political effect, in pulling prices up. Without question, it happened in 1973–74 and it happened again in 1979. We never seem to learn.

It is interesting to recall that 1977 was the year of the tight supplies scenario. Everybody was talking of an energy crisis, which was the conclusion of many studies. 1985 was going to be the critical year, though some said even earlier. We had the CIA studies and many others. Yet by 1978 the theme was complacency – the crisis is over – prices are going down – there is plenty of oil around. What happened was another accident. Iran came into the picture and then we were in tight supply once again. Just to remind you, this is not new, it has been happening for quite a while. But we did not focus on the problem from this perspective until fairly recently.

Where do we go from here, how might we deal with this situation? There would appear to be three possible scenarios, or alternative futures, although this is not a prediction. They are what might happen, but they are internally consistent in their elements. I will, with your permission, quote from my paper:[1]

[1] **'Energy and the Exercise of Power',** ***Foreign Affairs*****, Summer 1979.**

Looking beyond the current situation, what further developments might lie ahead in the energy and power relationship? Perhaps two, or even three, possible scenarios can be envisaged.

The first is a continuation of the present, with instability remaining for a number of years, particularly should Iran continue to experience internal political problems. Indeed, in the current "tight supply" situation, it would only need one other major OPEC nation to encounter difficulties for any reason. Certainly it must be expected that the consuming nations are currently re-evaluating their judgments of the long-term stability of each of the suppliers, and also their relationships with those whom they would judge to be the most reliable. However, it will take some time for any new assessment to be translated into political action, even given the necessary will to consider new accommodations with the suppliers.

Also, the political situation in the Middle East is far from resolved despite intensive efforts for an Arab-Israel agreement. As internal political pressures on Arab governments continue to build up, the prospects are that any available power sanctions might be used. These might well be cutbacks in production rather than outright suspension of supplies. They would certainly do little to improve stability.

For prices, the indications must be for continuing fluctuations as individual producers make their pricing decisions, with the overall outlook a rising trend. While power to control prices remains ultimately with the OPEC countries, the behaviour of the consumers may be the key factor. Thus, at times of shortage or major uncertainty, if consumers bid prices beyond OPEC price settings, then it will only be a matter of time before the awareness in individual OPEC countries of the value of oil to the consumers is translated into political pressures to raise prices, then into individual surcharges and finally into a raised OPEC price.

Accompanying such a situation would be a climate of accusations by consumers and particularly consumer governments of price gouging and windfall profits against the oil companies, with calls for sanctions against them and the suppliers.

The "instability" scenario seems likely to continue for the balance of 1979 at least. Against such uncertainties, economic prospects cannot be other than difficult. The climate for international trade and for investment must be influenced by concerns to limit exposure and risk. Essentially, despite a widespread desire to stabilize the economic and energy situations, the power to do so would be beyond that of any of the parties involved. The impact on the economic situation of the developed countries, not to mention that of the non-oil-producing developing countries, can only be serious, over a short time frame. For the longer term, it would plainly be most unhealthy for all concerned.

The second scenario would be a return to the status quo, to relative stability in world oil trade, and oil prices, at whatever higher price level the market eventually settles, be it $20 or $25 per barrel. This new plateau would require a degree of stability in Iranian production sufficient to ensure production at moderate volumes, although below historic levels, and no other major supply upsets.

Resultant cutbacks in consumption, both as a shock reaction similar to 1975 until users become accustomed to higher prices, as well as from true demand elasticity and an economic slowdown, would remove the strain on supplies and allow a return to order. Indeed, a short-term oversupply situation could appear and with it, within a year or two, a resurgence of some form of the complacency scenario to "show" that there is no energy problem.

The world's economies would eventually accommodate the changes and commence a new and probably slower growth trend, but short-term prospects, especially due to the one-time but much heralded inflationary influence of the price increase, will have been damaged.

This scenario would also be expected to display the all too familiar polarization between producers and consumers. As long as the price platêau remains below the cost of alternatives, their development will remain inhibited, and even if crash programs were initiated by political forces, their time scale is so long that it would be the end of the 1980s before they had much effect on volume of demand. Consideration of prices would most likely return to the familiar short-term debate on current year settings and their influence on consumer economies.

But once again the problem would not have been faced in its fundamental aspects, and, of course, the strains will start to build up again for the next cycle of the energy confrontation.

Or might a third scenario become possible? Might the world have finally learned enough from painful experience to make a "resolution" scenario credible? In such a scenario the energy power struggle would at last be resolved into a state of balance between the interests of the producers and consumers. Understanding the factors of power involved to achieve this is complex, yet never in the past did so many opportunities for analysis and understanding exist as there do today. Recent history has provided all of the necessary material – men should be able to draw the conclusions.

The consumer nations still do not realize that the OPEC nations are producing their irreplaceable resources much faster than they would choose based on long-term national considerations, or the concern in OPEC nations that their development needs have led them into this trap. Somehow the suppliers must back away before most of their oil is gone. Meanwhile, they must at least ensure adequate compensation in exchange for their oil, and they do not see that being achieved at prices well below the costs of alternatives. Instead they see OPEC subsidizing the developed world.

The producer countries have not helped their cause too much. The undreamed-of wealth after 1973 has, in many cases, not yet been well used to develop the infrastructure of these countries so that they can continue to produce wealth after the oil is gone; yet, for most, the years of financial surpluses are over. The producers' need was and is for long-term pricing and other contractual arrangements which would enable them to do a better job in developing their economies, but they remain far from certain what these are.

What is needed is a long-term understanding under which OPEC would now supply the volumes the world needs, at prices that consumers can afford in exchange for future considerations, financial assets, markets, prices, technology transfer and trade, in which the OPEC producers can have some confidence. There have been very few genuine efforts to understand or achieve such an accommodation.

On the one hand, the producers recognize their power to raise prices and to lower output – while at the same time they understand that their long-term interest lies in recognizing the need for the developed nations to continue their rate of growth, at a level that is not so low as to result in world stagnation. This could mean that some producing nations may have to make available their natural resources at a faster pace than required for the generation of the revenue necessary for the orderly economic development that will meet their political imperatives. And they need the cooperation of the industrialized consuming countries to achieve that kind of development.

On the other hand, the consumers must turn increasingly to alternative sources to conventional oil supplies, and must adjust their economics to the use of such sources, while guaranteeing to the producers a mechanism that will return to them in the long term the revenues generated but not required in the short term.

This scenario recognizes that a period of transition is required before the objectives of producer and consumer countries reach, if not a perfect match, at least a workable arrangement. In this situation, as far as price is concerned, surely the course of reason is to have the price of oil determined by long-term considerations. The only meaningful long-term price is the cost of replacement of conventional oil by other premium alternatives that are effectively not limited by the availability of natural resources. And the least disruptive development would be for the real price of oil to rise gradually to that level – some $20–30 (in 1979 dollars) – over a period of, say, ten to fifteen years.

Such a scenario would mean higher prices coming about over a period of time that would permit the consumers, developed and developing, to adjust to them. If it was clear that this was in fact the price trend, it would ensure that the new and otherwise marginal alternatives to oil would be ready when needed. And, more basically, it would open the way to a stable balance of power between producers and consumers.

Unfortunately, no consumer government has yet been able to encourage this kind of thinking. The pressures of short-term considerations, such as this year's trade balance, are too great. The world's political systems have not as yet developed that far. The power of the consumers is concentrated on the short term to keep today's energy prices down as far as possible. The world and its power interactions still function within a very short-term horizon.

For such reasons, the credibility of the resolution scenario must be suspect. However, without the necessary understanding and accommodation of long-term issues so that a balance of power is achieved, the alternative will be a continuation of the now familiar cycles of growing strain and then painful adjustment under a ratchet of increasing prices – until at last the level of premium alternatives is reached. The problem is that this could be a long, painful and dangerous time for the world.

Does anybody believe that the resolution scenario can be achieved? There may be some hopes, but I cannot expect so at this moment in time. My reasons are typified by a recent speech. I am sure that there are many such speeches but this one was made by the President of the United States. He is a rather important person in the world of energy today. A few quotes from the President's speech will explain better than I can whether we can believe in a resolution scenario; whether we should believe that inter-relationships and inter-dependence could be established because that is what we are talking about. I quote:

It is clear that the true problems of our nation are much deeper, deeper than gasoline lines or energy shortages; deeper even than inflation or recession. We can't go on consuming 40% more energy than we produce. When we import oil we are also importing inflation plus unemployment. We've got to use what we have. The Middle East has only 2% of the world's energy but the United

States has 24%. Our neck is stretched over the fence and OPEC has the knife. There will be other cartels and other shortages. American wisdom and courage right now can set a path to follow in the future. The real issue is freedom. We must deal with the energy position on a war footing. We have learned that piling up material goods cannot fill the emptiness of lives which have no confidence or purpose. All the traditions of our past, all the lessons of our heritage, all the promises of our future point to another path, the path of common purpose and the restoration of American values. That path leads to true freedom for our nation and ourselves. We can take the first steps down that path as we begin to solve our energy problems. Energy will be the immediate test of our ability to unite this nation.

And then he continues with regard to 'drain America first':

In little more than two decades we have gone from a position of energy independence to one in which almost half the oil we use comes from foreign countries at prices that are going through the roof. Our excessive dependence on OPEC has already taken a tremendous toll on our economy and on our people. This is the direct cause of the long lines that have made millions of you spend aggravating hours waiting for gasoline. It is a cause of the increased inflation and unemployment that we now face.

It continues in similar vein, but in particular I noted the reference to Mexico and Canada: 'We are and we will continue to be a good customer, a good neighbour and a good trading partner with both Mexico and with Canada.' They obviously must be doing something right. However, whilst this was being said by the President, Mr Schlesinger, who was also rather important in the energy field, said the following: 'The OPEC nations possess some 80% of the free world's proven oil reserves and that percentage is likely to increase. Moreover, market forces now control price trends inexorably driving the international price of oil upward.' And he also said: 'Indeed our neighbours Canada and Mexico and the North Sea producers have posted prices substantially above those of OPEC.' To quote Mr Schlesinger further: 'It is time for us to acknowledge reality however painful that may be.' However, you have to sympathise with Mr Carter because he is dealing with an internal political situation which is very difficult. But did anybody come out the next day and agree that Mr Carter had said the things that the American people wanted to hear? The *New York Times* had an Editorial on 'Riding Casually to War' and it ended: 'After 12 extraordinary days of deliberation, he [Carter] proposes not a war on energy or even the moral equivalent of war. His plan is accommodation to dependence. If he listens to the people as he promises to do, they need to tell him that this is not enough.' I don't know what enough would have been, but it wasn't enough. What about externally? Some of us here are believed to be members of a cartel.

Now, how would we react if we were a cartel and our biggest customer turned to us and said, in effect: 'You're killing us, we don't like you, we are almost at war with you, and furthermore, we are developing a strategy that's going to put you out of business. In 1985 or 1987 we are not going to import more than 3 million barrels of oil though we now import 8.' This is going to come at a time when most developing nations and producing nations need far more revenues than they do today. But yet the biggest customer is telling us that as far as it is concerned our role will end. But remember, it's not going to be tomorrow that our role ends. It's going to be in 1987 or 1988. 'But meanwhile don't stop the oil, provide 8 million barrels now so the US can then gradually and casually and effectively put you out of business.' So there you are. Do I believe in a resolution scenario? I don't think so. I wish I could.

To finish, I would like to return to the original article because I think there is an important aspect that has to be remembered:

> Energy remains a sensitive focus of power interaction. There are, regretably, few signs of long-term stabilization of the energy situation. The OPEC countries are still essentially developing countries and they have yet to understand fully how to use their power effectively. As long as the consumers are content to remain dependent rather than interdependent, they remain vulnerable and diminish their power. The present competition involving consumers, OPEC and the non-oil-developing countries is destructive to all.
>
> The world is producing wealth far below its potential – perhaps some 20 percent or $1,000 billion each year, below what would have been possible if growth had not been constrained since 1973. Interdependence is the only way out, but only the developed nations can really take the necessary steps leading to partnership with the developing world. This will be particularly difficult for the United States, whose whole history and culture have developed around independence rather than interdependence. However, the current position of the United States is close to a dependence on OPEC, and surely that is not something the United States will enjoy.
>
> What has to be understood is that there is no balance between today's energy requirements of the developed nations and the need, in social, political or economic terms, to produce these same requirements by the underdeveloped world. If the industrialized world were to reduce consumption of natural resources in order to keep pace with the long-term resource extraction policies by the underdeveloped producer nations, then world growth would come to a standstill and stagnation would result. If the underdeveloped world were to produce its resources at the rate required by the developed nations, it would exhaust its one element of power, natural resources, and would still sit at the tail end of the growth line in the future. So it seems that the time frame against which these fundamental social and political decisions are being made is wrong. As we can see, whatever the oil price, it is a symptom, not a cause, of differing needs. The producers need the understanding of the consumers, for whatever short-term benefits can be obtained will be short-lived, as has been demonstrated

by recent events.

Under current perceptions of power, long-term arrangements which correct the time-frame imbalances are difficult, since the representatives of current political systems live short political lives. There seems to be no answer but to change the perceptions of power — as a result of a recognition of the need to view them against a longer term time frame. Perhaps the institutions involved need not be changed. They could survive for a very long time if they were to change to a focus which is wide and long-term. For a long time the world has understood the need for a 'master plan' that would include the have-nots of the planet in the benefits of growth. The alternative inevitably will be social upheaval. Of course, every generation believes this will never happen in its own lifetime but, with increased desire for continuing growth, we may be approaching the fatal date more quickly than we think.

Part IV

WORLD ENERGY OUTLOOK

15 WORLD REGIONAL ENERGY MODELLING[1]

Wolf Häfele

There are numerous schemes and programmes for planning the energy future of certain countries. They provide a wealth of detailed measures but usually involve a fairly short (at best a medium-term) planning horizon of up to 15 years. After all, detailed measures cannot be undertaken outside a short time frame.

When innovations, that is major changes in an existing infrastructure, are involved much longer periods of time must be considered.

Compare, for example, the evolutions of various primary energy carriers in the energy market. Fig. 1 is a logistic representation showing the S curves of different energies penetrating the market (from share 0 to share 1) as straight lines. Their behaviour over a period longer than a century [1] is remarkably regular. This is just to underline the need for a long-range view once changes in the energy supply structure are at stake. If one takes into account that the life of a power plant averages 30 years, the time frame to consider for innovation may well be 50 years.

Looking at the energy problem from this angle, one is confronted with several difficulties:

- We are not equipped (neither analytically nor via the control of the market mechanisms) to come to grips with the interplay of short- and long-term aspects.

- International interdependence will have increased considerably 50 years from now. This will not come as a surprise, since already today 50% of the crude oil in the Federal Republic of

[1] This paper was published in *Futures*, February 1980, 12(1), 18-34. Reprinted by kind permission of IPC Science and Technology Press, Guildford, UK. The considerations discussed are founded on a five-year study of IIASA's Energy Systems Programme, with contributions from scientists from the US and USSR and 15 other countries in East and West. The study is being documented in a one-thousand-page volume on 'Energy in a Finite World – A Global systems Analysis', which is due to appear at the beginning of 1980.

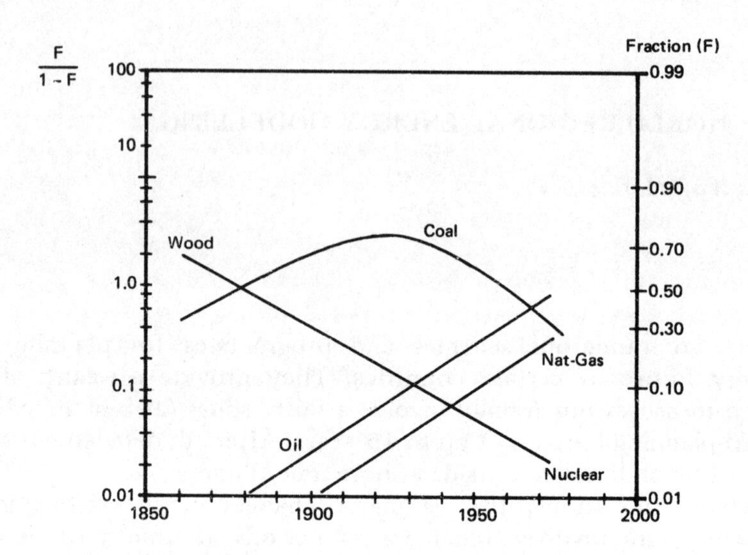

Fig. 1 World: primary energy substitution.

Germany, for example, originates from one single area, the Persian Gulf. Yet we will have to do better in viewing the world as a whole.

– There is a severe lack of input data, required for the standard tools of energy planning, with respect to the 50-year time frame. How will the world then look geopolitically? What about elasticities, i.e. the change in percent in the demand for a given secondary energy as a function of percentage changes in various determinants, such as prices or gross domestic product? Whereas most large-scale econometric models now in use in the US assume that elasticities are known in one form or another, the truth is that for long-range investigations no data exist for the relevant elasticities.

For these reasons, medium- and long-term strategies of energy supply are difficult to deal with, both in terms of substance and methodology.

An attempt at formulating long-term global energy supply strategies has been undertaken at the International Institute for Applied Systems Analysis (IIASA), Laxenburg, Austria. IIASA's Energy Systems Programme gave much thought to a qualitative understanding of the energy situation and a breakdown of the problem into feasible subsets. This enabled the Programme to proceed to a synthesis which depicts a global and long-term energy supply and demand picture.

The tool chosen is the writing of internally consistent scenarios: the greatest possible number of necessary conditions is identified and used to limit the scope of subjective judgment. To support the effort, a set of computer models is relied upon. The principle guiding this approach is plausibility. Two basic scenarios are constructed, makking the range between plausible upper and lower bounds. But since there is only *one* reality, the scenarios are construed as guidelines toward conceivable energy futures, and their outcomes are taken as indicators and not as predictors. Since the forces of the market have lost much of their regulatory power, such guidelines are indispensable. This is particularly true of the oil market, where prices and quantitites have come to be determined by political action, irrespective of the market mechanics. Market forces will work again properly, leading to the necessary investments, only when confidence and trust in the market are being restored.

Two Scenarios

For some steps in the description of the energy problem, it is useful to consider global overall figures like the following. The world today consumes about 8 Terawatt year per year of commercial energy. (Note that one TWyr/yr fairly accurately corresponds to one billion tons of coal equivalent, tce, per year.) The average per capita consumption then amounts to 2 kWyr/yr (see Fig. 2). About 70% of the world population, however, live on much less than the average, and a considerable number of them on only 0.2 kWyr/yr per capita. A conceivable addition of 0.3 kWyr/yr from burning wood and manure would mean much in this context. About 22% of the world

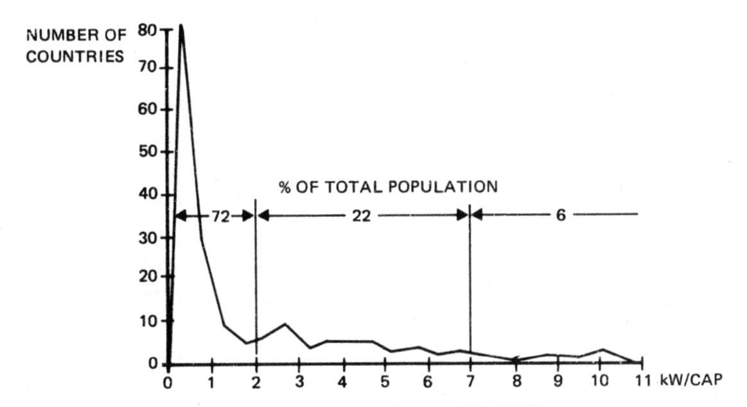

Fig. 2 Per capita commercial energy consumption world, 1975.

population including the Europeans use 2–7 kWyr/yr per person. The remaining 6% enjoy a per capita energy use of 7–12 kWyr/yr. If, as in Fig. 3, one assumes a doubling of the world population in the coming 50 years – a rather conservative assumption [2] – and an increase in the per capita average to 3 or 5 kWyr/yr, the world's energy demand would rise to 24 or 40 TWyr/yr, respectively. This simple consideration helps to assess the efficiency and the capacity required from future energy systems.

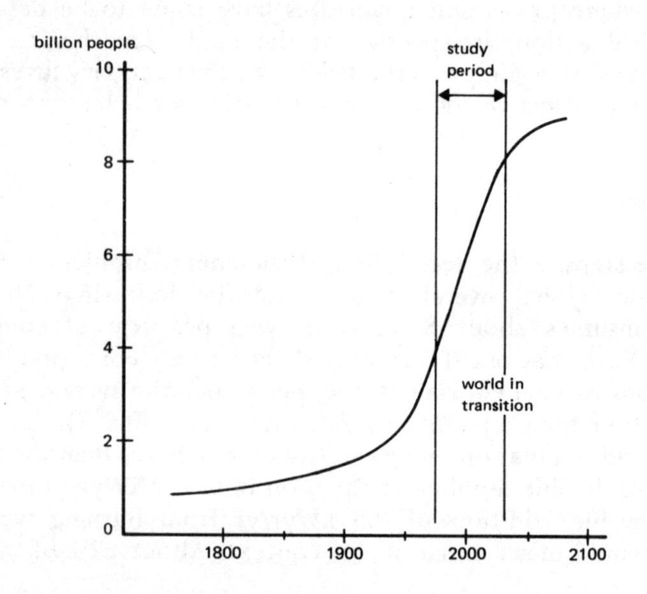

Fig. 3 **World population: historical and projected.**

Unfortunately, such rough guidelines are not sufficiently detailed for real-world decision making, and one is tempted to go back to the national framework. This cannot be done here, however, since it would impair the global vision of the problem. IIASA, in seeking a way out of this dilemma, has identified seven *regions* that describe the world as a whole. In this way, typical regional differences are accounted for and regional interdependencies identified (Fig. 4). These regions differ above all in their states of economic development and the availability of resources, and to a lesser extent in geographical conditions. Regions I II and III correspond to the so-called first and second worlds: the industrialized North. Regions

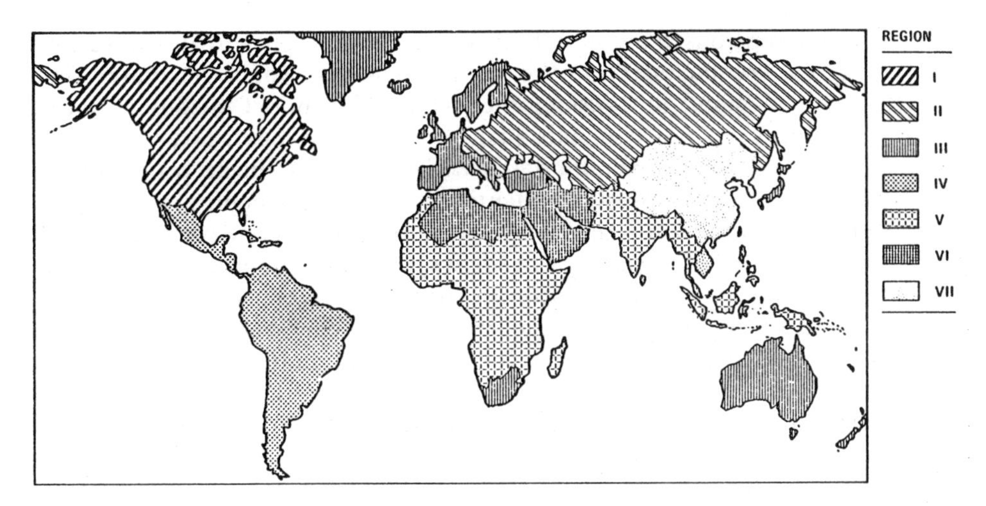

I (NA) NORTH AMERICA

II (SU/EE) THE SOVIET UNION AND E. EUROPE

III (WE/JANZ) W. EUROPE, JAPAN, AUSTRALIA, NEW ZEALAND, S. AFRICA, AND ISRAEL

IV (LA) LATIN MAERICA

V (Af/SEA) AFRICA (EXCEPT NORTHERN AFRICA AND S. AFRICA), SOUTH AND SOUTHEAST ASIA

VI (ME/NAf) MIDDLE EAST AND NORTHERN AFRICA

VII (C/CPA) CHINA AND CENTRALLY PLANNED ASIAN ECONOMIES

Fig. 4 Seven world regions.

IV, V, VI, and VII represent the developing third and fourth worlds, with widely differing national structures. A set of mathematical models is applied separately to each world region. This IIASA set of energy models is depicted in Fig. 5, with the larger computer models shown in boxes.

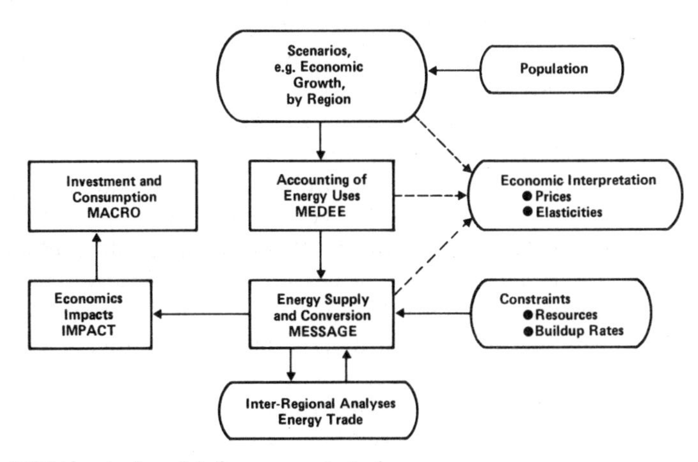

Fig. 5 IIASA's set of models for energy strategies.

Assumptions on population and economic growth in the various regions are included in the model MEDEE, which calculates final energy demand in considerable detail. The model output (a set of secondary energy demands such as electricity, heat, gas, etc.) is the input to the supply model MESSAGE. This linear programming model allocates specified quantities of primary energy, such as oil, gas, coal, uranium, etc., to the generation of secondary energy over a period of 50 years. It produces optimal discounted costs and, most important, takes into account various constraints, providing in this sense an optimal supply mix of primary energies in a region. The resulting requirements for direct and indirect investments in energy generation are accounted by the model IMPACT. The model uses an input-output approach to identify the effect these requirements may have on a given economy. The aggregated investments are then fed into the MACRO model, which helps assess the macroeconomic implications of changes in the ratio of energy consumption and investments. Using a quasi-formalized procedure which links international trade between the world regions, a first-order approximation of input data for MESSAGE is obtained. Prices and elasticities are also an output of the model.

While all the models in the set have been developed in their present form and applied at IIASA, their origins vary. MEDEE originates from the University of Grenoble, MACRO is based on work done in Canada and the US, and IMPACT and the world trade procedure come originally from the Siberian Power Institute at Irkutsk. The model MESSAGE has been completely developed at IIASA.

In the light of their explanations, it is crucial to emphasize the iterative character of the modelling procedure. The findings described below could in no way have been obtained from one run through the modelling loop.

Two scenarios have been constructed which are defined by two basic development variables, population and gross domestic product. Both scenarios, High and Low, are rather conservative, representing moderate departures from observed trends cases. In either scenario, population is assumed to grow to 8 billion in 2030 (and would then taper off to a sustainable level). The basic difference is in gross domestic product (GDP) projections: one scenario assumes a relatively low economic growth, fairly large advances in energy end-use technology, and a rather positive attitude towards energy saving of those concerned. The other scenario assumes a modestly high growth. The less conservative assumptions made on the supply side in both scenarios include effective and timely decision making and implementation, as well as due regard for the needs of the developing countries.

These assumptions are rather optimistic, marking the maximum bounds of what may be feasible, while the real world performance may turn out to be more modest.

Energy Demand

Let us now look more closely at how energy demand is dealt with in the two scenarios. To this end, it is useful to consider the economic development of the regions in terms of percent of per capita gross domestic product (GDP).

Table 1 1975 per capita GDP and growth rates for two scenarios to 2030.

		Growth Rate of Per Capita GDP (%/yr)			
		High Scenario		*Low Scenario*	
Region	*GDP Per Capita ($) 1975*	*1975-2000*	*2000 2030*	*1975-2000*	*2000-2030*
I	7046	2.9	1.8	1.7	0.7
II	3416	3.6	3.2	3.1	1.9
III	4259	3.0	1.8	1.7	0.9
IV	1066	3.0	2.4	1.6	1.9
V	239	2.8	2.4	1.7	1.4
VI	1429	3.8	2.8	2.4	1.2
VII	352	2.8	2.4	1.6	1.4

Table 1 gives the per capita GDP and its yearly growth rates used in the High and Low scenarios, 1975–2030. In the Low scenario, the per capita GDP growth for North America (region I) goes down to 0.7%/yr, and that for Europe (or more exactly, region III) comes to be only 0.9%/yr. Both values are meant to approximate zero economic growth. The highest growth rate, by contrast, is that for region VI (Middle East and Northern Africa) from now to the turn of the century (3.8%/yr). The Soviet Union and the Eastern European countries (region II) have generally high values but otherwise follow the decreasing trend.

Gross regional products (GRP) are obtained by multiplication of the GDP growth rates by regional population figures. (The population data here are from Keyfitz.) In the OECD countries, GRP annual growth rates, 1975–2030, range between only 2 and 3%; the evolution assumed follows the general decreasing trend.

More important than the economic data are the related values of energy demand. An adequate definition of energy flows from the source to consumption, differentiates at least between primary energy and secondary energy as well as energy use (see Fig. 6). The latter term in fact comprises what is called energy services, services which produce, for example, a fine piece of pottery, a warm room, or adequate illumination for reading. This energy service can be consumed (energy itself follows the law of conservation). The use of energy has to do with the negative entropy or negentropy (or information) content of energy. This rather abstract quantity, equivalent to the use of capital or work or to the impact of know-how, can completely or partially be consumed or substituted for: the piece of earthenware may break, the room may cool down, and the light photons be absorbed.

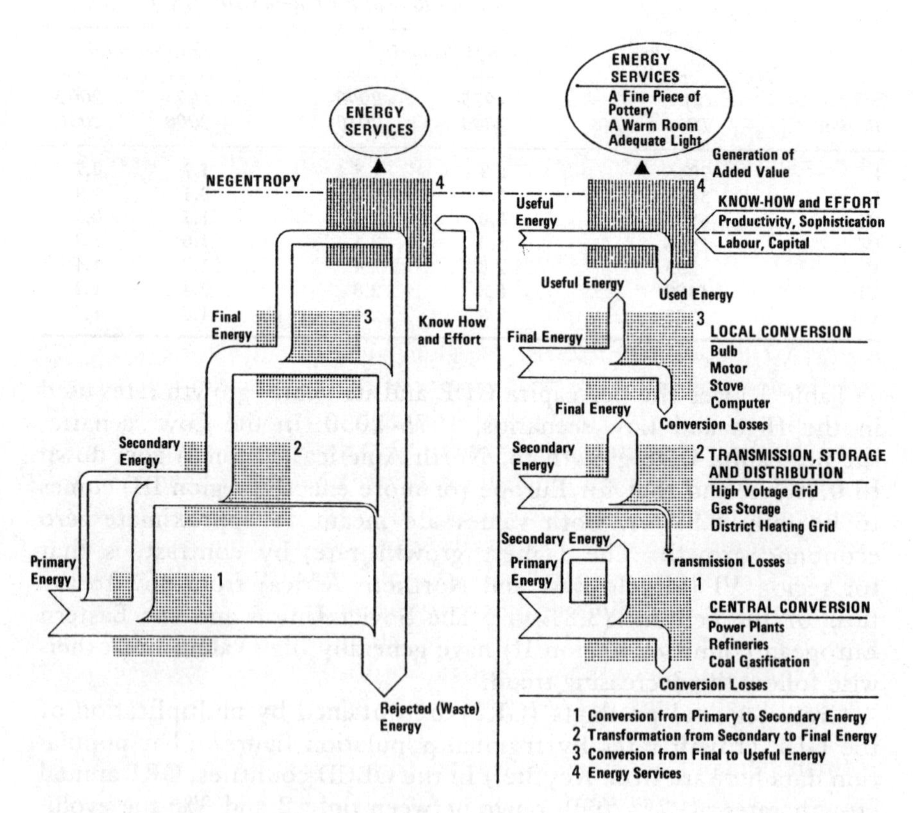

Fig. 6 Energy flows: definition of terms.

The point here is that the relationship between energy consumption — which depends on the level of a given economic activity — and the economic activity itself is not unambiguous and straightforward. No wonder, the issue is at the centre of a controversy today. It surfaces in the discussion on energy coefficients, that is the percentage of energy growth required for each percent of GDP growth. And, as we will see in a moment, the differentiation of final energy and primary energy is very important in this context.

The final energy-GDP coefficients for regions I, II, and III are estimated at around 0.8, but those for regions IV, V, VI, and VII at around 1.5, which demonstrates the need for the developing countries still to build up their infrastructures.

To argue this point in a convincing and credible manner, many details are necessary. MEDEE, the model for assessing long-term energy demand, is meant to provide these. It does in fact account for the great diversity of end-use categories and their interdependencies. Fig. 7 summarizes the relevant results. The final energy-GDP coefficient is assumed to go down to as low as 0.3 for the industrialized countries and to slightly less than 1.0 for the developing countries.

The respective coefficients of primary energy and GDP, on the other hand, may differ completely. (See Fig. 8, where ϵ_p for regions I, II, and III in 1975 was close to 1.0 but clearly above 0.8.) This is due to conversion losses at the level of secondary energy generation (electricity in particular). Considerable losses arise from coal liquefaction, which plays a major role in both scenarios for regions I and III, as will be seen later. This may well cause the primary energy-GDP coefficient to rise to values higher than 1.0.

It is not possible here to treat these energy demand calculations in greater detail. One point merits special attention, however. The demand for liquid secondary energy carriers, such as heating oil or gas, is shown to be a much more severe bottleneck than is widely assumed. Therefore, it is attempted in both scenarios to limit the use of liquid fuels to practically non-substitutable applications, such as transportation, feedstocks, and petrochemicals. Table 1 shows this use of liquids in percent. The shares vary between about 50% today and more than 90% in 2030. Therefore, it becomes more and more necessary to substitute district heat and electricity for heating oil and gas. As a consequence, there is a continuous steep increase in electrification (Fig. 9).

Such demand considerations lead to a per capita primary energy demand in the scenarios, as in Table 3: a world average of 3 or 4.5 kWyr/yr, respectively, in 2030 — as was noted above — instead

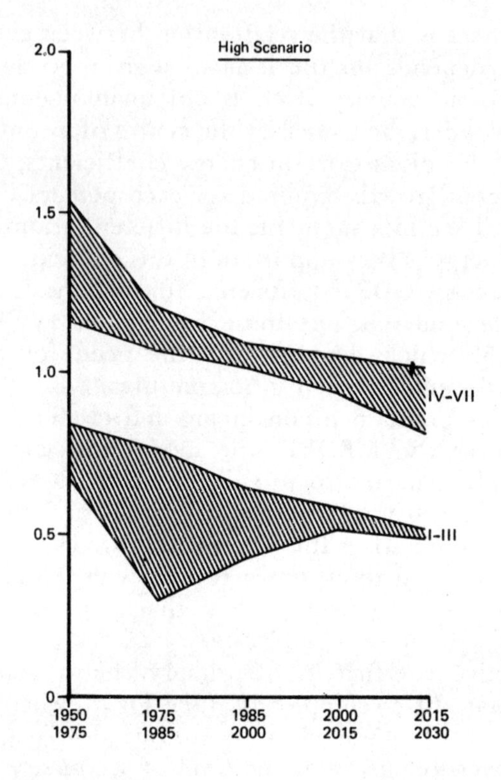

Fig. 7 Final energy-GDP coefficient.

Table 2 Use of liquids: percentage of liquid demand used for transportation and feedstocks.

Region	1975	*High* 2030	*Low* 2030
I	74	94	91
II	65	100	100
III	52	86	76
IV	69	90	89
V	58	91	88
VI	74	94	91

of 2 kWyr/yr per person as at present. While the per capita consumption ratio of regions I and III, compared to regions IV and V, improves by a factor of about two, considerable inequities between developed and developing countries remain, and the gap continues to be a problem far into the next century. The global primary energy demand is projected in Table 4, with 22 TWyr/yr in the Low scenario and 36 TWyr/yr in the High scenario.

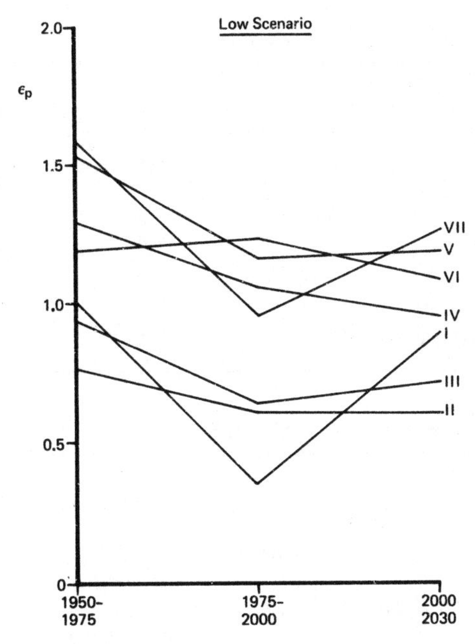

Fig. 8 Primary energy-GDP coefficient.

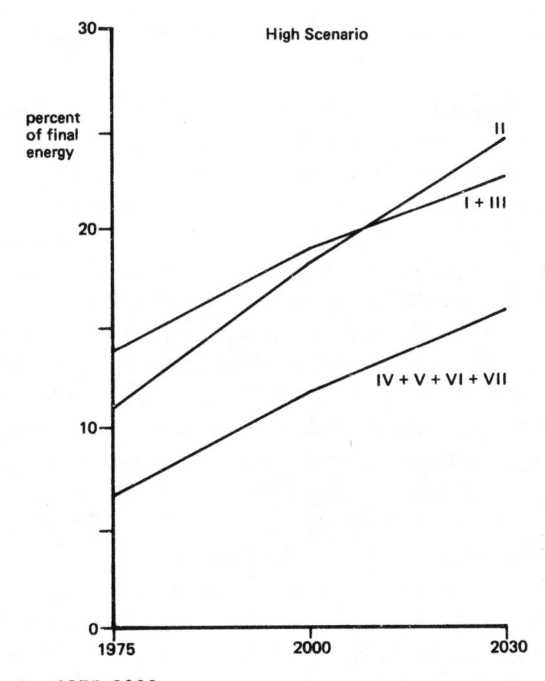

Fig. 9 Electrification 1975–2030.

Table 3 Primary energy per capita (kWyr/yr).

Regions	*1975*	*High 2030*	*Low 2030*
I + III	6.2	12.2 (2×)	8.2 (1.3×)
IV + V	0.4	1.9 (4.8×)	1.1 (2.9×)
World	2.1	4.5 (2.2×)	2.8 (1.4×)
Ratio $\frac{\text{I + III}}{\text{IV + V}}$	16.2	6.4	7.5

Table 4 Primary energy projections (TWyr/yr).

Regions	*1975*	*High 2030*	*Low 2030*
I + II + III	6.8	20.5 (3.0×)	13.9 (2.1×)
IV + V + VI + VII	1.5	15.2 (10.5×)	8.5 (5.8×)
World	8.2	35.7 (4.3×)	22.4 (2.7×)

However, it is easy to obtain even lower or higher values if the scenario assumptions are slightly varied. One method of comparison tried at IIASA is a 16 TWyr/yr scenario that retains the 2 kWyr/yr per capita average of world energy consumption. An increase in energy use in the developing countries must accordingly be offset by a negative energy consumption growth in the industrialized world. Table 5 describes this case for the seven world regions.

There is at present some agitation to promote a negative or zero energy growth for reasons other than energy supply difficulties [3]. Yet the impact this movement will have on our way of living cannot be grasped. At the other end of the spectrum, however, there are world energy consumption estimates of more than 40 TWyr/yr. The political concept of the New Economic Order, for example, pronounced by the UN group of the 77 at UNCTAD conferences, leads to such higher energy demand values [4]. But this is not inconsistent with an observation above that the link between energy and economy is not naturally a closed one. In this light, the energy demand figures of the High and Low scenarios fall well within the mid-range of today's projections.

Table 5 Per capita primary energy consumption, a 16 TW scenario, 1975–2030 (kWyr/yr).

Regions	*Base Year (1975)*	*2000*	*2030*
I	11.27	9.1	8.0
II	5.10	7.2	6.2
III	4.03	3.6	3.2
IV	1.06	1.8	2.8
V	0.23	0.5	0.7
VI	0.96	2.2	3.6
VII	0.51	1.0	1.2
World	2.1	2.0	2.0

Energy Reserves and Energy Resources

A clear distinction must be made between reserves and resources. Reserves, being resources that are proved, can be mined at economic rates. Resources then are considered to include reserves as well as a resource base: this resource base is presumed to exist from geological evidence, but its exploitation is not yet economic. Both categories vary continuously in quantity, the ultimate difference between them being technology. North Sea oil is a case in point. There, with the technology of floating platforms at hand, the resources have now become reserves, but only since 1973.

Estimations of resources traditionally differ from each other. Geologists apparently tend towards cautious estimates, while economists, guided by the role of price increases, proceed from a *de facto* unlimited resource base. Inherently different definitions apply to coal and gas, coal usually being estimated on a geological basis and oil with a view to maintaining a certain reserve-production ratio. Resource assessment is difficult, therefore, especially if assumptions of future conditions are involved, as in the present scenario.

While all this commands caution, one must produce numbers for the scenario definition. The estimates in Table 6 of ultimately recoverable fossil resources should be looked at this way. They are grouped in three price categories. Coal, oil, and gas of the cheapest category ($25/ton or $12/boe, respectively) make up about 1000 TWyr. Simple calculation shows that for 40 TWyr/yr this amount would be used up in 25 years, and public concern about resource scarcity originates from this. Realistically, categories II and III must also be included in the count, leading altogether to about 3000 TWyr. Also, it seems more appropriate to assume that world fossil energy use during the next five decades will average about

Table 6 Ultimately recoverable resources. Coal I: $25/t, II: $25-50/t; Oil, gas I: $12/boe, II: $12-20/boe, III; $20-25/boe.

Resource		*Coal (TWyr)*		*Oil (TWyr)*			*Gas (TWyr)*		
Cost Category		*I*	*II*	*I*	*II*	*III*	*I*	*II*	*III*
I	(NA)	174	232	23	26	125	34	40	29
II	(SU/EE)	136	448	37	45	69	66	51	31
III	(WE/JANZ)	93	151	17	3	21	19	5	14
IV	(LA)	10	11	19	81	110	17	12	14
V	(AF/SEA)	55	52	25	5	33	16	10	14
VI	(ME/NAF)	<1	<1	132	27	n.e.	108	10	14
VII	(C/CPA)	92	124	11	13	15	7	13	14
World		560	1019	264	200	373	267	141	130

15 TWyr/yr, which alleviates the situation. Of course, more details, in particular with regard to regional differences (see Table 6), were used in the modelling, as in the case of energy demand. But, even at this summary level, large differences are obvious between different qualities of coals, oil, and gas. For example, oil recovery in Saudi Arabia, Alaska, or the North Sea can not easily be compared. It shows that the 3000 TWyr in Table 6 are not a soft cushion to rest on.

Besides fossil resources, energy supply from renewable sources is getting much attention and raising great hopes. If we ignore the large-scale use of solar energy temporarily, we find the potential of the remaining sources to be limited. Much discussion would be needed to prove fully this statement. Since this is not possible here, a summary of the estimated potential of renewables is given in Table 7. It differentiates between what is theoretically or technically possible, and what may actually be feasible. The realizable potential of renewable sources appears to be about 10 TWyr/yr, a figure much lower than the expected demand. The average energy densities of renewables, on the other hand, are 0.1–1.0 W/m^2 (Fig. 10). With an indicative energy density of 0.5 W/m^2, this implies that 20 million km^2 of land will be needed to harvest the realizable potential of 10 TW – an area about as big as all the agricultural land in the world!

Another matter is the large-scale use of solar energy, the annual average density of which may be 20–40 W/m^2, and whose area requirement is relatively smaller. Extensive investigations [5] have shown that the need for land is bound to complicate solar use in some such cases, but might be resolved in general. A much greater long-term

Table 7 Estimated potential of world renewable energy supply.

Source	Potential: Technical	Realizable	Constraint
Forest and Fuel Farms	6.0	5.1	ecological climatological
Solar Panels Soil Storage Heat Pumps	5.0	1.0	economic technological
Hydropower	2.9	1.5	ecological social
Wind	3.0	1.0	economic
OTEC	1.0	0.5	ecological climatological technological
Geothermal	0.2	0.6	economic
Organic Wastes	0.1	0.1	balanced
Glacier Power	0.1	0	technological
Tidal	0.04	0	computational
Total	20	9.7 TW	

problem may be capital cost and energy storage. An overriding concern is the tremendous demand for material needed to cover such areas. The minimum material density is generally estimated to be 10 and 100 kg of concrete and iron respectively per square metre.

The solar power growth rates in Fig. 11 are estimated on this basis. With a hypothetical doubling of world concrete and iron production to be invested in solar power plants, between 250 and 1500 Gigawatt (thermal) could be installed per year. Therefore, and for other supporting reasons, the growth rate rather than the solar power potential appears to be a leading constraint on large-scale solar power application over the next 50 years. This, of course, assumes optimistically that capital cost and storage requirements can be met.

Now a few words about nuclear energy. In the present context, the relevant question is what magnitude of nuclear energy can be produced by 2030, leaving aside all institutional and societal issues. There is only a vague answer to it. Detailed studies – not discussed

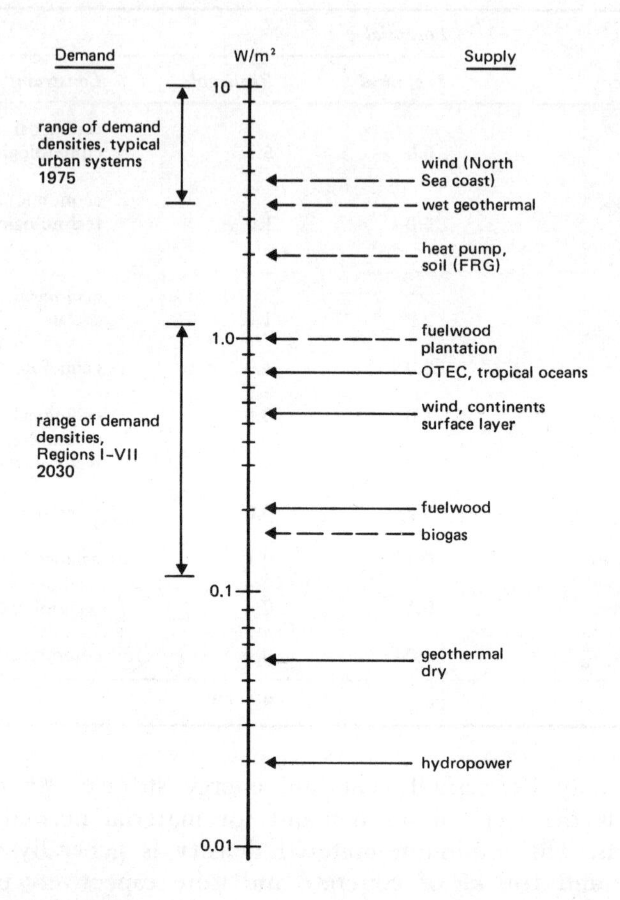

Fig. 10 Energies densities.

in the present context – indicate a worldwide realistic upper limit of 10 TW of installed electric capacity. It is trivial to show that the maximum could be less. To 10 TWe installed capacity would correspond a (commercial) primary energy consumption of 17 TWyr/yr. (Note that TWyr/yr always implies annual calorific input to produce power.) At such a growth rate, the sensitive parameter is the natural uranium (U nat) requirement, e.g. of light water reactors. 4.3 million tons at prices up to $130/kg U nat in the Western world are considered available today, for example, by the International Fuel Cycle Evaluation (INFCE). For the world as a whole and with worldwide prospection at the present US level, one could consider uranium resources of, say, 20 million tons – but

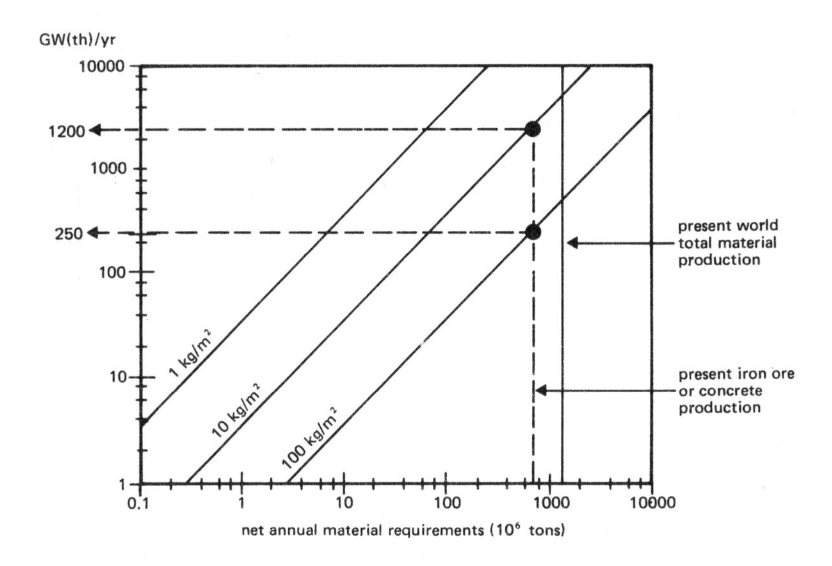

Fig. 11 **Material requirements: solar conversion systems of various net densities.**

this number should be taken as a working hypothesis rather than as an established fact. This large amount would be used up by about 2020, however, if light water reactors (LWRs) or other non-breeders were the only nuclear technology deployed. Therefore, breeders must play their part in time. For example, the plutonium from LWRs could be fed into fast breeder reactors (here LMFBRs, see Fig. 12) and be used as breeder inventory that is not consumed but breeds more fuel. A once-through of 20 million tons of U nat would lead to about 24,000 tons of Pu, which means that the 17 TWyr/yr in question could be produced for a virtually unlimited period of time. However, this possible reactor strategy presupposes an intensive build-up of fast breeder reactors now, which would become operative on a large scale by the turn of the century. There are, of course, various other possibilities besides this reactor strategy, but all of them require breeding.

In this respect it is useful to realize that the fusion reactor of the future, based on the present design, will also be a breeder reactor. Granting the central fusion process of energy release in the plasma to be typically different from the process of nuclear fissioning, there are nevertheless remarkable parallels between fusion and fission breeders in strategic energy planning and reactor operation

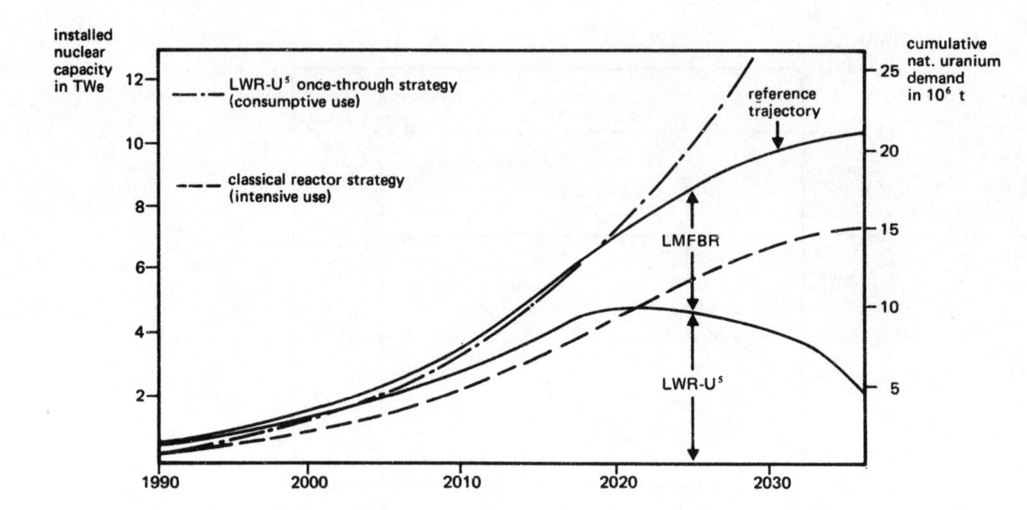

Fig. 12 **The classical reactor strategy.**

(lithium in fusion corresponds to U238 and Th232 in fission breeding, and tritium in fusion to plutonium and U233 in fission). In both cases, there are radioactive inventories and radioactive wastes. Both types of breeders today are geared to electricity generation. Lithium as well as uranium plus thorium resources are similar in size, either one yielding an energy output of about 20 kWh/g. In spite of these qualitative similarities, a technically mature fusion reactor could offer considerable quantitative advantages over the fission breeder [6], [7]. Technical maturation of fusion reactors, however, will continue far into the 21st century, and no more than 2–3 TWyr/yr from fusion are to be expected in 2030. Its share will possibly increase thereafter.

Table 8 attempts to summarize the world's resources, indicating what the potentials as well as the constraints are in producing and using these resources. The data are rather optimistic. Much more would have to be said if space permitted.

Energy Supply Strategies

How do resources fit into IIASA's set of energy models, attempting to simulate the energy demand-supply situation? Fig. 5 identifies resources in the lower right oval as an input to MESSAGE. Several

Table 8 Resources, production potentials, and contraints.

Source	*Production (TWyr/yr)*	*Resource (TWyr)*	*Constraints*
wood	2.5	∞	economy – environment
hydro	1–1.5	∞	economy – environment
total	6–(14)	∞	economy – (nature)
oil and gas	8–12 (?)	1,000	economy – environment – resources
coal	10–14 (??)	2,000 (?)	society – environment – economy
nuclear			
burner	12 for 2020	300	resources
breeders	⩽ 17 by 2030	300,000	build-up rates – resources
fusion	2–3 by 2030	300,000	technology – build-up rates
solar			
soft	1–2	∞	economy – land – infrastructure
hard	2–3 by 2030	∞	build-up rates – materials

extensive linear programming runs of the MESSAGE programme for the seven world regions lead to optimal energy supply strategies (within the defined context). For the purposes of this presentation, the aggregated global supply is of interest.

Fig. 13 shows the evolution of the primary energy mix by 2030. It is to be taken with a grain of salt, but note the slightly decreasing share of gas and the overall fairly constant share of oil together with synthetic fuels, e.g., methanol. Within this band, oil is increasingly replaced by synthetic liquids after 2000. One such source is autothermal coal liquefaction which is assumed to increase the demand for coal. Together with the traditional uses of coal, this leads to a rather uniform overall share of coal in the primary energy market, with the traditional share decreasing steadily. This decline is offset by a rise in nuclear energy for electricity generation. Among the various nuclear shares, that of the fast breeder increases quickly after the turn of the century. The rest of the primary energies remain fairly small until 2030, but let me repeat that solar and fusion could take on greater importance later. These results relate to the Low scenario. In terms of primary energy market shares, the mix for the High scenario does not differ significantly. But, of course, what we are looking for above all in this context is the absolute contributions of the various primary energies in the year 2030.

They are listed in Table 9. Indeed, oil production does not seem to decrease at all. It appears to be rather high, providing about 6.83 TWyr/yr in 2030 in the High scenario. So does gas production, contributing four times the value of today. All nuclear energy

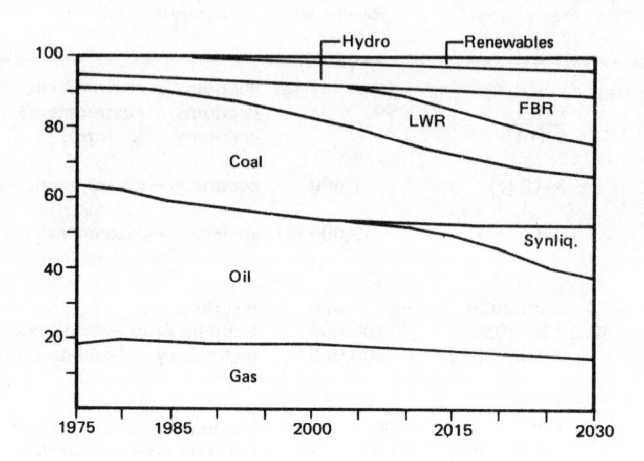

Fig. 13 **World: low demand, primary energy or equivalent.**

Table 9 **Two supply scenarios, global primary energy: 1975–2030 (TW).**

Primary source	*1975*	*High scenario* *2000*	*High scenario* *2030*	*Low scenario* *2000*	*Low scenario* *2030*
Oil	3.62	5.89	6.83	4.75	5.02
Gas	1.51	3.11	5.97	2.53	3.47
Coal	2.26	4.95	11.98	3.93	6.45
Nuclear 1	0.12	1.70	3.21	1.27	1.89
Nuclear 2	0	0.04	4.88	0.02	3.28
Hydro	0.50	0.83	1.46	0.83	1.46
Solar	0	0.10	0.49	0.09	0.30
Other	0.21	0.22	0.81	0.17	0.52
Total	8.21	16.84	35.65	13.59	22.39

production, of nonbreeders (nuclear 1) as well as breeders (nuclear 2), amounts to 8.1 TWyr/yr, that is about 23% of the total energy supply in the High scenario, and yet the number is far from the theoretical 17 TW potential. Most remarkably, these primary energies are supplemented by a coal production of almost 12 TWyr/yr, or about 13 billion tons of coal equivalent per year. Solar, which is just about 0.5 TWyr/yr in 2030, is a more or less *ad hoc* input to MESSAGE since the programme rejected solar contributions at the estimated cost levels. Solar may at best be 2 TWyr/yr, but certainly not more. Hydropower, too, may figure higher in 2030 than indicated by, say, 50%. In relative terms, the contributions of both solar and hydro appear rather small; absolutely speaking, they are enormous, and even more so are the absolute numbers for the other primary energies.

We may wish to react to these incredible quantities by reducing energy demand to values lower than those in the Low scenario. But, in doing so, we should have to face the problems that were discussed under the heading of a 16 TW scenario. In particular, a clarification would be needed as to the regions, the manner, and the extent in which the energy demand should be reduced.

The present scenario approach has the advantage of confronting us directly with the huge orders of magnitude that are required. This is not true of a national approach, which may allow an escape into imports if futures appear too unfavourable. When treating the world as a whole, as is done here, one must specify explicitly where imports for certain world regions could originate. As far as hopes for oil are concerned, extraction must be assumed to occur somewhere in the world. National difficulties can no longer be dismissed or transferred to the abstract notion of the global market when a worldwide perspective is taken.

Indeed, questions of oil imports were very important for our regional calculations. Thus region VI (largely though not fully identical with OPEC) appears to play a dominant role still by 2030. It cannot be expected that this region will continue to transfer its wealth of oil into inflationary capital, as would be the case if it simply complied with import requests as they are received. Rather, it is sensible to expect a limit on the region's oil production, here assumed to be 33 M b/d.

The block diagram in Fig. 14 illustrates the oil export-import situation for all regions in 1975 and 2030, according to the High scenario. One GWyr/yr, the unit given, corresponds to about 14,000 b/d. Region IV (South America) and region V (South East Asia and Africa) were also exporters in 1975, supplying region I (North America) and region III (Western Europe, Japan, and Australia). In 2030, the High scenario indicates that region I will produce sufficient oil to have no need for imports, in spite of the absolute increase in the region's energy demand; and that region III will show a reduction in imports, largely on account of autothermal coal liquefaction. The remainder should help alleviate the most severe energy needs of region V. However, it is clear that regions I and III will, on account of their purchasing power, still import more than is foreseen by the scenario, to the disadvantage of the developing region V.

One understands from the above that the main factors determining

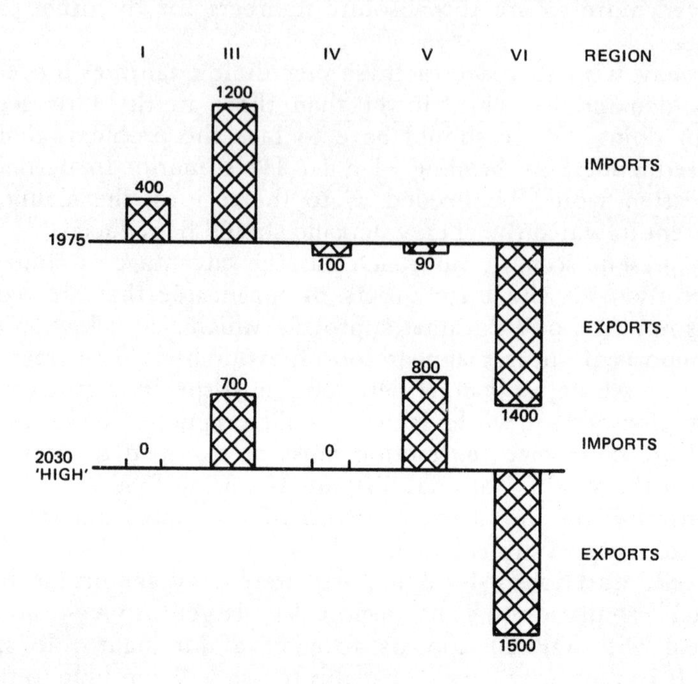

Fig. 14 Oil trading regions, 1975 and 2030 (GWyr/yr).

resource allocation and trade flows are first of all production ceilings and the possibility of increases in the production rate. The resources themselves are not actually exhausted by 2030. Compare the cumulative oil consumption by 2030 in Table 10: it is 68% in terms of price categories I and II but only 1% of highest-cost oil, which is oil shales and tar sands. For natural gas the ratios are 49% and 0%, and for coal 61% and 0%. In other words, out of a world total of about 3000 TWyr of resources, only about 900 TWyr will have been used up by 2030. This is, of course, the relatively cheap and clean resources having less impact on the environment than others. With an annual requirement of about 40 TWyr, for example, the remainder of approximately 2000 TWyr, would last for another 50 years. Given what we know today and what we can now anticipate for the future, the main constraint of the coming 5 decades will be production ceilings, with resources constraints coming to bear in the following half century. By that time a transition to nuclear and solar will be inevitable.

This exercise in how supply schemes affect resource allocation

Table 10 Cumulative uses of fossil fuels, 1975 to 2030, High scenario.

	Total Resource Available (TWyr)	*Total Consumed*	
		(TWyr)	*(%)*
Oil			
conventional (cat. I + II)	464	317	68
unconventional (cat. III)	373	4	1
Natural Gas			
conventional (cat. I + II)	408	199	49
unconventional	130	0	0
Coal			
cat. I	560	341	61
cat. II	1019	0	0

demonstrates the usefulness of the scenario approach, by which anticipated events are put into a chronologically meaningful order. Still, one should remember that it is scenarios we are dealing with here, and not predictions.

Coal in Europe

As shown in Table 9, the highest absolute shares in 2030 in both scenarios are those of coal production. They are worth looking into in more detail, given the present selling difficulties in coal and the at best short-term trend toward coal use, other than burning, for power generation. But at IIASA short-term and long-term are put into different boxes, which, in the context of coal use, means that one needs a long-term perspective to clarify matters. Although quantification is difficult, it is possible to derive to this end actual technological implications from the scenarios.

Calculations have been made for all regions, and in particular for Western Europe in region III. It appears that Western Europe does not have enough indigenous coal to meet an earmarked requirement of 14,000 million tce in 2030. Thus it would have to drive coal mining to the extreme of, say, about 500 million tce, and import the rest. This, in principle, could come from the US, but implies that they would have to mine 2000 million tce for their own needs plus 900 million tce for Europe!

In short, the insights gained from these calculations make coal a scarce resource after the turn of the century. It figures high in world trade and is likely to be processed by various new technologies, in order to substitute oil as a liquid secondary energy carrier.

Autothermal liquefaction, among the various coal conversion processes, is a process by which carbon atoms are transformed into hydrocarbons, such as methanol. A greater conversion efficiency than the present 25–29% would be desirable but this requires, for example, exogenous addition of large amounts of hydrogen (Fig. 15). Such an advanced process would require only one-third of the carbon needed in present autothermal processes, and the energy content of the resulting methanol would at equal parts be derived from hydrogen and carbon. This throws new light on the possible coupling of methanol production with nuclear and/or solar, serving to produce electrolytic hydrogen, for example. The overall constraining factor for such systems would be capital cost.

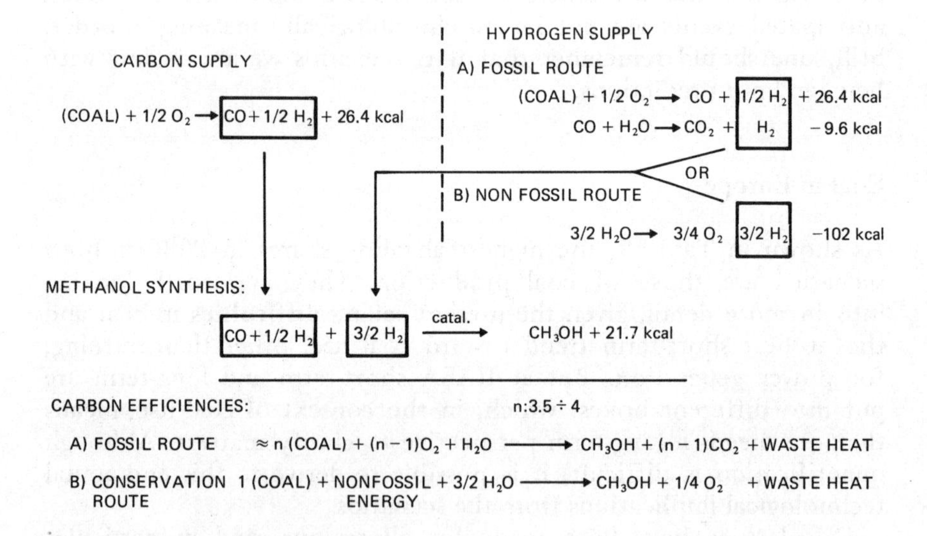

Fig. 15 Methanol production routes.

One cannot conclude a discussion on coal without touching on the CO_2 problem. The natural CO_2 content of the world's atmosphere compares to the release from burning about 500 TWyr of coal. With a fossil energy production of about 900 TWyr by 2030, suggested by the High scenario, a doubling of the atmospheric CO_2 content has to be expected, and an impact on the climate lasting for an indefinite period. Climatologists agree that an uneven warming of the earth's atmosphere would result in a minor variation of the average temperature but large changes (about 10 °C) in polar areas, and partial melting of the polar ice caps. At the present state of the art, there is no conception of what this will mean for actual climate patterns nor how certain the development is to occur. Experts can only appeal to decision-makers for highly flexible global energy strategies [8]. The prudent use of the carbon atom as discussed above may serve as an example.

Concluding Remarks

The foregoing considerations are an attempt to shed light on the complex interplay of medium- and long-term facets of the energy problem. It appears that, if due regard is given to both types of aspects, it is possible to point out ways to remedy the energy situation. The problems involved are only partially a matter of substance, and can be largely overcome if political and economic measures are guided by prudence and willpower.

References

1 Marchetti, C., Nakicenovic, N., Peterka, V., Fleck, F., *The Dynamics of Energy Systems and the Logistic Substitution Model*, vols. 1 and 2, AR-78-1A/B/C, Laxenburg, Austria, International Institute for Applied Systems Analysis, 1978.
2 Keyfitz, N., *Population of the World and Its Regions 1975–2030*, WP-79-, Laxenburg, Austria, International Institute for Applied Systems Analysis, forthcoming.
3 US Department of Energy, *Distributed Energy Systems in California's Future*, interim report, vols. 1 and 2 HCP/P7405-01/02, Washington, D.C., Office of Technology Impacts, 1978.
4 Leontief, W., *et al., The Future of the World Economy, A United Nations Study*, New York, Oxford University Press, 1977.
5 Häfele, W., *Der Beitrag der Sonnenenergie zur Deckung des gegenwärtigen und zukünftigen Energiebedarfs*, presentation, BMWF/ASSA Symposium on Solar Energy Research on the Occasion of Austria's National Holiday, Vienna, ASSA Information Service, 1978.
6 Kulcinski, G. L., Kessler, G., Holdren, J., Hafele W., Energy for the Long Run: Fission or Fusion? *American Scientist*, 67(1)1979, 78–89.
7 Häfele, W., Holdren, J. P., Kessler, G., Kulcinski, G. L., *Fusion and Fast Breeder Reactors*, RR-88-8, Laxenburg, Austria, International Institute for Applied Systems Analysis, 1976.
8 Williams, J. (ed.), *Carbon Dioxide, Climate and Society*, IIASA Proceedings Series Environment, CP-78-5, London, Pergamon Press, 1978.

16 WORLD ENERGY OUTLOOK: PERCEPTIONS OF AN OILMAN

C. C. Pocock[1]

It is possible that the consuming nations can muddle through the 1980s without an excessive pull on OPEC supplies, but this is contingent on two provisos. First, that OPEC lets its oil output rise gradually to its 35 M b/d potential; and, second, that consumer governments 'get things right'. In order to do this, they will have to encourage a crucial changeover from wasting oil by burning it for heating and power. If the consumer world is to survive, the proportion of oil used for steam-raising and heating in the non-Communist world outside North America must drop from its present 60% to 30% by the end of the century. Similarly, correct usage of oil for specialised purposes like transport, petrochemicals, and specialities would rise from 40% to 70%, representing a very large change.

This switch from burning would bring the rest of the non-Communist world closer to the current lighter refining pattern in the US, and the world must soon see that there is no other alternative. Ideally, no heating and boiler fuels should be oil, except in places where gas, coal, and nuclear power are unavailable or impractical.

Such consumer limitations are the only way for the industrialised countries to conform to OPEC's likely maximum oil production limit, which Shell estimates to be 35 M b/d maximum. That would be supplemented by non-OPEC crude output of 20 M b/d (slightly over present levels), rising by 1990 to some 25 M b/d, including 1 M b/d of synthetic oil.

Such a scenario would afford the time for a transition to alternative energy. But it also implies painfully slow economic growth and considerable input from coal and nuclear energy, as well as effective conservation – at least as effective as the Tokyo Summit (1979) guidelines. In that way, even if we are vulnerable to accidents in the near term, the consuming nations will manage throughout the 1980s without putting more pressure on OPEC.

[1] This paper expresses some of the views of the late C. C. Pocock as presented at the Seminar. This text appeared in *Petroleum Intelligence Weekly* as an interview with the Publisher, Ms Wanda Jablonski (*PIW*, October 8, 1979). Reprinted by kind permission.

The top 20 or so major oil companies still enjoy strong future prospects in the upstream producing sector, despite the fact that they are facing serious problems in the midstream sector from the growing activities of governments and national oil companies. In the downstream refining-marketing sector the major oil companies may face headaches in the future if they do not 'get things right' now.

Contrary to widespread views, I believe that the opportunities upstream for private companies are unlimited, over the next 10–20 years, for providing integrated management capabilities and expertise to national groups. Service agreements, partnerships, technical and managerial assistance, plus investment itself, all offer possibilities. Only international companies can integrate whole projects together, providing overall appraisal and a totality of effort, instead of bits of expertise from different hired contractors.

There is a lot of nonsense spoken about transfer of technology. The formula for Coca-Cola, for example, can be transferred, but without the right people and management the transfer of technology is meaningless. There are plenty of upstream prospects in OPEC areas, with exploration and development of difficult formations and new deeper reservoirs, as well as enhanced recovery, gas reinjection and help on gas export schemes. But the companies must realise that if they do not do the job properly, and if they have illusions about control, they will be thrown out in a minute.

In the midstream sector, the most significant market change of the past year has been the big upswing in new direct government-to-government oil deals, surprisingly, mainly to OECD countries. Such deals now cover more than 2 M b/d of crude taken out of conventional channels. The shift since the onset of the Iranian crisis is estimated at about 750,000 b/d to OECD states excluding Japan; another 750,000 b/d is being purchased for Japan under government sponsorship (though not strictly under government contract); and 500,000 b/d for non-OECD official purchases. In descending order of magnitude, the largest in the new wave of government-to-government direct buyers are Japan, France, Italy, Spain, and Turkey. This, compounded by additional activity by traders, has removed up to one-fourth of the oil in international trade from the usual oil-company circuit.

Such midstream problems do, however, provide an opportunity for the more versatile companies able to adapt to the drastic changes involved. Many companies have lost up to half their crude oil availability, forcing them out of third-party sales, and reduced supply flexibility and contract tolerances will squeeze them still further. On the other hand, all this can create new openings for groups which can balance the pluses and minuses and fill the gaps between the less efficient and more vulnerable direct deals. The majors, therefore, can provide an increasingly-needed service flexibility through processing arrangements, surplus products buy-back, backhauls, time or tonnage exchanges, ship-to-ship transfers, effective tanker management and so on.

Downstream operations still offer the major oil companies a position of strength. But the companies will probably be shifting out of bulk fuel oil markets in coming years, leaving them to coal and nuclear energy, while upgrading refineries to lighter products. They will be concentrating on selling facilities rather than commodities – for example, 'total comfort', not just heating oil, to home-owners and 'complete satisfaction', not only gasoline, to motorists.

Importantly, the shift means concentrating on profitable markets. If national security becomes a consumer government watchword, and unless some national markets become more remunerative, it will be necessary to turn elsewhere. Consumer countries cannot continue to expect equal treatment unless markets are equally remunerative. If company X had 150 barrels of oil for its affiliates and was then cut back to 100, these barrels would not necessarily be shared out equally. Oil's worth as a scarce commodity in different markets would have to be taken into account, although this carries a very serious public relations risk. Making higher profits, dealing in a scarce commodity and getting rich while the rest of the world gets poorer, obviously all requires a lot of explanation. The fact is, we are merely recycling the consumers' misery, and unless we do it well we will end up out of business.

Looking ahead to the middle of the next century, coal (either as fuel or feed for synthetics) is likely to be the main supplementary energy base for the US. But for Europe, there is no doubt that pipelined gas from the Soviet Union and the Middle East will be the important alternative to oil, with Europe's own depleted gas fields possibly supplying security storage. It is estimated that by the year 2000 the gas resources of the Soviet Union will be 155 billion barrels of oil equivalent, and that of the Middle East 280 billion. If the necessary incentives for timely exploitation are to be provided, gas prices must start being raised to equivalency with home-heating oil, which it should replace for space heating. Algeria and other countries who are refusing to sell their gas at fuel oil prices are absolutely correct.

17 WORLD ENERGY OUTLOOK: AN OPEC PERSPECTIVE

Ali Jaidah

My subject today is to speculate on the future of world energy from the point of view of oil-producing countries. I interpret this brief as meaning that I should identify the main issues, problems and opportunities that are likely to face the oil-producing countries in the coming years, and to relate these to the problems of energy facing the international community.

Probing into the future is an intellectually risky enterprise, and despite its many attractions not perhaps terribly rewarding. Yet as far back in history as is known man has always been fascinated by prophecy, horoscopes and fortune-telling. The reason is simple: the future is always a source of anxieties, especially in times of difficulty. Predictions, prophesies and horoscopes provide some relief to these anxieties, because we are less frightened by what we know than by what we do not know. The modern man, even when professing scepticism and rationality, is often an avid reader of horoscopes in his daily paper. Institutions, businesses and governments seek relief for their collective anxieties through forecasting models and planning scenarios. Of course, much sophisticated technology is now involved in model building and economic forecasting. But there are many variables that models cannot take into account, such as qualitative features of the world which resist being reduced to linear equations, and the possibility of accidental events about which the probability calculus does not necessarily make us any wiser.

Please do not interpret these remarks as criticism of energy models and forecasting scenarios. Rather, accept the remarks as a qualification to my own lecture, since I am also engaged in talking about the future, and may be carried away by speculation. Let me give you an example close to us; who has predicted events in Iran, and their implications for the world petroleum market? In the early months of 1978, had the oil companies or governments expected a crisis, they would not have allowed oil stocks to run down in the way they did.

When considering the future outlook of energy, the influence of the OPEC countries is limited to two aspects. First, the availability of a supply of crude oil and perhaps in the future of natural gas, and secondly the prices of crude oil. Thus the diagnosis of the future behaviour of OPEC has normally centred around how much oil OPEC countries are prepared to produce and at what price. There are multiple factors which influence the decision making process and I propose to divide these for convenience into three groups.

Techno-Economic Factors

I think that the most important new fact about the world petroleum market is its loss of flexibility. In the 1950s and 1960s available and potential supplies enabled a rapid growth in the consumption of oil, and the industry was able to cope with supply interruptions (whether brought about by conscious decisions like the nationalisation of oil in Iran under Mossadeq, or caused by technical accidents) by drawing on excess capacity available elsewhere. Most oil-producing countries were operating below capacity, and were witnessing an almost uninterrupted growth in output. Oil-exporting countries at the time were more constrained by their inability to administer freely the price of oil than by technical limitations on the level of output.

The world of the 1970s, and for that matter of the 1980s and 1990s, is one in which output flexibility seems to have gone. The rate at which available reserves of hydrocarbons are expected to grow does not match the rate at which consumption was growing in the 25 years following the Second World War. The world economy has adjusted to this new situation through a decline in the rate of increase of energy consumption. But this adjustment has been largely brought about by economic stagnation, rather than by structural and technological changes in the pattern of use. This may imply that every attempt at economic recovery, every upswing of the business cycle, would inevitably produce an imbalance between growing demand and constrained supply of oil. As I said earlier, the oil-producing countries do not have the amount of spare capacity required to compensate for these imbalances. The supply system cannot cope today as easily as it did in the 1950s and 1960s with accidents, whether natural, technical, or political, leading to interruption of production and exports, or even to a significant reduction of output in one of the major producing countries. In fact the recent decision of Saudi Arabia to increase allowable output from 8.5 M b/d to 9.5 M b/d, though providing short-term relief to a tight market, has further reduced the flexibility of the supply system. Should the situation get tight again, we would be in a worse position than before, because excess capacity available for emergency would have been reduced by 1 M b/d.

The oil-producing countries have unwittingly performed the role of buffer in the past. In a paper I delivered to another conference almost two years ago, I warned that OPEC countries neither wanted nor would be able to continue to serve the world as residual supplier. The role they have performed by providing a free stockpile at no charge at all, from which consuming countries chose at will to draw, or to let accumulate, will have to cease soon. I think that this is happening now. The OPEC countries as a whole are no longer able to act as a buffer. Their preferred minimum output, which is influenced by the revenue need of the countries' exchequers, is moving very close to the maximum allowable by technical factors.

Paradoxically, although the revenue needs of oil-producing countries have increased in a dramatic manner in recent years because of an explosion in government expenditure, many oil-producing countries still face the major problem of being unable to utilise their revenues in a productive and meaningful manner. Their spending is often characterised by waste. Their investment in productive assets at home suffers from a lack of complementary resources. Their placement in financial institutions are ill-ordered and affected by inflation and adverse movements in foreign-exchange parities, and they usually earn very meagre monetary returns.

Awareness is increasing in oil-producing countries of the fact that oil and gas are wasting assets, for which there are no immediate substitutes (development projects, which are one of the alternative assets are very slow to mature, and the financial papers tend to become worthless with the passage of time). We are particularly worried about this problem. In most oil-producing countries public opinion is increasingly bringing pressure to bear on governments, urging them to conserve oil, or at least to get sufficient economic rewards for the amounts exported. True, the dollar price of oil was increased significantly in 1973, very moderately thereafter, and once again in 1979. But by how much has the real price of oil, the price at constant dollar level, been increased? Inflation in the world economy has reduced the purchasing power of a unit of revenue. Inflation in the oil-producing countries, due to overcharging by the industrialised countries and to rapid overheating of their economies, has reduced even further the value of a unit of revenue. We are depleting our oil reserves faster than we would like; faster than our own people, concerned about their long-term future, would really like; and failing to obtain a monetary compensation that retains its value over time. We are blamed every time the dollar price of oil goes up, even when the rise falls short of the loss incurred by inflation and exchange depreciation.

To sum up: the techno-economic factors which have reduced the degree of freedom previously enjoyed by the world petroleum supply system, and which have increased the frustration of oil-producing countries, lead me to think that supply crises are likely to occur in the future.

It is impossible to tell in advance how frequent and how serious they will be, still less to predict their dates of occurrence. But if developments continue on the same lines as in recent years, it is almost certain that we shall all have to face a number of major disturbances on the world oil scene.

Socio-Political Factors

Political factors and conflicts may be the cause of instability in the future. We can distinguish three different contexts: domestic situations in the oil-producing countries, intra-regional political relationships, and the global environment which involves large blocs of countries, such as North/South or East/West. Domestically, most oil-producing countries face difficult socio-political challenges. The very sudden and very rapid increase in government expenditure has set in motion complex forces which can be difficult to control. Traditional values are being threatened; expectations have risen, but have not always been satisfied; new institutions have been established in an environment which is not always well-prepared to receive them, and in places social tensions have been exacerbated. Considering the difficulties, one is inclined to be more surprised by the resilience of social structures and political systems in most oil-producing countries, than by the upheavals and revolutions which have taken place in some instances. The strains are real, and we should never be so complacent as to forget their presence. I would like to add that our foreign friends do not always help in reducing the risk of instability. They burden us with new responsibilities; they interfere, sometimes discreetly, sometimes clumsily, in affairs which do not concern them. And the attempts of the superpowers to solve international problems to which we are a party have tended to produce nothing but further complications.

There are political problems leading to possible instability at the intra-regional level. I need not emphasise the significance of the Arab–Israeli conflict in this context. The recent effort of the US to bring partial peace in the Middle East may only have increased, if not the possibility of military conflict, that of internal instability. But the regional problems which concern oil-producing countries are not confined to the Middle East. Problems in Africa – with potentially dangerous conflict between the older sovereign states of that continent and the Republic of South Africa – also exist. If 1973 is known to us as the year of the Arab oil embargo, the 1980s might well turn out to be the years of African oil embargoes and sanctions, especially if Western governments and major companies continue with their present thinly disguised policy of supporting certain southern African regimes. Besides the Middle East and Africa, there is Latin America, which many observers tend to take for granted. Oil-consuming countries, especially in the Western hemisphere, may think that when everything else fails, then they will still be able to rely on stable petroleum supplies from Venezuela and Mexico. But is that so? There are potential sources of political instability in Latin America too.

Let us now turn to a third sphere, that of global relations between the big blocs. The North/South relation involves a host of issues – all important, all unsolved as yet. It is difficult to envisage an open conflict between South and North in the near future, but the future may be characterised by an embittered atmosphere which could lead to friction, localised disturbances, and minor but irritating acts of aggression. Finally, there are uncertainties about the relation between East and West. I am not only talking about the future of detente, but even of specific issues like the future energy balances of the socialist bloc. What would be the policies of a Soviet Union short of oil, if this were the case in 1985/90? What would be the implications of a Soviet oil surplus or oil shortage on the behaviour of Western consuming countries, and the OPEC producing countries, in the years to come?

Psychological Factors

The third set of factors we should take into account, both in analysing past events and in forecasting the future, can be referred to conveniently as psychological. When crises occur, or when the likelihood of crisis is perceived, various agents and parties involved do not always act or react rationally. This is one of the most worrying elements, should emergencies arise. All the parties are victims of pre-conceived ideas. Despite considerable progress in recent years I really doubt whether politicians, decision-makers and other influential agents in oil-consuming countries really understand the point of view of producing countries, and vice versa. Even if a high degree of intellectual understanding is achieved in times of relative calm, all seems to be forgotten when crises occur. One reason for this behaviour might be that politicians are sensitive to public opinion, and even if they know better will follow the lead of their constituencies.

But the role of psychological factors is not limited to situations of crisis. They intervene in the functioning of the oil market, and provide the motivation for a wide range of economic actions and economic policies. It is well-known for example, that buyers and sellers on a market (whether commodity, foreign exchange or stock exchange market) are often influenced by their individual interpretation of their collective psychology. When these factors are at play in the spot market for crude oil, we witness very sharp movements, sometimes up, sometimes down, in the spot price of oil, leading to disturbances of the whole price structure in the world petroleum market. Psychological, or if you prefer expectational, together with economic or commercial considerations, influence decisions about stocks in consuming countries. Recently we learned that the management of stocks can play a significant role in avoiding or precipitating short-term crises.

Psychological factors also influence the policies of oil-consuming and oil-producing countries. The desire to conserve oil for future generations in oil-producing countries does not entirely depend on cold economic calculations, but relates to a collective feeling of insecurity about the future, and to a value-judgement in favour of future generations. The failure of oil-consuming countries to adopt and implement a successful programme of energy conservation is partly due to the reluctance of governments to impose politically unpopular measures. But it is also attributable to a lack of conviction about the seriousness of the energy problem, and to the fact that the public at large is, probably for psychological reasons, almost totally incredulous.

The combination of techno-economic, political and psychological factors suggests that short-term crises are almost unavoidable, at least in the coming years. The real question in the short and medium term is not how to avoid a crisis, but how to manage it when it occurs, in order to avoid the development of a major disturbance. Skills and goodwill are required from all sides for the successful management of crises. I am not sure that an institutional framework involving producers and consumers could be successfully established in the immediate future. Institutions created to deal with emergencies at the international level do not have a good record of success, even when they were created by parties sharing common objectives. Take IEA for example, and its notable failure to provide relief during the 1979 mini-crisis. Attempts to increase mutual understanding between producers and consumers are of course welcome, and can contribute to better management of crises. I am only sceptical about efforts to institutionalise the dialogue too prematurely. There is more hope for the long term. Short-term crises are perhaps unavoidable, but long-term problems are more amenable to solution, provided we begin now. The paradox of the long term is that it is never there.

I would like to conclude by returning to the producers. Among the many issues which face them in the future, two deserve particular emphasis. First is the issue of wealth. Every time oil revenues increase because of a rise in prices or in the quantity exported, some complacency about the future creeps in. The people in oil-producing countries soon realise that wealth is not an end in itself. High revenues are needed for economic development, that is, for building up viable economies which can ensure high standards of living after the end of the oil era. How to bring about this type of development is the major issue facing oil-producing countries. We lack the necessary understanding of development processes, and we lack the adequate supplies of real factors of production which can speed up this process. In that area we get very little useful help from those who depend so heavily for their own welfare and development on our oil. Put in different terms, the economic issue facing oil-producing countries is how to maximise the real, as against monetary, returns from the production and export of oil. One aspect of this problem relates to the pricing of oil. There is no doubt in my mind that crude oil has been underpriced, even in recent years, and that the long period of price freeze, especially 1977–78, entailed an error of judgement. It is also clear that sudden rises in the price of oil after periods of price stagnation have adverse consequences on the world economy. The oil-producing countries should perhaps seek an improved method of administering the price of oil in the future. A second aspect of the economic issue is, how to translate revenue into productive assets. But this brings us back to the problem of economic development, which we have already dealt with.

The second issue is political. People in oil-producing countries, like people in the rest of the Third World, have aspirations which are not understood in the West. Governments are pulled in two opposite directions. Objectives of foreign policy exact from them concessions which sometimes contradict the aspirations of their own people. On the other hand, national objectives and aspirations on other occasions inspire intransigent policies and actions in the international arena. These conflicts of objectives are perhaps unavoidable, but they increase the risk of instability. A better international understanding of this political dilemma could help the governments of oil-producing countries in solving these contradictions, and in following a much smoother course.

Part V

ENERGY ISSUES AND POLICIES: OECD COUNTRIES

18 CRITICAL OVERVIEW OF US ENERGY POLICY[1]

Tom Stauffer

Summarizing US energy policy is like trying to summarize *Alice in Wonderland*. The story is quite unreal. The US is a country with abundant energy resources, but none the less over the next years the US will be haunted increasingly by the spectre of energy scarcity – truly potential poverty amidst real plenty.

First, let me emphasize the unbelievable – i.e. the US does indeed have an energy policy, or, more accurately, many energy policies. Each piece was conceived in isolation, unrelated to any broad concept of national need, and each piece was implemented independently of the conflicts and contradictions among the several pieces. These policies had many diverse objectives – redistributing windfalls or curbing pollution, for example – but most translated directly into reduced domestic energy production or increased domestic consumption. On balance they contributed little or nothing to domestic energy supplies, and they translated quite inadvertently but directly into rapidly increasing dependence upon imported oil.

Let me now survey the status of the four major energy sectors in the US, adding conservation and solar because of their central role in the postures of both Senator Kennedy and the major mass media, and then end by offering some interpretations of the principal themes in US domestic energy 'policy'.

Oil

Total oil imports into the US now exceed 8 M b/d and would exceed domestic crude oil production except for short-term stopgaps introduced by the Administration to disguise our rising dependency. The US was once the paramount oil producer in the world; today it is ranked lower than Avis Rent-a-Car – not even second; it is in third place, after Saudi Arabia and the USSR.

[1] This paper expresses T. Stauffer's views as presented at the Seminar but was written for *Petroleum Intelligence Weekly* (October 1, 1979). Reprinted by kind permission.

It is particularly important to recognise that US oil import levels today are understated, and understated seriously. First of all, we are no longer filling the Strategic Petroleum Reserve, which temporarily takes some 200,000 b/d off our import bill. Secondly, the Administration's short-run programme of substituting natural gas for oil has 'saved' somewhere between 400,000 and 600,000 b/d but will have to be reversed within about a year. This short-run expedient conveniently camouflages otherwise rising imports. The gas which backs out distillate and heavy fuel oil is 'bubble gas' reflecting a transient surplus of deliverability in an otherwise steadily declining system. Once that bubble bursts, imports must again surge upwards by at least that same amount, but the day of reckoning – with some luck – may be deferred beyond the 1980 election.

Price signals are no less distorted, stimulating demand rather than supply in the very best logic of the Red Queen. We still operate, at least in part, under historical prices, around $6 per barrel, and the issue of price deregulation has not yet really been resolved, because of the uncertain consequences of the windfall profits tax. Partisan debate rages, and it is quite unclear how the new windfall profits tax will interact with the new price structure. However, it is painfully clear that the net-of-tax US price for the US producer will be less than the world price (except possibly for special cases like California's heavy oils).

Thus incentives for domestic producers to produce, and especially the incentive for consumers to conserve, are both kept low by a price level well below the world market level. This theme of rejecting price mechanisms is part of the broader policy of rejecting market mechanisms. Moreover, the near-term prospects for US domestic production – we are the third largest producer in the world – are discouragingly dim. Our conventional crude oil supply is likely to fall off, and there will occur a slow decline over the next years. It would be a *tour de force* if we could manage even to hold our own. The more optimistic estimates of what price decontrol might mean, in terms of US supply, promised an additional 200,000–300,000 b/d. The real challenge in terms of US production is not more production, but whether or not we can forestall or slow the decline of the once prolific, but now very mature, domestic oil industry.

If we expect more domestic oil output, then we must seek recourse to unconventional deposits. We are particularly well endowed in that regard. The shale oil deposits of the American West contain well over 600-billion barrels of potentially recoverable oil. These are equivalent to some three times the known reserves of Saudi

Arabia, Moreover, some of those reserves occur in conjunction with deposits of aluminium-bearing material, and the US could simultaneously become self-sufficient in aluminium as a by-product of even a modestly scaled shale oil industry.

This is all well and good, but these resources are only potential resources, not ready reserves. Prospects for realization of this potential are singularly obscure at the present time. The cost of oil from shale has been a moving target for almost 20 years. The first estimate was about $3 a barrel in the mid-1960s, and the target varies between $30 and $45 today, but, I suggest, even these high costs exclude realistic allowances for cost escalation. The ultimate cost is uncertain, but it certainly will not be lower.

The environmental impacts are considerable and present further obstacles. It is not yet clear that the Western states will agree to fill their ravines and gullies with the talcum-like residue from such plants, in order to fuel the Eastern states. More and more, energy policy reflects fundamental regional political rifts in the US.

Unconventional oil sources do definitely exist; the resources are there. But the minimum lead time for construction of these plants is three to four years, with six or seven years being more likely. A massive programme would therefore be required to make even a remotely significant contribution by the end of the 1980s – 10 years from now. That is equal to five congressional terms, well in excess of the planning horizon possible in the US political environment of today, so we must discount this vast energy resource. The questions of price guarantees or the form of the subsidies are still open, the environmental opposition has only begun to mobilize; constructive action is distant.

Natural Gas

The natural gas balance is equally unattractive. Natural gas is a very important piece of US energy supply, but we still have price controls on most natural gas. Ostensibly, under the Natural Gas Policy Act of 1978, natural gas was decontrolled, but the date for full decontrol was in fact set in the mid-1980s. Gas prices are 'vintaged' – which is a very important concept peculiar to the US and which should be alien to all reasonable men. Under this system each source of gas is flagged and assigned a particular price, which remains fixed, irrespective of what happens subsequently in the market. Indicative of the ensuing distortions are some contracts which expired only a short time ago and which were pegged at 9¢ per Mcf – or 54¢ per barrel [of oil equivalent]! There are today 22 different tiers in the natural gas price structure, and the labelling with the proper tag is not yet complete.

One trend is clear, namely that the real price of new natural gas has been rising at the margin quite significantly over the last 6–8 years, even though the average price has risen only slowly. It was less than 30¢ in 1970, while today new gas, subject to court review, is priced at around $2.40. In spite of the steady rise in the real price of newly discovered natural gas, the response of supply to this price incentive has been disappointing, and I suggest that that implies that the gas industry, like the oil industry, is mature.

At the present time the gas discovery rate in the US is less than half of the production rate, so today's production is not self-sustaining; rather, it is production out of a dwindling inventory of proved reserves. We have not succeeded in making net additions to those proved reserves, in spite of the dramatic rise in prices for new gas and the marked increase in drilling activity, and the reserve-to-production ratio is still falling.

Unconventional sources of natural gas are abundant and still untapped as of yet, just as in the case for oil, but the question is quite muddled in the mind of the public. We note first that the media posture towards gas supply has been schizophrenic. When the media argue against decontrol of gas, they argue or emphasize evidence that there is so little gas left to find that the price incentives are meaningless. Hence, they posture, decontrol is wasteful and only generates windfalls for the producers. Conversely, the media equally emphatically report that there are vast but hidden reserves of natural gas, which the natural gas industry conceals in order to force higher prices.

There is a tremendously large potential gas resource in the US, and I use the word 'tremendously' advisedly. This resource occurs in unconventional deposits, among the most widely publicized of which are the geopressured deposits of the Texas Gulf coast. We have possibly 1,000-trillion cubic feet of gas in such deposits, at depths which are within the compass of conventional production technology. The problem, which is not widely discussed, is that the gas occurs in conjunction with water – about 30 cubic feet of gas per barrel of brine. Let us do some arithmetic: that is equivalent to some 30-odd barrels of brine per Mcf of gas, or 180–240 barrels of brine per barrel of oil equivalent. Thus 1 M b/d of energy in the form of geopressured gas is equivalent to creating a liquids-handling industry, akin in all technical features to oil, of 200 M b/d, or a tenfold explosion in the size of the domestic petroleum industry. I need not belabour that this supply is not imminent.

More interesting, but less publicised, are the tight gas deposits. These are roughly similar to conventional gas deposits but are differentiated by their very low permeabilities. Gas flows out of these reservoirs at a very low rate, so that many more wells are necessary for any given production level. The expected well density in these fields is of the order of 4 acres per well, in contrast to 640 acres for much conventional gas. The gas is accordingly expensive and still unproved hydro-fracturing techniques are required to free up that gas.

This gas is potentially producible at \$3–\$5 per Mcf, or \$18–\$30 per barrel of oil equivalent. At least 600-trillion cubic feet have been delineated; the Canadians appear to have another 600 tcf; and the ultimate resource may be at least four times this size. It is large, and it is known; many of those resources were discovered years ago, but no one was interested in the then-prevailing much lower prices. The discoveries were often not even booked. The questions today are drilling costs, requisite well densities, and the practicality of the massive hydro-fracturing programmes which will be needed. Again, this supply is not likely to be available until later in the 1980s.

Hence the prospects for both domestic oil and gas supply over the near term are dreary. The US oil and gas industries are both running 'open-choke', i.e. wide open. Except for the brief gas 'bubble' cited earlier, there is no spare deliverability in the system. Thus discoveries must equal production in order for us to hold our own. The only conceivable major offsets are Arctic and Alaskan gas, and it appears that both may not be sufficient to offset the basic decline until unconventional sources might be mobilized in the later 1980s.

Nuclear

Now, moving on to nuclear energy, I note once more that the resource base in the US is again very large – uniquely so among the industrialized states. We have very large deposits of low- and medium-cost uranium, large enough that they could serve even the earlier, larger-scale nuclear programme for a period of at least 30 years.

At the present time, US policy towards nuclear power is convoluted and contradictory. The Administration has chosen not to promote nuclear power actively as a transition to some more remote policy based upon renewables, such as solar or wind power. And Senator Kennedy, the ascendant Democratic candidate, as reflected in polls and current media campaigns, has gone even further and made a statement rejecting nuclear power even as a transitional option. The momentum of the US nuclear programme has been lost. Construction is slowing down; some plants have been cancelled, and others have been deferred.

The opposition to nuclear power is widely publicised but very narrowly based, drawn from groups neither elected nor representative. Here, as elsewhere on energy issues, one must distinguish carefully between reality and public perceptions. Truth in politics is whatever is believed, but some perceptions are real, while some are mere artifacts of the newsmaking process. The Union of Concerned Scientists is among the most widely quoted of the nuclear opponents, yet for practical purposes it has less than 10 active members – possibly as few as four. Decision-making within the Union seems to be confined to the latter cadre, yet this group is portrayed in the context of the national debate as if it reflected a consensus of academic or scientific opinion.

Well, the opposition has been successful. Our nuclear fuel reprocessing programme has been suspended, and the Administration has moved to discontinue the breeder development programme as well. It was in fact suspended by President Carter, but the Congress moved on its own initiative to override the President. It has authorized funds for the last two years to fund the construction of the prototype, and the LMFBR may actually be on stream later in the 1980s, only a year or so behind schedule, in spite of the Administration.

Otherwise, there is no significant forward motion left in the US nuclear programmes, and the electric utilities are increasingly reluctant to attempt to license nuclear plants. The delays are long, and the outcomes of the hearings are quite uncertain. The contradictions and perverse consequences of the quasi-moratorium are dramatically clear in the case of a recent development in California, where less nuclear power translates into more gas consumption, and hence more oil imports via substitution. The state, under another presidential hopeful, Gov. Jerry Brown, has suspended both coal and nuclear plant construction and, instead, has embarked upon the world's largest gas-turbine power plant – 1,200 megawatts. Moreover, it is expected that California, having rejected coal and nuclear power, will succeed in getting high priority claims on the declining supply of natural gas in order to fuel that station.

Coal

Let us turn finally to coal. Yet again, the US is more than amply endowed with coal resources. Our known deposits exceed 1,000-billion tons which have been reasonably precisely delineated. That is equivalent to 1,500 years of consumption at current levels, or looked at another way, it amounts to 750 Congressional terms laid end to end, a very long planning horizon, especially in the US today.

The route to coal-based plenty is beset with obstacles, the most obvious of which are the very real and very well-publicised environmental hazards associated with direct combustion of coal. Coal is sooty and carcinogenic, and thus socially unwelcome. Coal was indeed King during the industrial revolution, but it was deposed by oil. Now, the question is whether the Black Knight – the sometime King – will be allowed even to return, let alone to reign.

The central problem with coal, however, is not production – in sharp contrast with oil and gas. Rather, utilization is the bottleneck critically dominating coal as an energy source. Today the US actually enjoys – or suffers – surplus coal-producing capacity, in spite of an energy shortage. We have the coal capacity. We do not have the capability to consume it.

For an economy such as that of the US, any expansion in coal use is tantamount to creating some sort of synthetic fuels programme. Coal can compete with domestic uranium if used to generate electricity, and both compete with imported oil at the margin in that market. But more interesting is the question of how the US can use an old-fashioned fuel like coal and turn it into the forms of fuel which a modern industrial society is structured to use, namely gas or liquids.

The Administration has indeed recognized this basic truth: it has proposed an $80-billion programme to produce synfuels from coal, jumping the demonstration plant stages and proceeding directly and, most believe, rashly, to commercial-size plants, based upon technologies which, with one exception, have scarcely been proved at pilot-plant size. The programme presently is foundering in the Congress, a body which normally is not adverse to spending money.

The costs remain well above OPEC parity; coal gasification hovers around $35 per barrel, while the more remote liquids technology exceeds $40, except where coal is uniquely cheap, as in South Africa. Only two justifications can be advanced for a well-paced synfuels programme: (1) as insurance against an embargo, for which the rational premium is about $3 a barrel – the equivalent charge for a one-year's strategic stockpile; and (2) as pre-emptive capacity to reduce OPEC's market power. Neither argument has compelling political appeal – not enough to override the cost burden – and the Administration has failed in recent efforts to camouflage the costs.

Here, too, lead-times are an additional impediment. Even if the US were to move ahead, the plausible rate of implementation will be sufficiently slow that once more one can really exclude synfuels from making any significant contribution through the 1980s.

Exotic Sources

Lastly, let me touch upon one of the alternatives, 'exotic' energy sources – not because I can argue that these will make any greater contribution, but rather because two such 'exotic' sources have been advanced by Senator Kennedy. 'Conservation' and 'solar energy' have been designated as the bases of his energy programme for the US. Given the overwhelming media backing both for Senator Kennedy and his carefully researched energy placebos, these must be considered – even if not respected.

The first of the 'exotics' is conservation – negative demand as a form of supply. Conservation is an unassailable virtue – like truth, no one can oppose it. But, like truth, conservation evades convenient operational definition. The ultimate form of conservation, of course, is poverty; the Bangladeshi peasant is 'energy efficient'. But, beyond that dubious criterion for 'efficiency', any precision is elusive, and both the technologists and the economists have served the public badly. The physicist evokes the second law of thermodynamics to test efficiency, blissfully ignoring the inordinate costs necessary even to approximate such standards of efficiency. Conversely, the economists construct models, blithely ignorant of the second law of thermodynamics, and discover vast amounts of 'waste heat', not knowing its temperature is too low to boil even one egg.

Conservation is particularly meretricious as a policy target, because it is so virtuous. Yet 'waste' is not the opposite of conservation, and the trade-offs between technical, economic, and social tests of 'efficiency' are exceedingly complex. One cartoonist neatly epitomized that trade-off: 'What I want is a car small enough to meet President Carter's standards – but large enough to fit Tip O'Neill.'

Market-induced conservation reflects many responses by many millions of consuming units, ranging from householders who set back thermostats at night to large firms which install complex and expensive heat exchanges to integrate their plant processes and save fuel. Some of this already has occurred. In the US, residential fuel consumption is down some 10%, after all corrections for winter temperatures; and industrial fuel consumption, corrected for lower output, seems down by 5%–20%, with the biggest savings in the less fuel-intensive industries.

More such savings – perhaps as much again – should follow in time as older equipment is replaced by more fuel-efficient equipment designed to suit today's much higher fuel prices. Similarly, some retrofitting of older facilities will continue – storm windows, attic insulation, etc.

However, 'conservation' measures are acutely susceptible to the law of diminishing returns. Redesigning US cars to meet fuel economy standards, for example, while maintaining minimum comfort and performance standards, is equivalent to pricing the oil saved at over $60 per barrel, and most of the measures proposed are almost as capital-intensive at the margin as the more costly solar options.

At best, focus upon 'conservation' as an alternative to supply is an abdication of responsibility – politically tempting as it may be as an electoral expedient. Likely conservation may only offset the additional energy needed for growth. Any serious effort to force non-market conservation measures in the name of 'Conservation' may really be a euphemism for recession.

Solar technology is a beautiful example of the contrast between technological feasibility and economic viability. Solar energy works – of that there is no question – but the issue is cost.

Today, solar heating simply is not economical for any significant part of the US market, and the cost differential is large, even at current oil prices. Swimming pool heaters are indeed practical, but the demand is infinitesimal. Hot-water heating is economically feasible for newly constructed houses in some favoured parts of the country – but retrofits are far too costly, and the housing stock turns over only once in every 60 years or so.

The major market for which solar energy is targeted, space heating, involves a cost of $60–$120 a barrel for the crude oil displaced. Another test of the economics of solar energy is the capital investment cost per barrel of oil displaced – this amounts to $120,000 – $250,000 per daily barrel of oil equivalent. This is two to four times the capital intensity of nuclear power plants and illustrates neatly the theorem, 'There ain't no free lunch'. Or better, 'If there is cheap lunch, the taxi fare may be too high'. The capital costs of the equipment absolutely dominate the fact that the energy itself is free.

We end with a real stalemate. Our energy system in the US has been overdefined and overconstrained by the political process, and our 'crisis' highlights particularly painfully the point that economic analysis is only the beginning when one is trying to analyse large programmes. The political parameters dominate; economics and reason are subordinated.

Let me summarize with four propositions in which I shall try to isolate some central themes:

1. There is a fundamental paradox in the US energy imbalance: the US is uniquely capable of energy autarky, but that potential co-exists with the spectre of energy scarcity. The fact that such large, potentially accessible energy resources do exist in the US actually clutters and confuses the public debate. The public readily confuses potential resources with probable resources, and the prospective abundance in any one energy sector is used to justify inaction or paralysis in some other sector. 'A' negates 'B' etc., until one goes full circle to stalemate.

2. Secondly, the political process in the US has defaulted quite badly. It is the duty of a political system to reconcile conflicting priorities. The *Christian Science Monitor* summarized the dilemma quite succinctly, observing that US domestic policies were structured to guarantee an increasing dependence upon oil imports, while our foreign policy was also structured to ensure that we would not get those needed imports. It took some 10 to 15 years for that basic contradiction to surface, and, today, to mix metaphors, the chickens have come home to roost. The Arab–Israeli dispute, and the plight of the Palestinians, impinged not at all on the US public for some 25 years. US import dependence has brought the conflicts onto US shores in the form of gas lines and jeopardized jobs.

3. Energy supply has been totally politicized in the US. Previously we enjoyed the minimum of administrative intervention in energy markets in the US. Natural gas, however, was the first of the producing sectors to come under close regulation, and it was the first casualty. It looks as though oil may be the next.

4. The last proposition is the immediate consequence of the third. The political process, particularly in the US, concentrates only upon the short run. The politician thinks in terms of today – perhaps as far ahead as tomorrow morning's headlines, but more likely in terms of a pitch for the 11.00 o'clock evening news. Conversely, the energy supply industries are accustomed to plan over much longer time horizons; indeed they must. This is intrinsic and unavoidable; the gestation periods are commensurately long. Through this politicization, decisions which necessarily involve long horizons

are handed over to politicians who focus exclusively – and understandably – on the immediate short run. This, I suggest, is the most fundamental problem in the US today: this basic and painful conflict of time horizons. The 'bottom line' of US energy policy is a growing dependence upon imports over the short run, and perhaps well into the 1980s. Beneath the sound and fury of US rhetoric lies the austere reality that only a recessionary policy can curb imports. Having rejected price elasticities as policy parameters, the politicians now have recourse only to income elasticities. And thus the paradox: the spectre of real poverty amidst the promise of plenty.

19 ENERGY ISSUES AND POLICIES: THE UNITED KINGDOM

T. Philip Jones

Earlier this year (1979) we had a change of Government in the UK. The new Conservative Government has only been in office for a few months and is still in the process of shaping its energy policy. It undoubtedly will wish to make changes, and indeed in certain areas already has, to some of the detailed policies followed by the last Administration. But the nature of the energy problem which the new government faces has not changed and governments, whatever their political complexion, cannot avoid the long lead times and uncertainties which are inherent in the energy scene. Lead times are long whether it be for the development of a new coal mine or the construction of a new power station. And considerable uncertainties surround forecasts of both energy demand and supply. The UK Department of Energy will be publishing shortly our revised forecasts for the period up to the year 2000. I do not propose to discuss them in detail. First, because there is not time, and secondly because the only thing I would confidently say about them is that they are bound to be proved wrong.

One point, however, on which there is fairly wide agreement is that world supplies of oil will become increasingly scarce and expensive. The latest projections done by the Department of Energy suggest that by the end of the century oil prices could, depending on what assumptions are made about world economic growth, be between $1\frac{3}{4}$ and $2\frac{1}{2}$ times their present levels in real terms. As prices rise, we expect a general transition towards the use of fuels other than oil, with oil being confined increasingly to its premium uses – for example in transport and chemicals. The problem which all industrialised countries face – and we in the UK, despite the temporary respite provided by North Sea oil, face just as much as anyone else – is to manage this transition smoothly with the minimum disruption to growth prospects, employment, living standards, balance of payments and individual liberties.

It is worth recalling that in the UK it took 20 years, under highly favourable conditions, for oil to displace coal from the general industrial market. It will inevitably take some time to reverse that process. There is no easy alternative to oil as the mainstay of the energy economy. Nothing else is so readily handled, stored and transported. For a successful transition, much will depend on the stability of the climate against which we are working, and on economic growth which will itself help to generate the resources for new investment in conservation, supply and re-equipment.

Specifically, our task in the UK is to manage the transition from our present four-fuel economy – coal, oil, gas, and nuclear – to one where we can no longer rely on oil and gas from the North Sea and where international oil is likely to be increasingly scarce and expensive. We could, in the year 2000, depending on economic growth rates, be facing a demand of up to about 500 million tons of coal equivalent (Mtce) as compared to 360 Mtce now. To set against this possible demand we might have indigenous supplies of around 400 Mtce.

Our total UK Continental Shelf (UKCS) oil reserves are estimated to lie in the range of 2,400–4,400 tonnes. This implies that we may have already discovered between about half and two-thirds of the total. A proportion of what remains is believed to lie in deep water and, given the need to develop new technology as well as to find the oil, this is unlikely to be producible until towards the end of the century. Moves recently announced by the Government, for example in connection with its review of the role of the British National Oil Corporation, are designed to encourage companies to explore more widely and to invest more confidently in development. The Government wishes to encourage more investment both in drilling on already licensed territory and in deep waters on the UKCS. The Government's decision to examine with the industry the problems of the so-called marginal fields should also be of positive help.

On our present estimates, the UK would be self-sufficient in oil in net terms – while continuing, of course to trade North Sea crude for heavier crudes to obtain a refinery balance – for most of the 1980s, but could become a net importer again in the 1990s. While we can expect to remain a producer on a smaller scale well into the next century, we might need to import up to about a half of our oil requirements by the turn of the century.

Gas, almost entirely natural gas from the North Sea, currently provides nearly 20% of our total primary energy consumption. Total reserves of natural gas originally in place and available to the UK are estimated at 70 trillion ft^3. Gas supplies might build up from the present level of about 4,000 million ft^3/day (M cf/d) to about 6,000 M cf/d by around 1990, after which they would begin to decline. They will be reserved as far as possible for premium uses. Even so, on present prospects, there could be a need for substitute natural gas manufactured from coal and/or supplies of LNG before the end of the century.

Our oil and gas reserves therefore, while significant, are of a very limited life. And we believe that for the longer term we shall need, like most other OECD countries, to place increasing reliance on energy conservation, coal and nuclear power. This is not to neglect the renewable energy sources – wind, wave, tidal, geothermal, and solar energy. We have programmes under way on all of these but we do not see them making a significant contribution before the year 2000, although they might thereafter. But energy conservation, coal and nuclear power seem on most scenarios essential. Let me therefore say a brief word about each of them.

Energy Conservation

Views differ about the amount of savings that can be achieved by energy conservation. Indeed, there are those who believe that it is possible to achieve little or no growth in energy demand without impairing economic growth. Certainly we would expect – and indeed it is vital – that the previous relationship between energy growth and GDP growth should change. But I think it unlikely that we can have economic growth without energy growth. It would be running an unjustifiable risk to develop an energy policy on such assumptions. This is not to undermine the importance of energy conservation. Everyone would, I think, agree that energy conservation must be a central and permanent feature of energy policy. It certainly is in our's in the UK. We have an extensive programme covering all sectors of the economy and our forecasts allow for savings from energy conservation producing a 20% reduction in demand by the year 2000 – equivalent to some 60 Mtoe The trouble is that energy conservation is rather like virtue. We are all in favour of it in principle but we often find it jolly hard to put into practice. Economic pricing is, of course, essential, but I doubt if anyone is yet certain what, in addition to pricing, is the best combination of measures, regulations, or financial incentives, to achieve conservation. Perhaps this is not very surprising since conservation is after all a new area of policy. This is why the Government is currently reviewing the conservation programme to see whether the present balance is right.

Coal

We have very large resources of coal, estimated at 190 billion tonnes, of which perhaps about 45 billion tonnes might be ultimately recoverable. About 65% of coal production is currently used in power stations, but in the future coal is likely to be needed to replace oil for heating in industry, for the manufacture of synthetic natural gas and synthetic oil, and as a petrochemical feedstock. For this, we shall need a competitive, efficient, modern, coal industry. The National Coal Board is at present engaged on raising output from its present level of about 120 million tonnes p.a. towards 135 million tonnes by 1985 and sees scope for raising production to 170 million tonnes p.a. by 2000. This will, however, not be an easy task. It will require very substantial new investment. In the late 1950s and 1960s, as the use of oil and gas increased, the coal industry declined. In addition, many of our 19th century pits are now very close to exhaustion. In parallel with the creation of this new capacity, markets will need to be won back and new markets developed.

Nuclear Power

In the UK the 9 Maxnox and 2 AGR stations now commissioned have a total capacity of 7.4 GW and generated about 12% of our electricity requirements in 1978. We expect nuclear energy to be supplying around 20% of our requirements in the early 1980s when the first AGR programme is completed and by the year 2000 we could have built up to 40 GW of nuclear power, the equivalent of about 100 million tonnes of coal.

There is, of course, some public apprehension in the UK, as elsewhere, about the expansion of nuclear power. A balanced view, however, needs to be taken of the risks and benefits. Nuclear safety here, as in other countries, is of paramount importance and the Government will do all it can to promote it. The retention of public confidence is important to success. On any realistic scenario it is difficult to see how we can meet our energy demand at acceptable cost without a significant nuclear component. The Government has made it clear that nuclear power has a vital role to play in our energy supplies and that it intends to adopt a positive attitude towards its development in the UK.

I have concentrated on conservation, coal and nuclear power because we see these as the three essential elements in our policy in the longer term. But I would not like to leave you with the impression that we have a cut-and-dried policy or a detailed blueprint. We cannot predict the unexpected, but it will surely happen. The only way we can and do try to protect against this is by having a flexible policy and a broad framework which will enable us to adjust to changing circumstances, and which does not close options too soon. All forms of energy are uncertain in varying degrees. Every energy decision is likely to involve a wide range of other objectives – industrial, social, political and economic. We must seek the right balance between conflicting interests in each decision. There is no general formula. The question of balance is particularly important in terms of the interface between energy developments and conservation of the environment. The risk therefore needs to be spread over as many sources as possible and we need to have at our disposal the widest range of technologies. And finally we must recognise, as I hope and believe we do, that energy problems cannot be solved in isolation but only on the basis of international co-operation.

Part VI

ENERGY ISSUES AND POLICIES: THE SECOND AND THIRD WORLD

20 THE ENERGY POLICIES OF THE SOVIET UNION

Michael Kaser

Although only 6% of the world's population live in the USSR, the country produces one-quarter of the world's coal and natural gas and one-fifth of the world's oil, and has the objective and capability of generating eleven years hence more than 20% of its electricity from nuclear power. The Soviet Union's prospects as an energy producer are more assured than any other industrialized country because it possesses larger deposits of oil, peat and, probably, natural gas. It is also relevant that its timber stand (79 billion m^3) is greater than that of North America or of Brazil and constitutes one-third of the world total.

Table 1[1] Soviet energy resources

	Hard coal	*Brown coal*	*Oil*	*Natural gas*
	Thousand million tonnes		*Million tonnes*	*km^3*
Reserves	3,993	1,720	8,138	17,136
Production as percentage of Comecon	71	28	97	85

In 1978 the Soviet Union was the world's largest producer of oil (572 million tonnes) and coal (720 million tonnes), and was second only to the US in the extraction of natural gas (372 billion m^3). In that year its aggregate output of primary energy was 1,858 million tonnes of coal equivalent, and the pace of the expansion to this level can be judged from the fact that it passed the 1,000 million tonne mark scarcely 15 years ago. The Five-Year Plan to 1980 sets a production target of 2,809 million tonnes of coal equivalent (standardized at 7,000 kilocalories per tonne), including 1,053 million of coal, 940 million of oil, 599 million of natural gas, 82 million of hydroelectricity, and 135 million of other sources (nuclear,

[1] See notes on statistical sources at the end of this chapter.

firewood, peat and wastes). Assuming, as is likely, the fulfilment of the lower range of the Five Year Plan target for 1980,[1] energy production will have expanded over three Five Year Plans, that is since 1965, nearly 2.4 fold, while restraining consumption growth to little over double, 2.1 times. That excess of output over domestic use (largely achieved by the restriction of 'discretionary' uses such as private motoring) will have enabled its net exports of energy to have more than quadrupled (4.4 times to be precise).

For all this wealth, the pattern of energy supply and use presents numerous paradoxes. First, the Soviet utilization of natural resources per unit of output is higher than the average for industrial countries, whereas its consumption of energy on the same comparison is considerably lower. Secondly, the USSR before the Second World War exhibited a rate of growth of energy consumption exceeding that of total national product, but subsequently has achieved an energy rate lower than, or at most equal to, national product. Thirdly, the USSR is the world's largest oil exporter – thanks to the 1973 price explosion, oil exports (worth $6 billion in 1978, as Table 2 shows) constitute half of all Soviet hard-currency earnings – although domestic oil consumption per head is much lower than in the industrial countries to which it sells. Finally, one-fifth of Soviet household consumption is self-produced (by cutting wood or peat or by burning farm refuse), in contrast to the Western consumer who buys virtually all his supply commercially.

Table 2 Soviet exports in 1978 of oil and oil products by value

	Million valuta roubles	*Percentage of total export earnings*	*Value in US dollars (at 70c = 1 rbl.)*
To convertible currency partners	4,390	50.5	6,270
To socialist states	5,146	24.2	7,350
To developing countries	471	8.2	673
To all partners	10,052	28.2	14,293

Note: Estimation is involved since not all partners are listed.

The Soviet Union has nevertheless not played a major role in international energy politics. It was not invited to join either OPEC or IEA on their respective foundations, though in both cases considerations other than magnitude of the energy exports were involved. The low-key Soviet role may be attributed in part to the fact that

[1] Subsequent information shows that this target is unlikely to be reached.

the volume of the export balance has only recently become large and in part because most of the trade is with its fellow Communist countries. To take 1965 again as base, total oil exports were then only 43 million tonnes, of which almost exactly half (22.4 million) went to other Communist states. That same proportion – half to the West and half to the Communist world has broadly obtained ever since. As the USSR has now ceased to publish oil tonnages exported (see note on statistics at the end of this chapter), we have only values to rely upon: last year 51% of oil sales by value went to Communist partners, but if we allow for a rebate of some 15% on the price to them (it is reduced to 12% or so this year) the tonnage distribution was probably 55% to Communist states and 45% to the West – mostly the industrial West, with only a little to the Third World (see Table 2). Paradoxically, it is because the proportions of the Soviet energy balance are small that the matter is important – both in relation to home production and to the needs of customers in the industrial West. A relatively small reduction in Soviet output, a diversion of sales to fellow members of Comecon, or a steeper rise in domestic consumption could quickly eliminate the surplus and transform the USSR into a net buyer from OPEC.

This was the message of a widely disseminated, and as widely discussed, research study by the American CIA in April 1977.[1] Based on a peaking-out of the new West Siberian oilfields and the continuing decline of the old deposits, it forecast an output as low as 400 million tonnes of oil by 1985, that is 190 million tonnes below, or two-thirds of, the 1980 expected level. President Carter cited this report as a prime factor in the energy policy for the US he launched at that very time.

The CIA forecast was based on the rapid depletion of existing fields and the difficulties of bringing new deposits on stream within a time scale of 8 to 10 years (present Soviet experience in Siberia). The major Samotlor oilfield would reach its maximum in 1978 and could be maintained at a high level only until 1982. The fields that currently account for the bulk of Soviet production were experiencing serious water encroachment because a water injection technique was employed which maximized initial output per well. Although this minimized investment per unit of output during the early phase, the share of water in total fluid lifted could not fail to rise. (From 50% or so in 1975, the CIA forecast a share of 65% by 1980.) The large quantities of high-capacity submersible pumps and gas-lift equipment needed to stave off a fall in off-take were in excess of home production or import possibilities.

[1] ER 77-10270 *Prospects for Soviet Oil Production*, April 1977.

The CIA's second line of argument was that exploration drilling was insufficient to discover new reserves. The Agency's report to the President transformed the originally cautious estimate of a possible import deficit into a firm prediction that by 1985 'The Soviet Union and Eastern Europe will be importers of 3½–4½ M b/d'.[1] This shift of emphasis in presentation as well as the evidence itself were immediately noted. In the US one might mention particularly Dr. Marshall Goldman's critical evidence to the Department of Commerce Advisory Committee on East/West Trade in June 1977, and a report of the International Trade Commission in September. The CIA expanded its arguments in a further report[2] and in the space available to me in this paper I shall concentrate only on one issue, that of the reserves to production ratio. The CIA claimed that at the beginning of 1976 Soviet reserves classified as (a) proved and in exploitation, (b) proved and capable of exploitation, and (c) probable reserves, totalled 5,500 million tonnes:[3] that classification is approximately equivalent to proved and probable reserves in US practice.[4] As Professor Goldman pointed out, the CIA only two years before had estimated proved reserves alone at 10,000 million tonnes – 73 billion barrels against a revised 30 to 35 billion. Although analogies with US fields allowed the CIA to obtain a confirmatory estimate of 4,500 million tonnes (33 billion barrels), Soviet sources compiled by a French government agency[5] show 8,730 million tonnes (63 billion barrels[6]) for 1970, which agrees with the 8,138 million tonnes (59 billion barrels) in a 1978 Czechoslovak source. (This is the figure in my Table 1.) Unless the Soviet and Czech sources are deliberately false, one must set the CIA's pessimistic estimates aside. Even so the reserve/production ratio has been declining – from 25 years in 1970 to 16 years in 1977. This is not the 'over 20 years' which Peter Odell has recently advanced,[7] but it is above the US 1975 ratio of 7½ years. Moreover, one can perceive a reversal of the trend in the USSR. Thus in 1961–71 the annual rate of growth of reserves was only 5.5% against a production growth of 9.1%, but in 1971–75 the

[1] *The International Energy Situation to 1985.*

[2] ER 77-10425 *Prospects for Soviet Oil Production, a Supplemental Analysis,* July 1977.

[3] 40,000 million barrels. As a rate of barrels to tonnes, the CIA uses 7.18, 7.31, 7.38, 7.27, 7.33, 7.30, 7.29, and 7.33 (from conversions) in its *Supplemental Analysis.*

[4] 'Proved' reserves alone would be 4,100 to 4,800 million tonnes according to the CIA.

[5] Centre d'Études Prospectives et d'Information Internationales, *La Production Petrolière Soviètique à l'Horizon 1985.* May 1979.

[6] Assuming 7.25 barrels per tonne.

[7] In a working paper circulated by the Association of American Geographers. See also P. R. Odell and L. Vallenilla, *The Pressures of Oil*, London 1978.

rates were much closer together at 5.4 and 6.8% respectively.[1] In particular, exploration drilling in western Siberia, which dropped to as little as 367,000 in 1970 seems likely to meet its target of 1.13 million next year.[2]

My conclusion on the crucial issue of Soviet oil, is that an output of 680–700 million tonnes (reported as the 1985 Plan[3]) is more within the bounds of likelihood than the CIA's latest revision of 500 million tonnes as a maximum for 1985.[4] At worst, output will be 30 million tonnes above the likely 1980 output of 620 millions, the lower target of the Five Year Plan.

Technical data apart, it would seem inconceivable for the Soviet authorities to prejudice an export product which by itself brings in half its hard-currency earnings (Table 2). Oil accounts for a still higher share (56%) of its exports to the EEC. It will surely import the submersible pumps, gas-lift equipment, off-shore equipment, and rotary drills for deep boring in soft rock, which constitute its shopping list in the West. Any reasonable cost-benefit analysis in hard currency will show the appropriateness of such a course.[5] For us here, the question is whether Western support for Siberian development should be used as a political lever (as President Carter was on the point of doing a year ago), or should be given at the expense of expanding output in countries politically aligned to the West. An Exxon report last year pointed out that only 5% of exploratory drilling is conducted in developing countries, which have 50% of the potentially petroliferous configurations of the world.[6] On the other hand, the USSR has 37% of those potentially petroliferous regions, many of which could be explored and exploited sooner than those in remote parts of the Third World.

This leads me to the longer-term perspective. A report many of you may have seen by Petrostudies[7] goes to the opposite extreme from the CIA and forecasts a doubling of Soviet oil production by 1990. For my part, I stick with the forecast made to the Joint Economic Committee of the US Congress in 1976[8] of about a 50% increment (see Table 3). By then, the remote Arctic and Pacific off-shore deposits in the continental shelves of the Barents, Kara,

[1] Centre d'Études, op. cit.

[2] Ibid.

[3] Ibid.

[4] ER 79-10131 *Simulations of Soviet Growth Options KO 1985*, March 1979.

[5] The imposition of an embargo on sale of high-technology equipment by the US to the USSR in January 1980 patently changed the assumption of Soviet imports on which this statement was made to the Seminar.

[6] Cited by Odell, op. cit.

[7] Petrostudies, *Soviet Preparations for Major Boost of Oil Exports*, September 1980.

[8] *Soviet Economy in a New Perspective*, Washington EC, 1976.

and Okhotsk Seas, could have begun to be exploited (on an 8 to 10 year lead time). By then, too, natural gas – and here I use a CIA forecast[1] – should be 700 billion m^3 against 420 billion m^3 planned for next year. The urgency of promoting gas has been added to by the Iranian abandonment of IGAT 2 which would have delivered 17 B m^3 to the USSR against Soviet shipment to Western Europe and Czechoslovakia of 15 B m^3 on Iranian account (the margin of 2 B m^3 was the Soviet 'transit' earnings). The present cut of one-third of supplies through IGAT 1 is probably temporary.

The system of central planning in the Soviet Union has enabled energy consumption to be restrained, particularly by the low production of motor-cars, the absence of any major road-building (there is no East–West road at all across Siberia), the concentration of freight on to the railways, and generally by the tight rationing of energy to industrial users. With little consequent scope for conservation (or, save at high cost, substitution between sources), the USSR is committed to heavy investment for new production in remote regions in Siberia. All the output rise in oil, and almost all that in gas, is to come from that area, where special technologies for extremes of temperature, problems of drilling and transportation in permafrost, and absence of local settlements and manpower make for high capital outlays. The USSR is seeking industrial co-operation with Western corporations, which would share in the investment cost with repayment in products for 20 years after output begins.

I will pass quickly over the Soviet coal situation, where the vast, but difficult, Kansk-Achinsk fields[2] will be opened up by the Baikal–Amur Mainline – the second Trans-Siberian Railway – and should help achieve 1 billion tonnes output by 1990.

Save by errors of planning or of organization, the USSR should not in this century fall into deficit with respect to the domestic supply of energy (as the projection in Table 3 indicates). But it is insuring its future provision by diversification away from fossil fuels. Non-conventional sources (e.g. geothermal power in Kamchatka) are being explored.

I will look in detail at nuclear power prospects. Back in 1971 a target of 30,000 MW was set for 1980; this was trimmed first to 20,000, then to 18,500 and in fact only 10,000 MW have as yet been installed. This is certainly a slippage, but throughout the 1980s the vast Atommash works at Volgodonsk will be in commission,

[1] ER 78-10393 *USSR: Development of the Gas Industry*, July 1978.

[2] Easy to mine by open-cast shipping, but liable to self-ignition, and of low calorific content per tonne.

Table 3 East European and Soviet Energy Balances

		Production	*Consumption*	*Production as % of consumption*
		(million tons coal equivalent)		
East Europe	1965	359	370	97
	1970	418	459	91
	1975	475	552	86
Soviet Union	1965	925	829	112
	1970	1189	1055	113
	1975	1650	1411	117
	1990 (proj.)	3400	2950	115
East Europe & Soviet Union	1965	1285	1200	107
	1970	1607	1514	106
	1975	2125	1963	106
	1978	2510	na	na
	1980 Plan	2809	2410	117

with others undoubtedly to be added in the course of the decade. The political system allows no organization to lobby legislators or to mobilize public opinion against such an option, and the planned capacity has every chance of being built in the fuel-deficit regions of the European part of the USSR. The target for 1990 is said to be in the range 100,000 to 110,000 MW. With another 37,000 MW elsewhere in Comecon,[1] the total capacity in the group by 1990 should be some 140,000 MW. This compares with an EEC target of 160,000 to 200,000 MW by 1985, which – given similar failures in schedule – will only achieve between 70,000 and 80,000 MW.[2]

Suppose that the forecast in Table 3 is achieved: Soviet output will exceed consumption by about the same proportion (15%) as in 1975 (17%); but in absolute terms it will represent 450 million tonnes of coal equivalent, against 240 million in 1975. Comecon members in Europe that year had a net deficit of 77 million tonnes, of which the USSR supplied 126 million tonnes.[3] Doubling the Soviet contribution to that deficit in 1990 – say 250 million tonnes, still leaves 200 million tonnes for sale in the West, or three-quarters more than the 115 million tonnes delivered to the West in 1975.

[1] Deutsche Institut für Wirtschaftsforschung, (West) Berlin, as reported in *Financial Times*, 30 August 1979.

[2] Report to Energy Committee of European Parliament by the EEC Energy Commissioner, Guido Brunner, reported ibid., 12 September 1979.

[3] This figure comes from *Ekonomicheskoe sotrudrichestvo Stran-chlerov SEV* (Comecon's official journal), No. 3, 1979. The 1980 target was 180 million tonnes of coal equivalent. Soviet deliveries are more than the net East European deficit due to Polish exports of coal to the West.

Note on Statistical Sources

Soviet energy statistics are remarkably deficient for the second largest industrial power in the world; its political domination of the Council for Mutual Economic Assistance (Comecon) allows it to impose its preferences upon the Secretariat of that organization, which publishes an annual statistical abstract. Thus there are nothing like the comprehensive returns published by the OECD, EEC or their members.

Indeed, the volume of published material has been reduced by the Soviet abandonment of tonnage data for exports and the total suppression of import returns in the energy field. That this is contrary to the undertakings the USSR signed at Helsinki, I have argued recently in 'Monitoring East–West Trade for Madrid' (*The Contemporary Reivew*, May 1979), but we have no official returns on net trade to incorporate into an energy balance. The situation varies between admirable for Hungary to grossly inadequate in the GDR. 'The Communist Countries' for the present paper excludes China (which is the subject of Mr. Foster's paper) and minor countries, such as Albania, Mongolia, and Vietnam on which statistics are also deficient.

The USSR Central Statistical Office does not publish its own energy balances, and external estimates vary among themselves. Park (see references below) publishes five different estimates for six eastern European states. Table 3 uses the conventions of the Statistical Office of the European Communities (*Energy Statistics*) in converting energy sources into units of 'coal equivalent', viz. a standard calorific value of 7,000 kcal per kg. Soviet practice is the same (termed 'tons of conventional fuel'). The OECD unit (tons of a standard oil equivalent of 10,000 kcal per kg.) may of course be converted to (metric) tons of coal equivalent by multiplying by 0.7. The source of the 1965–75 series is the Statistical Office of the United Nations (Series J: *World Energy Supplies*); of the 1990 projection the Joint Economic Committee of the US Congress (reference below); and of the intervening series national estimates aggregated by the present writer. Soviet reports often utilize an energy balance limited to the major fuels (oil, gas, coal, hydro-electricity and, sometimes, peat) which is operationally useful in central planning since those sources are centrally allocated (on the system of 'material balances'), and some external authors compile estimates for every minor source, including privately-collected fuel wood. There are, furthermore, variations between estimates introduced by differences in conversion factors: Soviet compilations obviously use actual coefficients for each quality of fuel supplied, external estimates have to apply averages for each product. There is, for example, a wide variation between estimates (including private fuel wood) for the USSR by Campbell for 1965 (his last complete series) and those in Table 3 (million tons coal equivalent).

	Campbell	*Table 3*
Output	1016	925
Consumption	912	829
Net trade	104	96

References

1. R. W. Campbell, *Trends in the Soviet Oil and Gas Industry*, Baltimore, 1970.
2. D. Park, *Oil and Gas in Comecon Countries*, London, 1979.
3. US Congress, Joint Economic Committee, *Soviet Economy in a New Perspective*, Washington, 1976.

21 PETROLEUM PROSPECTS OF THE PEOPLES' REPUBLIC OF CHINA

John Foster

China's petroleum prospects are an issue of much interest today, particularly from the viewpoint of their potential impact on the international oil scene. Chinese policy toward the development of oil and gas resources has been supported by external procurement of equipment and new technology. China has sought to acquire a basic understanding of the technology underlying imported equipment and has then preferred to build up a domestic capability, rather than rely simply on foreign procurement. It has also preferred to build up its own technical competence gradually rather than rely on foreign technical assistance. Chinese policy appears to be continuing on this basis in the development of its onshore oil and gas resources, but it has shifted significantly for offshore exploration.

Offshore Exploration

China is now evincing considerable interest in offshore exploration. If substantial reserves can be found and developed offshore, they would be logistically well placed for delivery to domestic refineries as well as for export. To this end, Chinese policy has shifted significantly since 1977 toward an accelerated development of its offshore oil resources. These would help supply the growing need for oil both for domestic consumption and for export to finance capital goods imports, as China pursues a course of strong economic growth and industrialization. China is showing serious interest in having foreign companies participate in the exploration and development of offshore areas.

In 1978 the Chinese authorities began negotiations with Japan National Oil Corporation for a joint venture in the Bohai Gulf; negotiations were deferred in February 1979 but are reported to be resuming. Meanwhile, authorities have signed contracts in 1979 with foreign companies for eight offshore seismic programmes in the Yellow Sea or South China Sea. There is a host of foreign participants (including Petro-Canada) in one or other programme, and there is a foreign operator on their behalf for each programme. Two of these programmes are for adjoining areas of the Yellow Sea where Elf-Aquitaine and BP are the foreign operators. The other six programmes are for areas of the Yellow Sea. They include four surveys of adjacent areas of the Pearl River Mouth area, where the foreign operators are Exxon, Phillips, Mobil and Chevron; these areas are designated as a unified block, and participation in one survey requires participation in the other three, to be eligible for subsequent bidding. The fifth area of the Yellow Sea is south of Hainan Island (Atlantic-Richfield is the operator) and the sixth is westward in the Gulf of Tonkin (Amoco is the operator). Most surveys started in July–September and should be completed by late 1979 or early 1980. Each participant must provide summary interpretations to the Chinese authorities within 6–8 months of receiving data, i.e. during the last half of 1980. In these various arrangements, costs are borne by the foreign partner on a risk basis. If the results are encouraging, the Chinese authorities may allocate blocks for bidding for exploratory drilling. They have indicated their intention of nominating acreage in the Yellow Sea within 18 months of receiving interpretations, and in the South China Sea within 12 months.

Onshore Reserves

Some 23 basins are known onshore, and one (Bohai) is partly on and offshore. They contain 120 oil and gas fields, of which some 83 are in production. There are also large kerogen or oil shale deposits in China which are produced at the Fu-shun and Mao Ming mines; but operations have not been expanded in recent years. There have been no official estimates of Chinese oil reserves published since 1949, when exploration was very limited. But among external private estimates, Messrs. Meyerhoff and Willums have recently suggested that produced, proven, and probable reserves onshore may be about 14.4 billion barrels (Bb) excluding Bohai (Po Hai) offshore and shale oil, and that ultimate recoverable oil reserves may total 42 Bb. Other estimates appear to be of the same order of magnitude.

Regionally, the north and north-east basins are the most accessible and successfully explored areas of China for onshore development, with two-thirds of proven and probable onshore reserves being in these regions. In fact most Chinese oil production comes from there; predominant is the Daqing (Ta-ch'ing) field, and others of note are

Sheng-li and Dakang (Ta-kang). The discovery of a new field (Renqiu) was announced in 1975, though it may be an extension of the Dakang field. Much of China's future potential reserves are believed to be in the inaccessible and little developed western basins, some of which are close to the Soviet Union. Other basins with good potential lie in the southern region. However, an easier route in seeking new oil reserves looks to be offshore; and this seems to be the direction in which Chinese oil policy is headed.

Offshore Reserves

So far, there is no Chinese offshore production except from an extension of the Dakang (Ta-kang) and Sheng-li oilfields offshore Bohai (Po Hai) Gulf. But six basins are known offshore, and reserves may be as extensive there as onshore. Estimates of offshore reserves are still highly conjectural. A speculative indication of their ultimate potential is in the order of 30–40 billion barrels. Prospects are believed to be promising in the Chinese continental shelf, which stretches from the Bohai Gulf in the north-east to the Gulf of Tonkin in the south-west and includes the Yellow, East China, and South China Seas.

Oil Production

As told to us, oil seeps have been known in China since antiquity; as early as AD 1180 wells were drilled by bamboo 'drill pipe' to depths of 1,200 metres. The first modern field was discovered back in 1897. However, production was still insignificant (2,000 b/d) in 1949 when the Communist regime came to power, and it remained at a low level until the 1950s, when the three major fields now accounting for 80% of total output came into production: Daqing, Sheng-li and Dakang. These three fields are in the north and north-eastern regions. Relatively little comes from the far west and southern regions.

The main exploration thrust prior to the discovery of Daqing (Ta-ch'ing) in 1959 had been in the remote western region. Some of these basins have good to excellent potential. But they are in difficult terrain far from the main centres of consumption in the eastern and southern regions of China, and some basins are close to the Soviet border. Though major oil and gas fields in the southern region have been discovered, production there to date is relatively minor. There are major discovered oil reserves and very major reserves of gas in the Sichuan (Szechwan) basin.

Oil production, virtually all onshore, has expanded from some 100 M b/d in 1960 to 2.0 M b/d in 1978. It is officially estimated at 2.12 M b/d for 1979, only 2% higher than last year. The rapid growth in Chinese oil production took place despite adverse effects of the Cultural Revolution in 1969 and Gang of Four policies in 1976. It has been achieved basically by China's own efforts, supported by some Soviet assistance prior to 1960 and by judicious purchases of imported equipment.

China has announced targets of 4 M b/d for 1985 (growth of 10% p.a. from 1978) and 8 M b/d for 1990 (15% from 1985) and, most recently, 10 M b/d by the end of this century. These growth rates compare with those of 18% p.a. for 1970–75, slowing to 11% p.a. for 1975–78. Some external commentators have expressed scepticism about China's ability to achieve these targets, but there appears to be general agreement that production could rise considerably, particularly if offshore exploration efforts meet with success. While the production targets are ambitious and represent a difficult challenge, potential reserves appear to be large, and the new oil policies since 1977 heighten the prospects for a major break through in production levels.

Oil Exports

Oil exports are envisaged to be a main source of finance for the capital goods imports needed to accelerate economic growth. Chinese oil exports to its main customer, Japan, began in 1973. In 1978 they averaged about 120 M b/d. Exports to other countries add about another 60 M b/d.

The main export grade is Daqing (Ta-ch'ing) crude oil which is attractively low in sulphur content and quite light in gravity but is unfortunately low in its product yield of light fractions and high in paraffin content. In Japan it is to a large extent burned unrefined for electric power generation. The low sulphur content of Daqing oil is helpful in enabling compliance with Japanese air quality regulations. With a view to absorbing increased imports of Chinese oil, consideration is being given in Japan to the installation of hydro-cracking facilities, or alternatively to the possible availability of other grades of crude oil of better quality from new fields.

Some indication of oil exports from China in the medium-term future may be derived from the eight-year trade agreement which was reached in February 1978. It includes an expansion in Chinese oil exports from 136 M b/d in 1978 to 300 M b/d in 1982. Volumes for 1983–85 are to be negotiated in 1981, though political figures from the two countries have spoken of 600 M b/d for 1983. China will also export steam-coal and coking-coal to Japan in increasing yearly amounts. In return, China will import industrial plants and technology and construction materials.

Projections of oil exports from China nevertheless remain speculative. There is a general consensus of view that most production in the several years ahead is likely to be absorbed by the burgeoning domestic market, leaving a relatively small share for export. But China's offshore region is one of the world's large remaining unknown frontiers for exploration. Oil export prospects for the 1990s could be as limited as in the 1980s but are unpredictable.

Economic Prospects

Since 1949, China has shown good economic performance (perhaps 7% p.a.) in those years which were undisrupted: outside the periods of the Great Leap Forward, the Cultural Revolution, and the Gang of Four. But these latter periods were disastrous in economic terms. In February 1978, Chairman Hua Guofeng announced the well-known ten-year economic development plan which emphasizes steel-making, agriculture and infrastructure. It set ambitious targets which have this year been scaled down to more achievable prospects.

China's economic policy under Chairman Hua Guofeng (Hua Kuo-feng) and Vice-Premier Deng Xiaoping (Teng Hsiao-ping) is at a turning point. China is striving for a major expansion in economic development. While this has not meant a change from its basic philosophy of self-reliance, it is opening its doors wider to Western technology, equipment and finance, with a view to modernising its outmoded industrial and agricultural sectors.

The new commitment to economic development, announced at the National Peoples' Congress (NPC) in March 1978, still holds good, together with the new reliance on profit motive, market mechanism and managerial independence. But the targets and import requirements were sharply reduced at the NPC of June 1979. Deng Xiaoping himself seems to have steered this retreat or 'readjustment', acting on advice as well as pressure from political opponents.

In pursuing its policies since 1977, the new leadership is running a risk. And the possibility of political complications should be borne in mind. However, its hand is strengthened by the evident failures of previous policies. This all lessens the possibility that a new administration could scrap present policies. The immediate years ahead will be very revealing. The biggest constraints to the development plan lie in the economic infrastructure, management skilled manpower and foreign exchange. The plan envisages massive investment and other measures to overcome these constraints.

Foreign Trade

An important tool for achieving this capital expansion programme will be the publicised large-scale imports of technology and equipment. It will take a great effort to earn the additional foreign exchange needed to finance these imports. The ambitious buying plans of 1978 were running far ahead of resources, and some negotiations were accordingly delayed this spring while plans were being reassessed. Now that this appears completed, more moderate new contracts are being signed, and the earlier 'block-buster' proposals are no longer welcome.

To finance these imports, China is seeking a large expansion in exports of oil and manufactured goods. It also appears to be shifting its stance toward borrowing from abroad, which it has hitherto disfavoured. It is now considering a variety of purchase and credit schemes to earn or conserve hard currency. For some development projects China envisages product buyback schemes, compensating suppliers with output from the project. It is also seeking long-term low interest credit, perhaps commercial bank loans for direct profit financing, and even government-to-government loans. Known foreign lines of credit now amount to US $19 billion in medium- and long-term development credits, export credits, and commercial bank arrangements, together with US $6 billion short-term facilities from a Japanese banking syndicate. New loans are being negotiated. So China can probably obtain as much financing as it can absorb. And it clearly intends to steer a careful path.

After announcing their new three-year strategy this summer, the Chinese authorities (i) published their long-awaited code on joint ventures with foreign companies, and (ii) formally signed the framework for a trade agreement with the US which would open up most-favoured-nation treatment and trade credits. This agreement now requires approval by the US Congress.

Energy Sector

So what are the energy needs and exports in future years? Energy is the key to Chinese economic expansion. China relies on coal for roughly three-quarters of its domestic energy needs, though oil consumption is increasing rapidly.

Data are sketchy, but an indication of the amount of primary energy consumed in China in 1977 is in the order of 8.5 M b/doe (million barrels per day of oil equivalent). Of this amount, perhaps three-quarters comes from coal, just under 20% from oil, only 2% from hydro-electricity, and a little less from natural gas. There is also significant local consumption based on biomass energy, e.g. wood, roots and grass. Energy end-use appears highly angled towards industry and construction (estimated by some at 60%). The residential and commercial sector is next (28%), while transport (7%) and agriculture (5%) are relatively small energy users. UN data suggest yearly commercial energy consumption of about 4 barrels per head of population, compared with about 64 barrels or so for North America and 2 barrels for developing countries.

China's plans for economic and industrial development envisage a great increase in future energy consumption. This indicates the need to overcome constraints in energy production and related infrastructure, mainly in transport, oil-refining, and coal preparation. Oil demand looks likely to increase rapidly as China's economy develops. Particular end-uses of high growth include agriculture (as it becomes increasingly mechanised), railways (as they convert from coal to oil), highway construction, road transport, petrochemical, and fertilizer manufacture. There seems to be a consensus view that most oil production in the decade ahead will be absorbed by the burgeoning domestic market, leaving a relatively small share for export.

China is the world's third largest coal producer (behind the US and Soviet Union) and accounts for roughly one-eighth of world coal production. Output suffered setbacks during the Great Leap Forward and Cultural Revolution. Since then, however, it has evidently grown by about 4–5% p.a. in the last decade. Coal consumption is mainly concentrated in the industrial sector, while only a small part goes to electricity generation (in contrast to the US). The highest priority in coal usage is accordingly given to the industrial sector, where periodic shortages are reported to have acted as a brake in past years.

China is planning a massive expansion in the production of coal, which will continue to act as the main source of primary energy during the next decade. Though the railway dieselisation programme will lead to a diminishing rate for coal in the transport sector, this is expected to be offset by its increased use in electric power generation, particularly at mine-mouth. The plan calls for coal production to double from its 1977 level in 10 years and to double again by the year 2000 (7% p.a.), higher than in recent years (4–5% p.a.) Production targets are ambitious, bearing in mind logistical and infrastructure problems and the competition for investment resources from other sectors of the Chinese economy. Constraints may arise in the long-distance transport of coal to consuming areas; coal represents much of the total freight carried by rail, and the system is already congested.

Natural gas is the most under-used energy resource in China. It appears to contribute a mere 1% of total primary energy consumption. (Bad data but an indication.) The gas is mostly produced from the Szechwan basin. It is supplied to local industries and field operations, and it is not pipelined over long distances. I have no information on gas flaring or reinjection, though these practices may involve sizeable quantities. There were reports in 1978 that China was seeking assistance in building a plant for exporting gas, though it would seem economically preferable to use it domestically, thereby freeing oil for export.

Hydro-power contributes about 2% to China's energy supply. China's potential for hydro-power is immense, estimated by some to be 100 times the present installed capacity. But the hydro sites are far from major population centres, and a major constraint for China has been the technology of long-distance power transmission. Hence, China has favoured development of thermal generation. There are also reports that China is procuring two nuclear power plants from France. China is thought likely to push for faster growth in hydro-electric development in the future, in order to divert the use of oil from power generation into the export market. The rate at which hydro-electric capacity can be installed is hard to predict.

Now let us try our hand at some illustrative but helpful projections. If the targets for oil and coal are achieved, the production of primary energy could grow by roughly 8.5% p.a. from 1977 to 1985 to reach say 17 M b/doe. Suppose exports of oil were expanded to some 800 M b/d and of coal to some 500 M b/d o.e., to earn reasonable amounts of foreign exchange. There would then be some 15 M b/doe of primary energy available for domestic consumption, an increase of 7.5% p.a. from 1977. A very broad-brush approach suggests that this could be sufficient to fuel economic growth of over 5% p.a. if the energy coefficient remained roughly similar (1.4) to that in recent years. It could fuel economic growth at 6% p.a., if the coefficient were lower (say 1.2), more similar to that in other developing countries.

This conclusion perhaps sheds some light on the ambitious level of energy production targets. If expansion of energy output fell significantly short of these targets, there could be constraints of energy supply on China's overall economic growth objectives. By way of illustration, suppose that (a) coal production were to grow from 1977 to 1985 by 5% p.a. as in the last decade compared with the desired 7% p.a., and (b) oil production were to grow by 8.5% p.a. (to reach 3.5 M b/d), compared with the desired 10.5% p.a. Then there would in this illustration be only 12.5 M b/doe of primary energy available for domestic consumption, an increase of just under 5% p.a. from 1977. This would be enough to fuel economic growth of only 3.5% p.a. if the energy coefficient were to hold at its recent high level; and economic growth would improve to only about 4% p.a. supposing a lower coefficient (say 1.2).

These illustrations are based upon imprecise data and are all very rough and ready. But they do indicate the dimensions of the issues which confront the Chinese leadership today. They suggest why it is thrusting ahead with vigour (a) to develop all its energy resources, (b) to put particular emphasis on investment into both coal and oil, and (c) to bring in foreign participation in offshore oil exploration and development.

22 ENERGY BALANCES AND PROSPECTS OF DEVELOPING COUNTRIES

Staff, OPEC Special Fund

The Third World accounts for over 70% of world population but consumes only 16% of the world's commercial energy. The consumption of 94 developing countries having 40% of the world population represents 8% of world commercial energy consumption. By commercial energy we mean oil, gas, coal and electricity generated from hydro and nuclear sources. We do not include therefore the renewable traditional energy sources such as fuelwood, and agricultural and animal waste, which represent as much as half of the energy supply in developing countries.

Commercial Energy

In the field of commercial energy alone, developing countries are, as a group, net energy exporters to the developed world. The surplus of energy produced over energy consumed is to be found mainly in petroleum and, to a lesser degree, in natural gas. This fact, however, does not reflect adequately the differences that exist among developing countries. Apart from OPEC Member Countries which exported in 1978 a total of 27.9 M b/d of oil,[1] there are twelve countries which are oil exporters: Angola, Bahrain, Bolivia, Congo, Egypt, Malaysia, Mexico, Syria, Oman, Trinidad and Tobago, Tunisia, and Zaïre. The volume of exports of the latter countries reached 2.1 M b/d in 1977. The rest of the developing countries are net oil importers.

Commercial energy consumption of net oil-importing countries totalled an equivalent of 8.06 M b/d in 1975, of which 4.33 million was consumed in the form of oil. Their aggregate oil production, on the other hand, was estimated at about 1 M b/d and projections show that by 1990 production is expected to reach 2.5 M b/d. The above production levels represent an increase in oil-sufficiency

[1] This figure includes both crdue oil and refined petroleum products.

from 21% in 1975 to 30% in 1990 (see Table 1). However, this production of oil will remain limited to a small number of the oil-importing countries (Argentina, Brazil, Chile, Columbia, India, Peru, and Turkey) and represents only 2% of world oil production.

Table 1 Oil imports of oil-importing developing countries (86 countries)

	1975/6	*1990*	*1975/76–1990 Growth rates (%/Yr.)*
Oil Production, M b/d	1,048	2,550	6.6
Net Oil Imports, M b/d	3,835	5,908	3.1
Total Oil demand, M b/d	4,883	8,458	4.0
Self-sufficiency	21%	30%	
Approximate value of Net Oil Imports (US $ billion 1976)	17.5	27.0	

Source: Parra, Antonio R., *Overview of the energy situation in oil-importing countries,* March 1979.

After accounting for their production of oil, the net oil imports of oil-importing countries reached 3.8 M b/d in 1975 and are expected to increase at an average rate of 3% per annum in the next 5–10 years. Only eight countries (Brazil, India, the Philippines, South Korea, the Netherlands Antilles, Taiwan, Thailand, and Turkey) import, individually, more than 100,000 b/d and thus account for roughly 70% of the total net oil imports of oil-importing developing countries. Only two of the above-mentioned eight countries have per capita incomes below US $500. The net oil imports of all non-OPEC developing countries, on the other hand, was estimated at 1.9 M b/d for 1975 after taking into account the oil exports of non-OPEC member developing countries (see Table 2).

Commercial energy demand of developing countries will experience an increase in the order of 4% to 5% per annum as the structure of their economies changes and they concentrate more on the industrial sectors. This will bring about a growing pressure in their energy supply situation. The problem of securing additional energy supplies will be particularly severe in those countries not endowed with oil, coal and gas reserves, and which have an adequate level of foreign exchange earnings to meet their oil import bill.

Coal and natural gas do not present ready alternatives to oil in most developing countries. Coal geological resources in developing countries are estimated to account only for 2.3% of the world total. Almost 90% of these resources are to be found in the US,

Table 2 Total oil import bill at current prices:

– Oil import volume (1975/76):		3.8 M b/d × 365 = 1.387 billion b/year
– Total oil import bill at current prices:		1.387 × $21.41 = $30.514 billion
– Incremental bill (assuming rise of $7.45 per barrel):		1.387 × $7.45 = $10.3 billion
– Excluding:	Brazil	823.7 (000) b/d
	S. Korea	356
	N. Antilles	314.3
	Taiwan	295
	Turkey	256.9
	Argentina	78.5
	Chile	64.6
	Panama	60.4
	Antigua	16.1
	Fiji	6.9
	Colombia	1.9
	Belize	1.3
		2,275.6 (000) b/d or
		$18.27 b. per year
– All other developing countries		
– Oil imports 1.53 M b/d		
– Oil imports bill 1.53 × 364 × 21.41 = $11,956		
– Incremental bill 1.53 × 365 × 7.45 = $4.16 billion		

Source: Parra, Antonio R., *Overview of the energy situation in oil importing countries*, March 1979.

the USSR and China. According to a World Bank report, coal resources in the developing world are concentrated in Botswana, Columbia, Mexico, Brazil, India, Korea, Vietnam, Turkey, Swaziland, and Yugoslavia. The increases in coal production in these countries will be mainly consumed locally and are not expected to meet the increases in energy demand of other developing countries.[1] Moreover, there are restrictions in increasing the supply of coal mainly due to mining, handling, and transportation difficulties.

Nor does natural gas present a viable alternative to the developing non-oil producing world. Most of the reserves of natural gas are located in the oil-producing countries. Furthermore, its transport is costly and requires heavy capital investments in both the exporting and importing countries.

Hydropower is still largely unexploited in developing countries and supplies only a small percentage of total commercial energy. Electricity from all sources represents roughly 8 to 10% of total commercial energy consumption in developing countries and the share of hydro-electricity in total electricity production is estimated

[1] IBRD, *Coal Development Potential and Prospects in Developing Countries*, August 1979.

to be 66%, 30%, and 54% respectively for the regions of Africa, the Far East (excluding Japan) and Latin America. In fact, the share of electricity is only about 10% of the world's total commercial energy consumption. It should be noted, however, that production of electrical energy is only 200 kWh/person in developing countries as compared to 4,300 kWh/person in the developed countries.

The rise in energy demand in developing countries, including OPEC members, will continue to increase their share in world commercial energy consumption which rose from 6.2% in 1955 to 8.2% in 1970 and to 10% in 1975. These countries' consumption is estimated to reach 25% of world total by the year 2000. This increase in their share of energy consumption will cause competition for energy supplies with the industrialized world, particularly in the oil market. The fact that present technology is petroleum-based lessens the flexibility of developing countries in reducing their oil consumption and in increasing the utilization of alternative energy sources. This is less true in the case of developed countries where there is greater scope for conservation.

Non-conventional energy sources, such as solar and geothermal energy are not, at present, viable alternatives to oil in meeting growing energy demands. They will require technological breakthroughs to make their usage economically feasible. Nuclear energy may not be a practical alternative for the large majority of developing countries, due to the minimum size required for feasible nuclear plants in addition to the high cost and sophisticated technology. Therefore, when considering future energy supplies in developing countries, it is probably oil which will have to cover most of the gap between energy supply and energy demand in non-oil producing countries.

Non-Commercial Energy

Increases in energy demand will also affect the non-commercial energy supplies in developing countries. Although the share of non-commercial energy in total energy consumption declines as countries move into more advanced stages of economic development,[1] population growth, particularly in rural areas increases non-commercial energy demand. The larger the proportion of rural population, the higher the share of non-commercial energy consumption is likely to be.

The amount of non-commercial energy used in the developing

[1] In India, for instance, non-commercial energy accounted for 75% of total consumption in 1953/54 and only for 58% in 1970/71.

world is hard to estimate due to its decentralized usage and unorganized market. Nevertheless, a series of studies have been carried out on the subject and have indicated that the share of non-commerial energy in the total energy supply could be as high as 67.6%, 49.6%, and 29.8% in the regions of Africa, the Far East (excluding Japan), and Latin America, respectively (see Table 3).

Fuelwood is the main source of non-commercial energy. Its wide usage in an unorganized manner is causing serious environmental problems particularly in Africa and Asia. The percentage of wood production used as fuel in developing countries is 80% as compared with 10% in developed countries. The irregular usage of wood in developing countries for other than industrial purposes does not promote reforestation and results, therefore, in problems of deforestation and erosion. More could be done in the field of planning and management in order to assure both an efficient use of wood and the future availability of adequate fuelwood supply.

Agricultural waste, on the other hand, although used as fuel, is cumbersome to use and to collect. A large part of it is used in the respective processing industries such as bagasse in the sugar factories or oil-seed waste in oil mills. Otherwise agricultural waste is burnt, thereby producing serious damage to the land, since in this way nutrients, which should return to the soil, are destroyed.

The third renewable non-commercial energy is animal dung. Through a process of anearobic fermentation this waste can provide both fuel and fertilizer. This process produces a gas which is 60% methane and the leftover can be used as fertilizer. This also applies to other organic waste. At present, however, the direct use of animal and agricultural waste prevails, and bio-gas plants are still at the experimental stage. Continued efforts in developing bio-gas technology would be to the benefit of developing countries since it can provide both fuel and fertilizer to the rural areas from renewable sources. The amount of energy which could be provided in this manner would be suitable, however, only for small-scale and decentralized consumption.

OPEC and the Oil-importing Developing Countries

Oil-importing developing countries will continue to depend on oil imports to meet the bulk of their increases in domestic energy demand. In this regard, they will face a twofold problem in the coming decade: securing oil supplies in a highly competitive market and meeting the increase in cost as a result of higher oil prices.

Table 3 Regionwise energy consumption (Mtce) – 1973[1]

Region[5]	*Commercial Energy*	*Fuelwood*[6]	*Agricultural Waste*[7]	*Total*
Africa[2]	66 (32.3)	116 (56.8)	22 (10.9)	204
Middle East	109 (86.5)	5 (4.0)	12 (9.5)	126
Far East[3]	247 (50.4)	143 (29.2)	100 (20.4)	490
CPE Asia	502 (83.0)	66 (10.9)	37 (6.1)	605
Latin America	144 (60.5)	72 (30.2)	22 (9.3)	238
Central America	160 (82.1)	26 (13.3)	9 (4.6)	195
West Europe	2568 (98.5)	18 (1.1)	6 (0.4)	1592
CPE Europe	1771 (96.3)	43 (2.3)	24 (1.4)	1838
North America	2723 (99.7)	7 (0.3)	0	2730
World[4]	7885 (92.2)	498 (5.8)	167 (1.9)	8550
Sum of first 5 regions	1228	430	137	1858
% of world	15.57	86.34	82	21.73

[1] Figures in parentheses are percentages.
[2] Excluding South Africa.
[3] Excluding Japan
[4] Includes all countries (i.e. Japan, South Africa, Australia, etc.).
[5] Regions as defined by World Energy Study, UN, 1975.
[6] *Source: FAO Production Yearbook,* 1973.
[7] *Source*: J. K. Parikh, *Energy Systems and Development*, International Institute for Applied Systems Analysis, Laxenburg, August 1979.

OPEC Member Countries could play a significant role in assuring oil supplies through long- and medium-term contracts with individual countries. However, their margin for action is limited to the degree in which OPEC countries control the marketing of their products. Consequently the co-operation of oil companies is a necessary and determinant factor. To this effect, the OPEC Oil Ministers in their recent conferences have asked both the industrialized countries and the private oil companies to play a more constructive role in guaranteeing oil supplies to developing countries. Developed countries are called upon to control their energy consumption while oil companies are requested not to engage in price speculation practices.

The stringent market situation in oil, on the other hand, and the consequent increases in oil prices have resulted, and will continue to result, in additional financial outflows from oil-importing developing countries. The annual oil import bill of this group of countries based on the oil import volumes for 1975/76 amounts to $30 billion per year if calculated at current prices averaged at $21.41 per barrel.

As indicated earlier, the bulk of oil imports remains concentrated on a few countries, six of which are high income developing countries which, with the exception of Turkey, have relatively good prospects of meeting their oil import requirements without recourse to major external financing. The remaining oil-importing developing countries, which form the majority, require external assistance for the importation of an aggregate volume of at least 1.5 M b/d. The total cost of these imports is calculated to be $12 billion per year at current prices; the last price increase representing on average one-third of the total oil import bill, i.e. about $4 billion per year as of July 1979.

Projections for the next 5 to 10 years show an increase, in absolute terms, in the volume of net oil imports to 4.3 M b/d in 1985 and 5.9 M b/d by 1990. Both these figures assume an average annual growth rate of net oil imports between 3% and 3.5% based on high oil production growth rates of 8.4% and 6.6% and an increase in oil demand of 5.2% and 4% respectively. Thus, even if we assume no further increases in oil prices, the oil import bill of oil-importing developing countries (OIDC) will grow at an annual rate of 3–3.5% in the next 5 years, that is to say, by US $1 billion per year.

According to a World Bank report the consolidated net oil imports of all developing countries, excluding only OPEC Member Countries, will, however, decrease from 1.9 M b/d in 1975 to 0.75 M b/d in 1980 and to 0.08 M b/d in 1985. This reduction will be principally due to increases in oil production of non-OPEC oil-exporting countries. The capital requirements of the oil sector of both oil-importing and oil-exporting countries (other than OPEC) is estimated to be in the order of $5.65 billion per annum for the next 6 years, of which 25% to 30% will be for exploration activities. The investment of OIDC alone is estimated at US $3.39 billion per year of which US $1.07 billion represent the investment requirement of those countries with per capita incomes below US $625[1] (see Table 4).

The magnitudes of investment finance for oil projects and the risks involved in the stages prior to oil production, require that

[1] IBRD, *A Programme to Accelerate Petroleum Production in Developing Countries*, January 1979.

Table 4 Investment in oil and gas by non-OPEC developing countries – estimated annual requirements 1976–85

Developing Countries with per capita Incomes in 1976 of:	*Annual average*	
	Oil *(1977 $ m)*	*Gas*
$1051 and above		
Net Oil Exporters	1,000	450
Net Oil Importers	1,575	225
	2,575	675
$626–1050		
Net Oil Exporters	200	152
Net Oil Importers	750	100
	950	252
$252–625		
Net Oil Exporters	930	50
Net Oil Importers	420	100
	1,350	150
Below $250		
Net Oil Exporters	100	10
Net Oil Importers	650	138
	750	148
Sub-total: Net Oil Exporters	2,230	662
Sub-total: Net Oil Importers	3,395	563
Grand total	5,625	1,225

Source: World Bank, *Staff Working Paper 289*, April 1978 (Per capita income limits are expressed in 1976 $).

Notes: 'Oil' includes investment requirements in oil and gas exploration, development of oil, production of oil and associated gas, and crude oil pipelines in all non-OPEC developing countries. The exploration stage is assumed to account for 25 to 30% of total investment requirements in the upstream phase. The relative costs of the various exploration activities are approximately 5 to 10% for geological surveys, 15 to 30% for geophysical prospecting and 60 to 75% for drilling.

'Gas' refers only to investment in development of non-associated gas and gas pipelines: excludes investment in liquefied natural gas (LNG) projects except in Malaysia.

Investment Requirements for oil relate to the projected output in the non-OPEC developing countries of 8.40 M b/doe (of which 2.85 M b/doe in OIDCs) by 1985. They are not comparable with the estimates made in July 1977 because: (a) they are expressed in 1977 rather than 1975 dollars; (b) they cover only upstream investment (including crude oil pipelines); (c) the real costs of petroleum development are now estimated to be significantly higher; and (d) the earlier estimates were related to a level of output that was considered feasible if maximum efforts were made.

funds be forthcoming from various sources, both private and public. The programme of the World Bank is an attempt in this direction. It mainly entails the partial financing of the development stage of oil and gas and it is estimated that by 1983 the Bank's operation in these sectors will rise to $1,230 million of which 60% would be to finance production facilities covering 20% of total investment. Emphasis is also being placed on the energy sector in the activities of other development finance institutions.[1]

The OPEC Special Fund, as the collective aid facility of OPEC Member Countries, is also giving high priority to the energy sector.[2] Thus far, the Fund has been involved mainly in electricity projects. Power generation and electrification have traditionally received the bulk of development project assistance to the energy sector from other sources as well. OECD statistics show that both these sub-sectors accounted for as much as 85% of the total lending in energy from DAC countries[3] and multilateral institutions (both private and public) to developing countries in 1977.

In addition to the OPEC Fund, OPEC Member Countries are in a position to support the energy sectors of developing countries bilaterally and through other institutions. The full extent of OPEC's role in the development of the energy sector in the developing world is yet to be determined. Joint-ventures in the oil and gas industries, technical assistance and participation in pilot projects of alternative sources are all sound possibilities.

OPEC Member Countries have contributed to the development efforts of Third World Countries long before the events of 1973. The Kuwait Fund, the Arab Fund, and the Abu Dhabi Fund are examples of this. However, it was only after the increases in oil prices that these countries were able to intensify their aid programmes and seek other ways of assisting developing countries.

OPEC aid has not, however, been extended as a compensation for the higher cost of oil imports. If it had been, it would have benefited mainly the more advanced developing countries which

[1] For instance, the Inter-American Development Bank, in particular, is considering the establishment of a special guarantee fund for fuel and non-fuel minerals with the view of complementing existing national insurance and guarantee programmes. Such a mechanism would provide additional security to investors by extending insurance against political and commercial risks for both equity and debt capital. In addition to this project, the Bank will participate directly in the financing of oil projects thereby encouraging the mobilization of resources for this sector.

[2] Although emphasis is placed on projects which reduce energy dependence, the Fund does not limit its activities to the energy sector. Such a policy would result in a reduction of its lending to the degree that beneficiary countries do not have energy projects ready for immediate implementation.

[3] The DAC is comprised mainly of OECD aid donors.

Table 5 Selected developing countries' net oil imports and total receipts from OPEC sources

Region	*Net oil imports*			*Increase since 1973*		*Total receipts from OPEC sources*		*Total receipts as a share of increase in net oil imports*	
	(millions of dollars)							*(per cent)*	
	1973	*1974*	*1975*	*1974*	*1975*	*1974*	*1975*	*1974*	*1975*
North Africa[1]	55	227	221	172	166	17	113	10.0	68.2
Other developing Africa[2]	364	882	996	518	632	522	702	100.7	111.0
(excluding Sudan)	(340)	(797)	(900)	(457)	(561)	(159)	(406)	(34.9)	(72.5)
Asia: Middle East[3]	6	16	17	10	11	89	148	893.0	1,347.3
Other developing Asia[4]	2,228	5,563	6,572	3,335	4,344	1,168	1,532	35.0	35.3
Developing America[5]	1,281	4,612	4,569	3,331	3,282	231	606	6.9	18.5
Total[6]	3,934	11,300	12,376	7,366	8,441	2,027	3,102	27.5	36.7
of which:									
Land-locked countries[7]	53	115	139	62	87	10	48	16.9	55.2
Least developed countries[8]	127	334	372	207	245	574	617	277.3	251.9
MSA countries[9]	1,298	2,480	2,827	1,182	1,529	1,685	2,023	142.5	132.3
LAS members[10]	86	328	334	243	248	468	557	193.0	224.0

Source: UNCTAD, Document TD/B/685/Add.1.

[1] Morocco.

[2] 16 countries: Burundi, Central African Empire, Ghana, Ivory Coast, Kenya, Liberia, Mauritius, Malawi, Senegal, Sierra Leone, Sudan, Togo, United Republic of Cameroon, United Republic of Tanzania, Zaire, Zambia.

[3] Yemen Arab Republic.

[4] 12 countries: including Bangladesh, India, Malaysia, Nepal, Pakistan, Philippines, Republic of Korea, Singapore, Sri Lanka, Thailand.

[5] 13 countries: Argentina, Brazil, Chile, Costa Rica, El Salvador, Guatemala, Honduras, Jamaica, Nicaragua, Panama, Paraguay, Peru, Uruguay.

[6] 43 countries. [7] 5 countries. [8] 8 countries. [9] 19 countries. [10] 3 countries.

are in a better position to resort to other types of finance. In fact, the larger part of OPEC aid has been directed to the least developed and the most seriously affected countries, which account for a relatively small share of oil imports. Engaging in compensatory financing would also have meant subsidizing oil, thereby altering the comparative cost of energy sources which could then result in an inadequate and inefficient allocation of resources.

Given the fact that OPEC aid is geared towards development in general and is not a compensation for higher oil prices, OPEC Member Countries have contributed in some years amounts which, in aggre-

Table 6 OPEC and DAC countries' official financial flows to LDCs and Multilateral agencies, 1974-77 – a comparison.

Total Disbursements	*1974*	*1975*	*1976*	*1977*
1. *In US $ million*				
OPEC	7,557	11,463	9,161	7,588
DAC	11,517	13,506	13,401	14,138
2. *In per cent*				
Ratio of OPEC to Total Disbursements	65.6	45.9	40.6	34.8
OPEC Flows as share of own GNP	4.40	5.62	3.78	2.65
DAC Flows as share of own GNP	0.33	0.35	0.32	0.30

Source: UN Document TD/B/C.7/31 (Summary).

Notes: Data for OPEC and throughout the period 1974-76 are taken from the summary of the UNCTAD Report 'Financial Solidarity for Development', UN Doc.TD/B/C.7/31. Data covering the year 1977 for OPEC and the whole period 1974-77 for DAC are taken from the OECD/DAC source.

OPEC here means 10 countries only; Ecuador, Gabon, and Indonesia are not included. Percentage of flows to GNP would be much higher if Nigeria and Algeria are also excluded. All OPEC donors are developing countries; most of them suffer from deficit in their current accounts.

OPEC financial flows are derived from depletable natural resources, unlike DAC flows which are overwhelmingly drawn from renewable sources of wealth.

All OPEC financial flows are untied, unlike most of the DAC flows which are inevitably tied to procurement from DAC countries as a group. Thus, both DAC and OPEC financial flows to LDCs impart benefits to DAC countries; benefits that are translated into a higher level of employment, lower unit costs, exports of spare parts and services, follow-up contracts, etc.

OPEC disbursements at ODA terms represented in 1976, 2.21% of donors' GNP while concessional disbursements from DAC countries represented 0.33% of their GNP. If the data are restricted to the major four OPEC donors, the ratio of ODA disbursements to GNP would range between 5% and 10%.

DAC disbursements exclude Southern Europe, Israel, and dependent territories, which do not receive assistance from OPEC sources.

Table 7 Flow of workers' remittances and its share in total imports and exports of goods in selected labour exporting countries.

Country	*1973 Remittances*[1]	*1974 Remittances*	*As per cent of Exports*	*Imports*
Algeria		390	9	9
Bangladesh	22.8	36	13	2
Egypt	85	189	11	5
India[3]	183	276	8	5
Jordan	45	75	48	12
Morocco		356	21	17
Pakistan[3]	147	151	15	6
Syrian Arab Republic	51	62	8	4
Tunisia		118	13	9
Turkey		1,425	93	33
Yemen Arab Republic	108	159	1,325	69
Yemen PDR	33	41	410	23

Sources: IMF consolidated balance of payments reports.
[1] In current prices, million dollars, gross figures.
[2] Estimates.

gate, exceeded the amount of the increase in the cost of oil imports for many recipient countries. This has particularly been the case in most of the least developed and most seriously affected countries (see Table 5).

In the past years official disbursements in concessional ODA terms[1] have represented on average 1.5% to 2.7% of the aggregate GNP of OPEC Member Countries. In the more significant individual cases of the major OPEC donors, the ratio has been much higher (see Table 6). The surplus of OPEC countries as a group has declined and their growing domestic capital needs may not permit them to increase and/or engage indefinitely in long-term concessional aid programmes. Therefore, when considering future grounds for co-operation between OPEC and other developing countries more attention should be given to the role of non-concessional flows.

OPEC countries have channelled non-concessional funds to developing countries mainly through multilateral institutions and investment companies purchasing bond issues. Bilateral non-concessional assistance has been limited mainly to deposits in Central Banks and oil credits. Direct capital investments in developing countries contain a high risk element and require an infrastructure which have made them less attractive to OPEC Member Countries and to surplus developing countries in general. Efforts to increase direct investments in other developing countries would, however, provide

[1] Official Development Assistance.

1975 Remit-tances	*As percent of Exports*	*Imports*	*1976 Remit-tances*	*As percent of Export*	*Import*	*1977 Remit-tances*	*As percent of Export*	*Import*
466	11	7	245	5	4	246	4	3
35	9	1	36	10	1	83	18	9
367	23	7	754	47	18	1,425	66	27
490	12	8	750[3]	17	12	n.a.	n.a.	n.a.
167	109	18	396	198	34	425	186	38
533	35	18	548	43	16	577	44	18
230	22	8	353	31	12	1,118	88	40
55	6	3	51[3]	5	2	n.a.	n.a.	n.a.
146	17	8	135	17	8	142	16	8
1,312	94	25	982	50	17	982	56	17
221	1,556	72	525	4,269	137	1,013	5,449	139
56	373	32	115	261	40	179[4]	352	49

[3] Fiscal year ending June of the indicated year.
[4] Preliminary.

these countries with risk and loan capital for their development projects while helping those OPEC Member Countries with available liquid assets to diversify their investment portfolio.

OPEC countries also represent a growing potential market for the exports of other developing countries. While imports from developing countries increased eightfold during the period 1971–77, from $1.2 to $9.8 billion, the trade gap between OPEC countries and the rest of the developing world widened from $2.3 to $19.4 billion. There is a need to encourage trade among developing countries and this will require among other things, better flows of information, infrastructure in communications and transport and other trade related facilities.

The accelerated development programmes undertaken by OPEC Member Countries have also generated additional employment opportunities to migrant labour of other developing countries. This has been particularly significant in the Gulf countries with small populations where the demand for managerial, skilled and unskilled labour largely exceeds the available national labour force. Thus, the deployment of immigrant labour in OPEC countries has resulted in an important source of foreign exchange earnings to many developing countries as workers' remittances. In fact, the flow of remittances may represent as much as 50% of the total export value of some developing countries (see Table 7). The areas for co-operation among developing countries abound and constitute an important element for the achievement of higher levels of economic growth in the Third World.

Part VII

ENERGY ISSUES AND POLICIES: OIL-EXPORTING COUNTRIES

23 CRUDE OIL: ISSUES AND POLICIES FOR OIL-EXPORTING COUNTRIES

René Ortiz

The tight supply situation and the panic that occurred during the first half of 1979 represent only the symptoms of the underlying oil supply/demand dilemma. In its broadest scope, the world energy problem reflects the limited nature of world oil resources, and although the world is not running out of oil, current worldwide reserves are not increasing to any appreciable extent. If the present trend of consumption continues, we shall be producing at a greater rate than reserves are being added to, and this will cause a fall of output within the coming two decades, as most experts expect. Limited oil reserves have already forced a fall in US production, and the same thing is expected to happen soon in the USSR. This is an alarming situation, considering that these two countries account for one-third of world oil production, and one which is compounded by the fact that the amount of discovered reserves in both countries has fallen sharply in recent years. Thus, contrary to the views that became popular during the temporary, supposed oil glut of 1977/78, the world does not have years in which to make a smooth transition to alternative energy sources unless substantial efforts in increasing the resource base are made, and consumers take the necessary steps for adjustment, whether in more efficient use of the limited hydrocarbon resources or through incorporating into the system other alternatives.

The three decades prior to 1973 constituted the great era of the petroleum-based world economy. Huge discoveries in the Middle East, Africa, and the USSR during the 1940s, 50s and 60s brought rapid increases in world productive capacity. During the 1950s, the bulk of the increase in oil supplies came from the US, Canada and Venezuela, but by the 1970s the Middle East and Nigeria were providing virtually the entire increment. Fig. 1 gives total world production, together with the contribution of OPEC, the Middle East and US; while Fig. 2 shows the evolution of oil production for the world, OPEC and OPEC Middle East.

It can be easily seen that while OPEC's and OPEC Middle East's

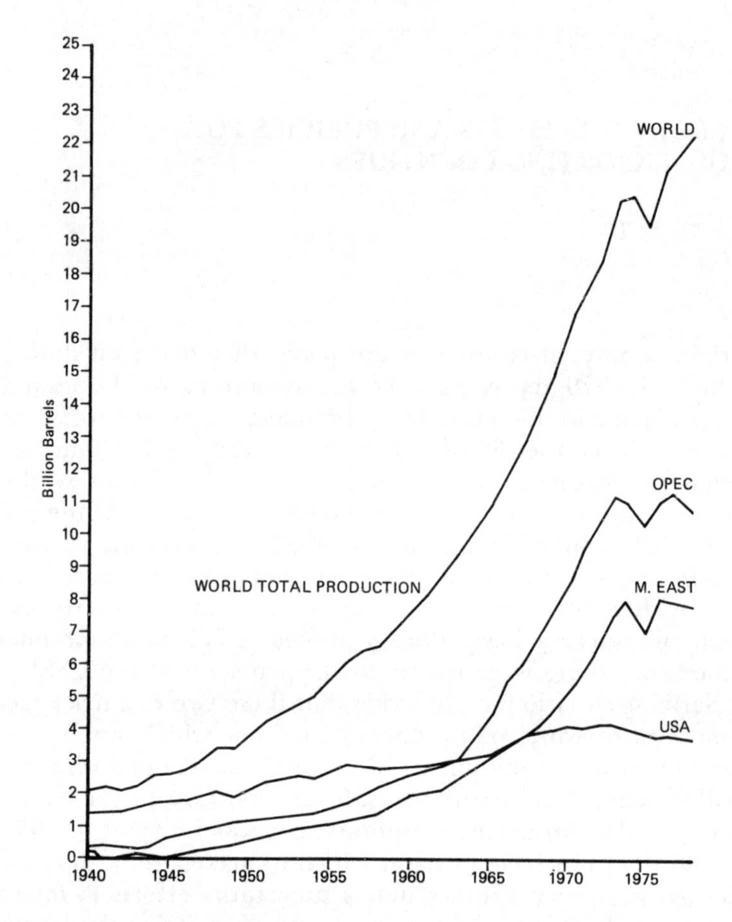

Fig. 1 World crude production.

shares in total world production were 4.8% and 24.8% respectively in 1960, these shares increased to 55.5% and 36.9% respectively in 1973. The proven reserves of OPEC, on the other hand, did not increase in the same manner: while OPEC's production was growing at an annual rate of more than 8%, reserves were growing at a rate of about only 4% per year.

Looking at the development of the reserves-to-production ratio of total OPEC (or, if you prefer, life in years at respective production rates) it can be easily seen from Fig. 3 that, while the ratio was about 65 years in 1960, it decreased sharply to about 32 years in 1977. The situation is more acute for the Middle Eastern Members of our Organization, where this ratio declined from about 85 years in 1960 to about 30 years in 1977. This trend clearly reflects the

	1960	%	*1968*	%	*1973*	%	*1978*	%
1. World	7,674,460	100	14,146,318	100	20,367,981	100	22,525,540	100
2. OPEC	3,130,584	40.8	6,857,218	48.5	11,310,802	55.5	10,878,830	48.3
3. OPEC Middle East	1,903,229	24.8	3,972,713	20.1	7,518,423	36.9	7,583,748	33.7
4. OPEC Middle East/ total OPEC	=	60.8		57.9		66.5		69.7

Fig. 2 The evolution of oil production: World, OPEC, OPEC Middle East (000 barrels).

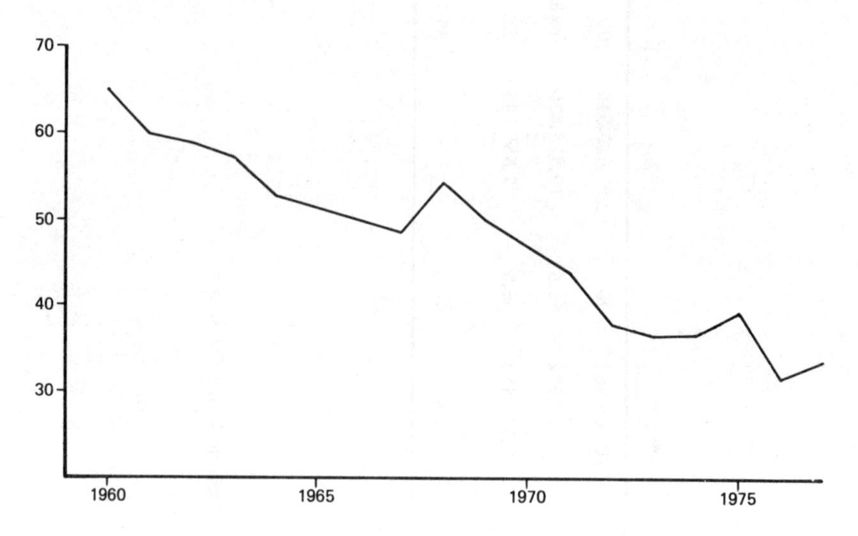

Fig. 3 Ratio of estimated proven reserves of crude oil to production of crude oil in OPEC Member Countries.

exploitation policies of previous concessionaires and foreign operators, under which minimum investment was made in exploration, the dominant trend of activities being the search for conventional petroleum deposits of large size, located in accessible areas. Additionally, development and production technology was directed towards extracting free flowing crudes, thereby, and without appropriate evaluation, leaving out as uneconomical small discoveries, and refraining from making investment in reservoir maintenance or in increasing the recovery factor.

Until 1973 this traditional approach was considered by these operators as justified by low oil prices, with the exploitation of other petroleum deposits and the making of heavy investment in exploration being regarded as uneconomical. Further, no plans were made for investment to utilize the associated gas; hence, most of it was flared, resulting in the further dissipation of valuable energy resources.

In 1973, our own Member Countries started to play an increasing role in the development of their own natural resources, and began to manifest a keen awareness of the necessity for sound conservation policies. As oil represented for our Member Countries the main foreign exchange earner around which their social and economic development centred, and as oil is depletable, extending the life span of the reserves became an important national objective. Further, the demand for a volume of revenues compatible with our needs started to play a predominant role in our production policies. Hence, it is not surprising that some of our Members are monitoring carefully their existing productive capacity, increasing it modestly or refraining from making the investment necessary to expand it.

This trend towards cautious policies is the symptom of a strong preference for a production profile that stretches our reserves over a longer period, of an aversion to even the smallest risk of impairing ultimate oil recovery. There is concern about the disruptive economic and social effects of a massive inflow of oil money, and the keeping of a large share of that wealth in the form of financial assets subject to inflationary impacts, currency variations or the guaranteeing of those assets abroad. Further, although the prime objective is to transform these depletable assets into fixed ones, through development and industrialization, some of our Member Countries with substantial financial assets at their disposal are encountering obstacles to the achievement of their economic development goals.

The number of countries that have imposed political constraints on production has grown markedly over the past several years and now includes countries with roughly 60% of the world's total crude reserves. Some of these countries are outside OPEC and include, surprisingly, some developed countries. Norway, for example, has established conservation policies with regard to the rate of development and capacity expansion, due to the fear of the disruptive effects of the inflow of oil money into other sectors of the economy. Canada (Province of Alberta) is adopting a tight production policy, whether due to environmental constraints or in order to cater for the needs of future generations. Mexico also has a conservation view about the kind of reserves-to-production ratio it wishes to maintain in the years ahead.

Many major producers, both within our Organization and outside it, are not only restricting the development of new capacity, but are also holding free flowing production below the maximum sustainable, through setting a production ceiling on their output. Saudi Arabia, for example, has had a production ceiling of 8.5 M b/d since 1974; Kuwait's ceiling of 2 M b/d expresses its strong view in this regard; the United Arab Emirates limits its output to 80% of capacity for similar reasons; and the latest Iranian government action has set the ceiling at 3.5–4 M b/d. Nevertheless, these production limits have been, and could be, relaxed in goodwill gestures. This goodwill was shown during the first half of 1979, when a 7.7% increase of OPEC's output came to help offset some of the shortfall caused by the Iranian disruption which had induced a temporary uncertainty in the market. Therefore, the practice of setting a ceiling will be a dominant and major factor in the supply and availability of hydrocarbons, unless greater economic incentive and assurance can be given for such producers to utilize their revenue, and increase their resource base in an optimal and logical manner.

Oil production in OPEC Member Countries, outside the Gulf, is limited by productive capacity, which is unlikely to expand appreciably during the next few years. Thus, if production in the Gulf stays at the announced ceiling, world output will remain nearly constant. Outside OPEC, likely changes in production and capacity will tend to offset each other. It is expected, in particular, that there will be a marked increase in North Sea oil production, reaching a probable peak in 1982–83; a decline in US production; a decline in the net exports of oil to the socialist countries, as Soviet production peaks and begins to drop; an increase in production in less developed countries (LDCs) outside OPEC, especially Mexico and Egypt, most of which, however, will be offset by the rise in oil consumption of the non-OPEC LDCs themselves.

With oil supply from traditional sources thus restricted, increasing the resource base and developing alternative sources of energy become important if the required world economic growth is to be sustained. Except for natural gas, the resource base for alternative energies is sufficient to allow a large expansion of output, but there are severe cost and environmental factors which will have to be overcome. Even with the enhanced productivity resulting from current oil prices, large scale development of some of these resources will take many years. During the next 3–4 years even the optimistic projection of production of non-oil energy resources in the OECD, which assumes an increase of 2 M b/d of oil equivalent in coal supplies and no further delays in nuclear power programmes, would result in only a 1–1½% rate of growth in total energy supplies for OECD countries. Furthermore, if energy conservation continues at the rate of recent years, OECD energy demand will exceed energy supply for annual rates of OECD economic growth above 2½%, which is well below the desired rate of growth. Therefore, the industrialized countries, somehow, will have to adjust to a slow growth of energy supply and to a stable or declining oil supply. The adjustment will take the form of increased energy conservation, reduced economic growth, or, most likely, some combination of both of these. However, holding energy demand to projected supply levels without lowering OECD economic growth targeted below the 3–3½% annual rate of growth normally considered acceptable, would require unprecedented rates of conservation, substantially higher than in recent years and higher even than those achieved in 1973–74 and the subsequent adjustment of the oil prices. The inadequacy of the energy-mix balance to be supplied to meet demand at acceptable rates of economic growth could cause chaos in the oil market and would lead naturally to more pressure on the oil price.

In order to alleviate the situation, therefore, it is only logical that efforts should be directed to expanding the oil resource base, as the supply problem is not only related to the ability to sell, the ability to pay, and the value of the unit export, but is also closely connected with, and predominantly affected by, the available resource base. I am sure that you will all agree with me that the attitude of a producer would be relaxed greatly if his resource base were increased, and if an energy-mix balance could be achieved. The principle of adding new discoveries of crude has always been recognized, at least at the same pace of production flow, out of the deposits. Unfortunately, it has been kept as a principle rather than adopted as a practice.

It has always been maintained that exploration represents the very heart of the petroleum industry. Exploration is where it all starts, and all other segments of the industry, which include, but are not limited to refining, marketing and transporting, owe their existence to the success of exploration. It has been estimated that areas currently productive contain up to two-thirds of the future potential, with the mature producing areas containing 30% and the developing productive areas 40%, of the remaining commercial undiscovered potential. Intensive exploration, whether through applying advanced geophysical techniques, or increasing drilling activity to delineate extensions to known fields, deeper targets, stratigraphical traps, etc., are the most likely sources of quick additional reserves.

With this factor in mind, and in the light of the situation that I have outlined about the attitudes of foreign operators in the OPEC area, it is very clear that exploration activity in OPEC Member Countries represents a tremendous opportunity for increasing the world's oil resource base and a logical step towards stretching the energy transition period confronting us.

Our own Member Countries are intensifying their exploration activities with large investments to support their reserves. However, I am sure that all of you realize the difficulties and problems with which we are faced in setting the required pace, whether due to lack of manpower, cost of equipment and technology, etc. Nevertheless, one cannot expect a country to utilize and invest a large proportion of its oil income, an income that is needed for development and diversification of its economy, in increasing its oil resource base without positive collaboration from those who possess the technology and expertise required, and which should be provided in a fair and equitable manner. In this regard, it has to be emphasized that the time of concessions and other similar types of agreements is past. It is no longer possible to visualize the use of these contractual bases in carrying out exploration activities, since as the international community has recognized, new conditions now apply and have to be dealt with.

As a result of the situation outlined in the foregoing, the proven recoverable oil reserves within OPEC Member Countries now stand at 450 billion barrels, and are, based on a recovery factor of about 30%. Increasing the recovery efficiency will add a substantial amount to the availability of hydrocarbon resources at a comparatively low cost. In this regard, those possessing the needed expertise and technology for increasing such resource base have the opportunity to co-operate with our Member Countries, or their national oil companies, in the implementation of such steps. Furthermore, our Member Countries possess a large reserve of low-grade deposits which, if developed, will contribute a substantial amount to the supply of hydrocarbons.

Turning our attention now to gas: OPEC's gas reserves are presently estimated at 138 billion barrels of oil equivalent, or about 40% of the total world gas reserves. Gas production from OPEC Member Countries is, however, in the order of 4 million barrels of oil equivalent, which represents 18% of the total world gas production. It is important, therefore, to note that in spite of the low production rate which OPEC's gas production represents *vis-à-vis* the rest of the world, when compared to its gas reserve position, a considerable amount of OPEC's gas production is flared. The amount of gas that is flared as a proportion of total gas production, ranges from 35 to 92% according to the different Member Countries' potential for either domestic consumption or export. This malpractice of gas flaring, introduced in the past for well-known reasons, took no account of its future utilization and its attractiveness as a clean primary energy source. Therefore, another logical approach to extend the energy-transition period, is the gradual incorporation of gas into the energy-mix balance, bearing in mind the price incentive to encourage development in the downstream stage.

Although our Member Countries are embarking on large investments in implementing gas projects, nevertheless, various problems constrain and hinder such developments. It is a well-known fact that gas projects are highly capital-intensive and gas utilization needs the necessary market penetration to take its appropriate share and alleviate the problem of satisfying the total primary energy demand. It is also a well-known fact that in order to penetrate any market you have to deliver the commodity at a price competitive with alternatives. Due, however, to the low prices of refined products with which gas is competing in the major consuming markets, the need to give an economic incentive for conversion so that gas may be utilized; the high cost of its transportation *vis-à-vis* crude or petroleum products; and the cost of regasification – all these factors make the economic viability of gas projects doubtful and reduce the incentive of Member Countries to embark on the huge investments required, especially bearing in mind that this gas is mostly produced as an associate and is tied to the production of crude. The problem of c.i.f. *vis-à-vis* f.o.b. equalization is a well known dilemma as, when you netback the gas to the point of production, the f.o.b. value will be much lower than the crude value on a Btu basis, and, sometimes, when you netback to the wellhead, you might have a negative result. I am sure you will all understand that the economic incentive for the production and utilization of gas is of vital importance if we want this clean source to take its rightful place in the total energy demand balance. Therefore, besides the need for various competent companies with the required technological know-how to extend the necessary co-operation to the appropriate authorities in our Member Countries, the economic incentive for enhancing the market penetration of gas is also required.

I have tried here to highlight the huge opportunities in the upstream operations of the oil industry that are available in our Member Countries, the pursuit of which is in the interest of both producers and consumers, emphasizing the exploration developments and utilization of gas. The mode and type of agreements to implement such projects can take various forms, depending on the requirements and needs of the various parties concerned. However, it has to be borne in mind that the conceptual framework under which these investments can be made must be based on the following:

1. Respect for the national sovereignty of the owner of these resources.
2. A fair price for equipment and services provided.
3. Involvement of nationals in all phases of implementing such projects, so that adequate technological and operational capability can be acquired.
4. Extension of technical assistance for the efficient and continuous utilization of such projects.
5. Enhancement of the viability of such projects through opening the necessary markets to them.

24 REFINING AND PETROCHEMICALS: DEVELOPMENTS IN SOME OIL-EXPORTING COUNTRIES

Aziz Al Watari

I. PETROLEUM REFINING

Crude oil by itself has little use in direct application. Its full value is realized only when it is processed into refined products for specific end uses. In addition, refining is a necessary first step for the downstream development of some liquid petroleum fractions. In a series of processes, including primary distillation and sometimes secondary conversion, fuel and non-fuel products are manufactured to meet energy requirements not only for specific fuels, but also for lubricants, asphalts and waxes. Feedstocks, in the form of naphtha and gas oil, are also produced for the petrochemical industry. This latter function of the refining industry confers upon it additional importance whenever significant petrochemical activities exist alongside, or are being planned. The prudent industrial planner can ill afford not to take advantage of the inherent potential economics of joint or integrated refining/petrochemical processing, especially in growing markets. A refining industry creates value-added and has the potential to provide input to linked industrial activities.

The Arab Refining Industry

Currently, there are 40 refineries in the Arab world ranging in capacity from 5,000 to 565,000 b/d. Their nominal combined capacity is 3,366,000 b/d.[1] Assuming an overall utilization rate of 80% (and fuel consumption plus losses at 7.5% of throughput), their combined product output is 2.5 M b/d. Of this, OAPEC nominal capacity at 2.98 M b/d represents 3.2% of the world total. These figures compare with OAPEC's crude production in 1978 of some 20 M b/d and OAPEC states' average consumption of

[1] Against 900 refineries in the world with a combined capacity of 80 M b/d (1978).

refined products of 886,000 b/d. That for the Arab world as a whole was 1.17 M b/d (the last two figures exclude ship bunkering use).

Arab Demand for Energy

Commercial energy demand in the Arab world is basically a demand for petroleum (oil and gas), which perhaps meets more than 95% of its total energy requirements. The balance comes largely from hydroelectricity and, to a lesser extent, from coal. The Arab region is perhaps more dependent on oil and gas for its commercial energy needs than is any other region. A review of the emerging state of Arab energy demand provides an appropriate background for estimating its requirements of refined products.

Recent demand growth rates for oil products in OAPEC member states have been rather high. Statistics show a rise of from 440,000 b/d in 1973 to 847,500 b/d in 1977, representing an arithmetic average rate of increase of 17.8% over the 4-year period (see Table 1). Per capita annual energy consumption remains low at 465 kilograms oil equivalent (Kgoe) in 1976, against an average of 1407 Kgoe for the world and 4346 Kgoe for the developed economies of Western Europe.

Demand estimates for the long term are difficult to make on the basis of sectoral analysis, because, in many instances, statistics on specific consumption of oil products are not available. Such constraints and qualifications notwithstanding, a recent OAPEC study using a macroeconomic methodology (energy/GDP elasticities) came out with the forecast shown in Table 2. In terms of per barrel oil equivalent, the energy demand estimates for the Arab world are 2.42, 3.3 and 5.95 M b/d for the target years 1985, 1990 and 2000 respectively. Given the low degree of accuracy expected of such forecasts, it is worth noting that the study's figures come rather close to the technical projection of oil products demand.[1] Finally, one independent consulting firm estimates Arab demand for refined products for the years 1985 and 1990 at 2.3 and 3.3 M b/d respectively.

[1] This could imply that these studies did not properly account for the increasing use of flare gas whose potential is quite important, being flared currently at about 1.3 M b/d of oil equivalent.

Table 1 Total world commercial energy consumption – annual growth rate (%).

	1960–65	*1965–70*	*1970–75*	*1960–75*	*1960–73*	*1970–73*	*1973–75*	*1975–76*
World	4.2	5.7	3.0	4.3	4.9	4.8	0.4	5.6
Developed Countries	4.6	5.7	2.1	4.1	4.9	4.2	(0.01)	5.6
Western Europe	4.9	5.4	1.5	4.3	5.0	4.5	(2.5)	6.9
Centrally Planned Economies	2.8	5.1	5.2	3.7	4.2	5.0	4.9	4.8
Developing Countries	6.3	8.2	5.4	6.6	7.2	7.3	2.7	7.8
Arab Countries	5.8	8.9	11.3	8.6	7.9	,10.1	13.1	14.2

Sources: UN, *World Energy Supplies*, Series J, No. 2.

Ali Sadik *et al.*, 'Energy Demand Forecast for the Arab Countries (1985, 1990, 2000)', paper presented at the First Arab Eneryg Conference, Abu Dhabi, UAE., March 1979.

Table 2 Total commercial energy consumption forecast for Arab countries (most likely outcome of three scenarios) (10^6 Mtoe)

Country	*1975*	*1985*	*1990*	*2000*
Algeria	5.42	12.41	17.54	34.0
Bahrain	2.10	4.23	5.02	7.47
Egypt	10.72	20.70	28.7	55.5
Iraq	5.33	8.89	11.11	21.0
Jordan	0.75	1.02	1.27	
Kuwait	5.83	12.9	16.65	21.7
Lebanon	1.81	3.51	4.69	7.0
Libyan Jamahiriya	1.82	3.37	5.22	12.57
Morocco	3.13	5.48	7.3	12.73
Mauritania	0.097	0.29	0.38	0.584
Oman	0.247	0.355	0.45	0.81
Qatar	1.53	7.19	9.51	17.6
Somalia	0.103	0.14	0.2	0.31
Saudi Arabia	10.51	20.8	29.7	46.7
Sudan	1.78	3.93	6.35	12.9
Syria	3.51	6.9	9.53	18.23
Tunisia	1.7	3.21	4.4	7.9
UAE	2.07	4.57	6.8	13.22
Yemen AR	0.22	0.45	0.65	1.34
Yemen PDR	0.38	0.68	1.5	3.1
Total	59.057	121.025	166.970	297.254

Sources: UN, *World Energy Supplies*, 1972–1976, Series J, No. 2.
Ali Sadik *el al.*, 'Energy Demand Forecast for the Arab Countries (1985, 1990, 2000)', paper presented at the First Arab Energy Conference, Abu Dhabi, UAE, March 1979.

Refined Products Supplies

Despite the current outlook for crude oil supplies, world refining capacity rose during 1977–78 by 4% to stand currently at about 80 M b/d. The yearly average rate of capacity growth during the period 1973–78 was markedly lower than that during the decade 1968–78. The slowdown from 6.17% for 1968–78 to 4.17% for the last 5 years marks the beginning of a new trend in the post-1973 era.

Capacity utilization is a different matter. Whereas the industry traditionally operated at an average 90% of nominal capacity, in 1978 it stood at only about 70%. Capacities as low as 46% were recorded in Italy (1975) and at around 60% in most major European countries. The US has not suffered such capacity-use decline, and has in fact kept above the 80% level for the past five years. Despite the over-capacity, 3 M b/d of new world capacity is under construction (excluding the centrally-planned economies) with another 10 M b/d being planned.[1]

[1] *Oil and Gas Journal*, April 24, 1978.

Additional refining capacities of 1.27 M b/d may be completed between now and 1985, which could bring OAPEC's total to 4.26 M b/d. Further planned capacities until 1990 may well boost the total to 5 M b/d. Capacity of the rest of the Arab world is given as 381,000 b/d in 1977; 225,000 b/d are scheduled for completion by 1985, bringing the total to 606,000 b/d. Additional capacities planned for completion by 1990 may bring the total to 766,000 b/d.

Markets

Apart from a few isolated cases, petroleum products markets are interrelated and highly integrated, particularly in Western Europe and the US. Until recently, opportunities for trade within and between regions existed to varying degrees, but future patterns of such trade are far from certain. Mathematical models (LP transportation), for example, provide a picture of an optimal world system of supplies and requirements equilibrating the marginal cost of refining in all important markets. However, we live in a divided world where criteria other than pure economics or current commercial feasibilities and conveniences may be decisive factors in refining investments. There are, for instance, the longer term developmental aspects of the resource-rich Third World countries, such as those of OAPEC, and the policies (and politics) of energy in the major oil-consuming countries. These and other factors, notably technological advances in the production of various forms of energy and/or the devices and appliances that use energy, make for a very complex and unpredictable situation. The energy/economic patterns that will emerge during the next 25 years therefore defy any real understanding at present. This is because we are currently going through a transition, and the possibilities of likely substitutions and complementarities are infinite.

Because of the interlinking of refined products markets, major developments in all major markets must be of interest to OAPEC states, albeit those nearer to them more so. But the global character of the refining industry will remain an important feature of this industry for the foreseeable future.

OAPEC. By all accounts, a relatively rapid growth of inland oil products demand has been forecast for most OAPEC states. The reason is high economic growth and ambitious development programmes in all the major oil-producing states. Naturally, the first priority for individual OAPEC members is to satisfy their own demand for petroleum products. From the various available surveys and studies, there seems little doubt that they will be able to meet demand in the foreseeable future. Moreover, projections of refined products balances for OAPEC states show an exportable surplus for the region as a whole of between 1.0 and 1.2 M b/d between 1980 and 1985.

Other Arab Countries. Products demand for the Arab world outside the OAPEC region is also expected to continue growing at high rates, although somewhat lower than OAPEC's. These states benefit from Arab financial assistance, labour exports, and increasing inter-regional trade. It is estimated that their oil products demand will be about 840,000 b/d by 1990. This, however, is only one-fourth that of OAPEC states' total expected demand for oil products the same year. Total products demand (excluding refinery fuel consumption plus bunkers) in non-OAPEC Arab states exceeded their refinery output by 34,000 b/d in 1976. Deficits will appear for all products in 1980 and shortage in total production of between 79,000 and 267,000 b/d for the years 1985 and 1990 is forecast by some studies.

It should be noted once again that the balancing exercise given above does not imply the existence of actual markets (mechanism and infrastructure) for the purpose of regional trade. A policy of developing an inter-Arab market is required if OAPEC states are to take advantage of the opportunity to develop markets in the Arab world and to benefit from developing their oil products trade in a co-ordinated manner. Assuming the geographic distribution of oil resources to remain unchanged over the next 10 years, it would be of interest to both oil and non-oil states to agree on refinery product supply arrangements, and on terms of trade and payment, so as to relieve the non-oil states of the heavy investment requirements for additional refining capacities required by the mid-1980s. The OAPEC states would, in return, acquire a natural market for their expected surplus. Present arrangements between Kuwait and South Yemen, and between Iraq and Somalia are examples of co-operation. However, a more comprehensive approach that rationalizes the economics of supply and distribution over the long run, and covers most Arab states, would be highly desirable to stabilize expectations of energy supply to the non-oil Arab states, whose number is likely to rise. The problem here, however, is the preference of these states, like nearly all crude-importing countries, for self-sufficiency in refining facilities for reasons of security, balance of payment, and perhaps expected benefits from industrialization.

Few national oil entities in OAPEC states have discovered, on their own, the benefits of co-ordination to achieve better product balances, improved capacity use and, consequently, better overall operating economics in their marketing efforts. Here OAPEC can play a role to promote and develop such efforts, and, in fact, it is currently working towards this end.

Export Refining. The trend towards market-located refineries, rather than source-located ones, seems to have started after World War II. In 1949, more than 50% of world total refining capacity was located in crude-producing areas such as the US and elsewhere. By 1959, while the US share remained the same, a shift away from the Middle East was becoming pronounced. In that year only 5.7% of world capacity was still located in the Middle East, while the shares of Western Europe and Japan, considered to be the Middle East's natural markets, rose to 14% of the world total. By 1977, the Middle East's relative refining importance, had declined to 4.1% while Europe's increased to 34.1% of the world total. The arguments in justification of the shift are well known – security, foreign exchange savings for the crude importers, and lower capital and operating costs for the multinational oil companies (MNCs). In addition, crude is cheaper to transport than are refined products, especially in large tankers.

Current export refining capacity in the Arabian Gulf and Aden is estimated at some 2 M b/d and the world total (outside the CPEs) is about 8 M b/d. Output of Middle East refineries has been directed primarily to markets in Asia and the Far East, with Japan absorbing the bulk, and also for bunkering use. Figures on products exports from the Middle East are incomplete, but the pattern of almost all Middle East refineries (including OAPEC's) emphasizes heavy oils, i.e. diesel and fuel oils, which amounted to almost 70% of total products exports in 1971 (excluding bunkers). The export-refining centres – Kuwait, Bahrain and Saudi Arabia – have produced anywhere from 34 to 47% of OAPEC's fuel oil output in recent years. These percentages are higher than the world average crude yield (including the other major export-refining centres). It is noteworthy, however, that the 200,000 b/d Shuaiba refinery in Kuwait has different features: it produces a rather high proportion of 'white' products (70% + of crude throughput) using hydrocracking and intensive hydrogen processes for product treatment including 'H-oil' processing of some heavy ends.

Future Market Development. The present uncertainty regarding energy policies of the major industrial nations, especially as regards target-setting to reduce oil imports and future rates of economic growth, renders demand forecasts for refined products in major consuming markets highly uncertain. A decline in crude oil production obviously implies, among other things, a lower refinery utilization rate in the short and medium terms. If this persists, it should curb investment in new capacity. On the other hand, if steady demand for refined products is coupled with a decline in crude production (predicted for the early 1980s, if not yet the case as some believe), then crude oil availability becomes a key factor for the refining industry's future in all major centres.

Other current developments will surely change the pattern of refined products supply/demand. Developments such as the increasing proportion of heavy crudes in OPEC's crude output, the rising cost of energy required to process them for meeting the relatively high white products demand and environmentally acceptable product qualities, and the response of decision-makers, are also of significant importance in shaping the future of the refining industry on both sides of the trade line for crude and products.

A review of refined products supply/demand balances for the future set against the changing background of the aforementioned developments might clarify the desirability of various investment possibilities in the mid-1980s. At the moment, there is little indication as to how the major oil producers are viewing such developments in their investment decisions. Delay in committing investments to export refining might just be that indication.

The exportable surplus of OAPEC states to markets beyond the Arab region is mainly directed to some of the major oil-consuming nations in the OECD. The OECD countries account for 80% of world refined products consumption and the same percentage of imports. Imports of refined products into the OECD (disposal) amounted to 134 M tons in 1978 or 2.6 M b/d. Three main petroleum-importing areas are considered below, namely, Western Europe, Japan and the US. Between them they represent a demand for products imports or additional refining capacity of 4 M b/d in 1985 and some 12 M b/d by 1990 (assuming constant real prices for crude oil). A few remarks on each is therefore in order.

WESTERN EUROPE. This area currently has a large surplus in refining capacity which is ostensibly adequate to meet primary distillation requirements until 1985, with only a marginal deficit thereafter. Paradoxically, Western Europe continues to add to primary and conversion capacities. The latest figures for capacities under construction give 436,087 b/d of primary distillation and

423,500 b/d of catalytic cracking. It is also expected that Western Europe will add 1 M b/d of new capacity by 1985 to bring the total to 21.6 M b/d. A further 3 M b/d by 1990 may be needed, according to one independent consulting firm. Such developments would seem to provide an opportunity to absorb part or all the exportable surplus expected from OAPEC in the late 1980s, assuming that the current products imports from Eastern Europe (currently at 50,000 b/d) will cease,[1] and further assuming that the rate of capacity utilization in Western Europe continues below historic levels and stabilizes at between 69 and 78% for the next decade, i.e. at a throughput of 14.8 to 16.8 billion b/d.

JAPAN. With 49 refineries, a crude distillation capacity of 5.9 M b/d (March 1978), a relatively small cracking capacity of 325,885 b/d, and a reforming capacity of 557,000 b/d, capacities in Japan are expected to be adequate until 1982. The Japanese refining industry is tightly controlled by the government, which authorizes and allocates capacity, negotiates bilateral agreements with other governments for crude supplies (this activity will increase as MNC suppliers withdraw from Japan) and, finally, sets prices and regulates imports. From November 1973 until 1977 no licences were granted for new capacities. Japan may therefore be a favourable market for future Gulf-based OAPEC capacity given its historic import trends and continuing relatively high demand for heavy fuels (although declining percentage-wise relative to total products consumption). Both middle distillate and gasoline percentages have been rising in the last 10 years.

THE UNITED STATES. According to *Oil & Gas Journal*, the 289 refineries currently operating in the US have a combined capacity of 17.17 M b/d, with 5.05 M b/d of cat-cracking and 3.84 M b/d of reforming. The US is the largest market in the world, and its refining industry runs at the highest nominal capacity use rate of all major markets, some 90%. Imports, essentially of fuel oil, come from export refineries in the Caribbean and run at over 2 M b/d.

Beyond 1980, new US capacity looks rather uncertain. In addition to the environmental factors, the gap between domestic demand and domestic capacity will continue to widen, unless new investment can be economically justified. Furthermore, inflation has made

[1] *Petroleum Intelligence Weekly* reported on August 20, 1979, that the EEC countries imported an astounding 16.3 million tons of products from the USSR and 2.2 million tons from Romania in 1978.

grass-root plants non-competitive with old ones, according to *Oil & Gas Journal* (April 29, 1973). There is also a question of energy policy uncertainties, especially regarding self-sufficiency in refining capacity. The 2 M b/d gap between US domestic refining and demand of 2 M b/d is largely due to the preference of major US refiners to minimize domestic residual fuel production in favour of producing more white products. This trend is continuing, despite the emergence of new capacity should be built between 1980 and 1985, or whether

It is thus up to the US Government to decide whether 1–1.5 M b/d of new capacity should be built between 1980–85, or whether the equivalent products imports will be allowed. Most analysts predict continued products imports and no more grass-roots refineries after 1980. With these conditions, plus unfavourable crude-product freight differentials, the opportunities for significant OAPEC product penetration do not seem to be encouraging. Caribbean (including Mexican) and European refineries are better located for competitive trade.

Refining Economics

The total cost incurred in the production (refining, transport and terminalling) and c.i.f. delivery to export markets of a refined barrel of oil is a key parameter, because the criterion for ascertaining profitability of an export-refining project is the price less the build-up cost. In order to estimate the competitiveness of an export-refining industry located in the Gulf, one study[1] has adopted a project basis whereby built-up cost economics are compared between one idealized refinery in the Gulf and four similar ones located in the US East Coast, North West Europe (Rotterdam), the Mediterranean (Europe), and in Japan. All these refineries are of the same configuration, i.e. of simple hydroskimming design (no conversion processing), and produce an identical product slate. Allowance for scale economy is made only for the Gulf-located refinery, assumed at 250,000 b/d capacity, while the rest are assumed at 175,000 b/d. All costs are based on 1976 prices. All refineries are also assumed to take Arabian light crude and produce a full product slate, i.e. C_3 and C_4 LPG, gasoline, naphtha, kerosene, jet fuel, gas oil and fuel oil. The analysts emphasize that the purpose of their study is 'to examine the economic effects of size and location on refineries'. A comparative costs summary is shown in Table 3 which clearly shows that, on the basis of production cost for 1980,

[1] The Economist Intelligence Unit, London.

Table 3 Cost of supplying a refined product barrel to major consumer areas, 1980 ($/barrel Arabian Light)

		US East Coast		*Japan*		*NW Europe*		*S. Europe*	
		E	*C*	*E*	*C*	*E*	*C*	*E*	*C*
Refining	(1)	1.07	1.24	1.07	1.02	1.07	1.09	1.07	1.11
	(2)	0.69		0.69		0.69		0.69	
Freight									
– crude				1.51		0.66		1.11	0.97
– products		2.95		2.24		2.40		1.98	
Terminalling		0.03		0.03		0.03		0.03	
Total	(1)	4.05	2.75	3.34	1.68	3.50	2.20	3.08	2.08
	(2)	3.67		2.96		3.12		2.70	
Differential									
	(1)	1.30		1.66		1.30		1.00	
	(2)	0.92		1.28		0.92		0.62	

Cost of supplying a refined product barrel to major consumer areas, 1985 ($/barrel Arabian Light)

		US East Coast		*Japan*		*NW Europe*		*S. Europe*	
		E	*C*	*E*	*C*	*E*	*C*	*E*	*C*
Refining	(1)	1.08	1.24	1.08	1.02	1.08	1.09	1.08	1.11
	(2)	0.70		0.70		0.70		0.70	
Freight									
– crude			1.60		0.81		1.34		1.02
– products		3.07		2.34		2.50		2.05	
Terminalling		0.03		0.03		0.03		0.03	
Total	(1)	4.18	2.84	3.45	1.83	3.61	2.43	3.16	2.13
	(2)	3.80		3.07		3.23		2.78	
Differential									
	(1)	1.34		1.62		1.18		1.03	
	(2)	0.96		1.24		0.80		0.65	

E = Middle East export refinery C = Consumer country refinery (1) Liquid refinery fuel (2) Associated gas as refinery fuel a Middle East-located refinery using natural gas for fuel enjoys a cost advantage over those in US East Coast, Japan, North West Europe and Southern Europe of 55, 31, 40 and 42 US cents per barrel, respectively. But this advantage is more than offset, according to the study, by the crude/product freight differential, leaving the Middle East-located refinery at a delivered cost disadvantage to the other locations of anywhere between $0.62 and $1.28 per barrel, depending on transport distance and ignoring canal dues when relevant. The study concludes that the relative profitability of an export refinery improves by some $0.25/barrel for every 1,000 miles of movement closer to the market refinery. However, because of its view of global supply/demand, the study's final words were 'market opportunities will in each individual case be a function of the type of individual refinery, its location and the company(ies) involved in its operation'.

Another study,[1] basing its cost analysis on similar but smaller refineries (100,000 b/d hydroskimming units) in the Gulf, North Africa and Southern Italy, gives the results shown in Table 4. It shows OAPEC locations in North Africa to have a cost advantage over Italy's, and even the Saudi Arabian's location cost penalty is much lower, at 18.1 ¢/b. This is in contrast to the former study, which estimates the minimum cost penalty of Gulf-produced products at 62 ¢/b *vis-à-vis* the Southern Italy location.

It should be noted that both studies essentially use opportunity cost pricing of all major inputs, but a more relevant and realistic approach to the costing of capital and utilities would change these results significantly. One can question the fuel costing in the first case, where in the fuel oil-fired refinery of the Gulf fuel cost was shown to be higher than in the other locations, i.e. 33 ¢/b against 24.98 ¢/b. The second case, too, can be questioned on the basis of its use of 53.5 ¢/b as the fuel oil cost in all refinery locations.[2]

One can also argue that the economics of current product tankers may change as expansion of this sector of petroleum transport takes place to meet the rising requirements for additional production of refined products expected from the OAPEC area. Larger product tankers would almost certainly contribute to narrowing the gap between crude and product freight rates – a factor so crucial to both analyses. But here, too, the situation may not be so simple as it

[1] *Petrochemical Educational Course*, Vol. 1, Study prepared by Comerint S.P.A. (ENI) for OAPEC, Rome 1977.

[2] These assumptions ignore freight cost savings on fuels used on OAPEC sites which market-located refineries gain having to import an additional quantity of some 7% of their crude throughput as fuel.

Table 4 Operating costs of a Hydroskimming Refinery (1) (cents/barrel)

Site	*Libyan Jamhiriya*	*Libyan Jamhiriya*	*Algeria*	*Saudi Arabia*	*Italy*
Capacity: M t/yr	2.5 + 2.5	5	5	5	5
Item:					
Depreciation and Interest	95.0	87.7	92.2	106.0	74.0
Interest on Working Capital	5.3	5.3	5.3	5.3	5.0
Total Financing Costs	100.3	93.0	97.5	111.5	79.0
Fuel Oil	53.5	53.5	53.5	53.5	53.5
(Natural Gas)	(16.5)	(16.5)	(16.5)	(16.5)	
Cat. & Chem.	0.6	0.6	0.6	0.65	0.5
Tel.	3.6	3.6	3.6	3.95	3.4
Labour	17.0	17.0	19.5	15.3	13.2
Maintenance Material	5.7	5.4	5.6	6.5	3.1
Insurance	2.8	2.6	2.7	3.1	2.1
Overheads	4.2	4.2	4.9	3.8	3.3
Direct Costs	87.4	86.9	90.4	86.8	79.1
(Natural Gas)	(50.4)	(49.9)	(53.4)	(49.8)	
Transportation	20.0	20.0	8.0	30.0	15.0
Total c.i.f. End Market	207.7	199.9	195.9	228.3	173.1
(Natural Gas)	(170.7)	(162.9)	(158.9)	(191.3)	

(1) Excluding taxes, infrastructure costs, land cost.

seems, given the persistence of the present depressed market for crude oil carriers, which are operating at rates so low as barely to recover variable cost (and not even all can do that).

The hydroskimming idealized refinery used for comparison may not be the typical design of future refineries. They will more likely be designed to optimize high-value white products. This implies extensive use of hydrocracking, deep desulphurization, and reforming – all more energy- and capital-intensive than hydroskimming. With available flare gas and favourable capital finance, OAPEC states will improve their given advantages, everything else being the same.

In both studies the cost of associated natural gas used in OAPEC-located refineries as a refinery fuel was reckoned at gathering cost. But the use of associated natural gas as a refinery fuel for products exported by oil-exporting states produces a value for that gas close to its crude oil price equivalent. For example, $1.70 (1975 prices) of additional national revenue accrues for each 600 ft^3 used in refining a barrel of crude, or $17/b equivalent. Refining has been shown to offer the highest opportunity value for associated natural gas use, next to re-injection into reservoirs.

In both studies, the financial component in the built-up cost is shown to be higher for source-located refineries, based in part on the assumption that capital will be obtained by loan at interest rates higher than those offered in the industrial countries. Yet, even on the basis of opportunity cost calculations, one may assume that part of the investment capital may be in equity form obtained possibly at concessionary rates of interest or on terms involving no financial loss to OAPEC governments; but at lower cost than the case in a typical industrial country by an amount equal to the bankers' spread and/or underwriting cost. There is also the likelihood of advantages from income tax exemptions, which are normally chargeable to equity capital in similar plants located in developed countries. Since the host governments are simultaneously the taxing authority and the beneficial owners of the petroleum resources, the distinction between oil income revenue and tax revenue is not important; it is the sum of the two which is financially significant. As a consequence, capital financial charges for plants sited in the Gulf may be around 11%, while that for a European, US or Japanese plant is 16%.

Finally, a crucial element in cost comparison is the possibility of subsidization. This may take the form of governments bearing the cost of infrastructure such as ports, roads and housing. Since such facilities are developmental in nature, they should logically be charged to the development budget. Other subsidies may take the form of low interest loans, grace periods on government loans, subordinated government equity capital, and lower prices for crude and products, as used to be given to the MNCs during the concessionary era. Once some of the commercial criteria more relevant to developed economies are suspended, the competitive potential of Gulf-based refining is clearly limited only by the availability of oil and gas.

Marketing Options

The marketing efforts required to sell refined products directly to consumers can be substantial. There are alternatives, however. OAPEC export projects output can be marketed by the national oil companies (NOCs) through long-term arrangements with major consumers and/or large independent distributors in consuming areas. They can also start or acquire existing marketing organizations, or enter into joint ventures with integrated distribution.

It is not intended here to analyse each of these approaches, but rather to remark briefly on the options they provide. The first approach involves minimal effort, but it is not so secure. The second offers security but at a prohibitive cost amounting to an investment expenditure of some 70% of the size of investment in the export refinery itself. The third may be the most convenient and least risky, being backed by the expertise of the partner, and may be the most likely to succeed in the medium term under normal conditions. No one policy is generally satisfactory for all states. OAPEC can be the kind of institution required to co-ordinate the marketing strategies of its member states, which have acquired a significant stake in export refining.

Policies and Strategies

Project analysis is fundamental to an understanding of the comparative economics of refineries sited near petroleum sources relative to those located in major markets. In the idealized first case, it was shown that the economics based on the opportunity cost reckoning is not favourable. But when relevant factors concerning fuel and capital financial cost are introduced, a refinery project – given the existence of market opportunity in major, more favourably located, markets – becomes competitive at an advantage which is inversely proportional to the distance between source and markets. The economic perspective of an export refining industry has to include, on the benefits side, linkage, feedback and complementarity effects, especially with regard to petrochemical projects taking feed from the refineries. We think that the prime objective of supplying Arab domestic markets using economic size units (but not necessarily risky maximum scale just to capture the largest scale economies available) may be a reasonable target until some of the uncertainties are removed.

With capital equipment costs giving rise to a 30–40% higher installed plant cost in the OAPEC area over a similar one in the major OECD countries, an initiative must be taken to raise the level of direct technology transfer. This implies increasing indigenous input of skilled manpower in the conception, design, procurement and installation of refineries, especially in countries such as Iraq, Algeria, and even Saudi Arabia. Such an approach can in the long run contribute significantly to a reduction in capital cost, as shown by the latest expansion of the Basra refinery in Iraq. In the longer term, specialization in the hydrocarbon process industry calls for industrial R & D to help develop a technological base, know-how and the efficient management of those component technologies that make up the bulk of processing units. This is one crucial factor that has contributed to the success and technological dominance of the typical integrated oil company.

Although the OAPEC producers cannot afford to integrate downstream as far as the garage distribution of motor fuels and of heating oil depots (nor is it politically prudent for them to do so given the risk and enormous investment required to own such real assets), they can still capture the benefits of integration by arriving at negotiated agreements with governments in the major markets to allow, without restrictions, the incremental quantities of refined products expected between now and 1990.

The last 5 years have seen a lengthening in the OAPEC export-refining capacity outlook of from 1.18 M b/d of additional capacities for completion by 1980 (envisaged in 1975) to a mostly unrealized 1.27 M b/d for completion beyond 1980, closer to 1985. The production and refining world of the next 20 years will be quite different once oil production has peaked. Processing locally the entire crude production of OAPEC states is, in the judgement of this writer, neither feasible nor desirable given Arab economic priorities. But processing 25–30% of Arab crude can and should be expected at this time. How much crude to refine is not the most important question. What product pattern to use and how efficiently this can be done are the more important issues in the longer term. Only an able, indigenous technological base can tackle the problems involved – witness Mexico, for example.

II. PETROCHEMICALS

Petroleum-derived chemicals are numerous. Not only do they include most organic chemicals, but also a significant proportion of inorganic chemicals such as ammonia,[1] sulphur, and carbon black. Organic chemicals dominate with 90% of total world output produced from petroleum hydrocarbons.[2] No simple classification can encompass all petrochemicals, but there are four main large tonnage end-use petrochemicals, namely, plastics, fibres, synthetic rubber (elastomers) and detergents. Together with basic and intermediate petrochemicals they make up the industry's output, which last year amounted to some 70 million tons.

[1] 80% of world ammonia production is derived from petroleum sources, mainly natural gas.

[2] This includes naphtha, refinery gases, natural gas and NGL, and fuel oil.

Until the early 1970s, hardly any OAPEC country directed investment to this industry, despite the abundance of cheap raw materials and a potential market. Apart from the institutional arrangements governing petroleum production and refining industries in the member states (or maybe because of them) the economic parameters were not favourable. For one thing, most of the hydrocarbon feed was not of crucial significance in the manufacturing cost of most bulk petrochemicals. Secondly, the prevailing commercial plant size was larger than the size of the domestic market in most, if not all, member states. Thirdly, and perhaps most importantly, the required investment finance in most OAPEC countries was not available. Finally, infrastructure – whether physical or human – was inadequate.

As for joint-venture type arrangements, the conditions demanded by technology and know-how owners and controllers for co-operating with host governments or national entities, were too restrictive and the cost of royalties and licence fees was considered too high to be acceptable.

With the 1973–74 oil price rise, and the subsequently higher costs of energy and feedstock, the manufacturing costs of many basic and intermediate chemicals became favourable in those OAPEC states with large petroleum reserves and production. Furthermore, being volume exporters of crude oil, sufficient additional funds were available to meet the huge investment requirements of this capital intensive industry. With the accelerating rate of economic growth in the OAPEC and Arab region, market size is at, or can soon catch up with, the economic plant size required for a number of petrochemical commodities. Also, specific opportunities for exports to certain markets either exist or will develop in the future.

The industry worldwide is undergoing changes. For example, its historically high growth rates have declined and will perhaps settle at much lower levels. Raw material prices are raising production cost (and prices), and thereby contributing not only to lower demand growth for certain products but also to providing an economic rationale for shifting location to areas of abundant reserves of low-cost fluid hydrocarbons. Because of these changes, some industrial relocation is expected. Such developments might lead to certain industry-wide technological and organizational (degree of integration) changes, and are likely to be influenced by the evolving relationships between petroleum producers and the actors in major petrochemical markets, i.e. governments, MNCs, national companies. These developments provide new opportunities for different interests on the petrochemical scene and pose problems, old and new. I shall attempt to delve into these matters later.

Production Costs and Demand Conditions

A comparative production cost analysis for a number of petrochemicals has been computed by several consulting firms. Some studies show that there is a clear cost advantage for the OAPEC states in the production of petrochemicals except benzene and, surprisingly, PP, LDPE and HDPE. Of significant importance are favourable cost differences in producing the highly important bulk chemicals – ethylene, VCM, ammonia and urea. Their cost advantages are likely to increase with the rise in the real cost of energy (utilities) and feedstocks, making such advantages decisive in the selling of these materials in certain export markets.

Changes in the Cost Structure. A plant built in 1972 enjoyed a 9% manufacturing cost advantage over one built in 1977. In the case of the all-important ethylene, the proportion of the fixed cost part (of the total manufacturing cost) declined from 39.5% in 1972 to only 19.3% in 1977, while variable costs, essentially raw materials and utilities, rose from 44.4% in 1972 to 72.9% in 1977. The implication is that countries with large, cheap petroleum reserves are favoured in the production of basic petrochemcials.

The production cost rises, owing to energy and feedstock price increases, are leading the industry to adopt energy conservation measures that use raw materials more efficiently. Furthermore, a significant part of the increase in construction costs is the result of stringent environmental controls.

Changes in Demand Conditions. The general decline in economic activity is another important factor in the slowdown of petrochemicals demand growth rates which historically rose at twice that of the GNP. Also of significance is the influence of public awareness in developed countries about the industry's environmental consequences – a factor leading in the direction of lower consumption for certain important end-products. These trends could lead to a structural shift in demand for petrochemicals in many OECD markets.

The industry today operates at a relatively low capacity rate. This is causing a reduced investment rate in the industry and increasing price competition at market outlets. The number of projects worldwide has already declined from over 1,000 in 1975 to 750 as currently reported. It is widely predicted that no major investment for new petrochemical complexes will be forthcoming before the supply/demand balance is restored, a situation which, with the continuing economic slowdown, may last until 1985.

During the 1960s, demand grew at a fast rate, but began to decelerate in the early 1970s (1976 demand equalled 1974), as pointed out earlier. In the developed countries the growth rate slowed but then steadied as markets stabilized after reaching saturation levels. But in the light of emerging economic conditions, the slowdown may actually have been the start of a shift in the attitude of consumers and producers of petrochemicals. However, for the time being, the effect of rises in the prices of petrochemicals on the level of their demand has been limited by concommitant rises in the prices of their substitutes. This new trend will perhaps set demand at a level and growth rate more moderate than it had been hitherto.

As for the developing countries, the demand growth rate has typically been higher, albeit more irregular. It is observed that demand growth was less affected in 1974–75 than was the case in the developed countries. This can be explained by the continuing higher economic growth rate in the developing countries, where potential demand remains very large.

International Trade

Value-wise, US exports of synthetic fibres, thermoplastics and synthetic rubber represented 1% of its total exports in 1976. For the EEC, total internal and external exchanges of similar petrochemical end-products were worth $5 billion that year. In Japan, synthetic fibres constituted 2.5% of total exports. International trade in basic and intermediates is lower in value. US exports of BTX were worth $250 million in 1976. International trade patterns for the main end-petrochemicals for the year 1973 are shown in Table 5. The major part of the trade is in end-product exchanges in the following order: plastics, synthetic fibres, synthetic rubber and detergents.

In terms of tonnage, plastics represent 70%. Plastics growth in the world trade peaked in 1973 at 12 million tons from 3 million tons in 1965. The EEC, Japan and US were the main exporters of end-products with 8.8, 1.6 and 1.4 million tons (5.4 million tons in internal exchange), respectively. Other West European nations accounted for 1.3, Asia 1.7, and Latin America 0.6 million tons.

It can be concluded from Table 5 that the EEC is a net exporter, the US imports mainly butane and butadiene, and Japan is a net exporter of all products except the xylenes and methanol. As regards the trade in those basic and intermediate petrochemicals of primary

Table 5 World trade patterns for the main end petrochemicals by region, 1973 (percentage)

Exports *From*	*EEC*		*US*		*Japan*		*Other countries*		*Total*
To	*Import*	*Export*	*Import*	*Exports*	*Import*	*Export*	*Import*	*Export*	*Import*
EEC	89.05	60.72	5.34	21.38	2.10	7.48	3.50	35.07	100
Other Western European countries	91.21	17.04	3.80	4.17	1.49	1.45	3.50	9.62	100
Eastern Europe	81.52	6.56	0.55	0.26	6.34	2.67	11.59	13.70	100
United States	40.63	1.71			29.55	6.49	29.82	18.43	100
Canada	7.90	0.39	87.36	25.23	3.61	0.93	1.13	0.82	100
Latin America	35.76	3.00	39.87	19.62	20.93	9.15	3.44	4.24	100
Japan	35.72	0.61	56.49	5.67			7.79	1.96	100
Other Asia and Oceania	26.00	5.69	17.27	22.17	52.92	60.41	3.81	12.24	100
Africa	61.27	4.29	3.65	1.50	31.27	11.42	3.81	3.92	100
Total export		100.00		100.00		100.00		100.00	

Source: *First World-wide Study on the Petrochemical Industry: 1975–2000*, UNIDO/ICIS. 83, December 12, 1978.

concern to the future OAPEC industry, the following represents the main trade flows:

(a) Olefins. Ethylene accounts for 80% of world olefin demand, primarily in Western Europe and Japan. Little international trade actually exists in ethylene and propylene as commodities, except for minor amounts within the EEC. There is, however, a large amount of trade in the C_4s from Europe to the US.
(b) Aromatics. The US exports to Europe benzene, toluene, and xylenes. In recent years Eastern Europe has also been exporting to Western Europe.
(c) Intermediate products. The US exports methanol, styrene, and cychlohexane to the EEC, while raw materials for synthetic fibres and plastics are exported from the US, EEC and Japan to LDCs which have acquired polymer processing facilities.
(d) End-products. In 1975, 1.1 million tons of end-products were exported from the EEC, the US and Japan, the last being the biggest exporter.

The major trends in international trade as identified in a UNIDO study sketch the following picture: expectations of toughening competition, a substantial increase in the trade of basic and intermediate products, and a geographical broadening of the whole chemical trade. For the EEC, internal trade will continue at the present level. The EEC countries will export more propylene to the US because of their ethylene surplus. Butadiene will also continue to flow to the US barring any feedstock shift for basic petrochemicals production. Conversely, aromatics will continue to flow at current levels from the US to Europe. Finally, Japan's export tonnage is expected to remain constant with a decrease in the ratio of exports to consumption due to competition from the Middle East and South-east Asia.

The major changes predicted by the UNIDO study regarding production and price exchanges are as follows:

(a) Regional self-sufficiency is on the rise as a trend.
(b) The position of the developed countries is diminishing.
(c) Despite (a) above, trade will be maintained volume-wise at current levels owing to the rise in consumption and the appearance of new production flows. The latter will take the form of more speciality products from developed to developing countries and a reverse flow of commodity chemicals. The latter will require increasing amounts of resources for the development of marketing and distribution facilities from the new exporters.

OAPEC States' Projects and Markets

Against the background of the supply/demand balance for the target years 1982, 1987 and 1992 in six major producing and/or consuming areas of the world, the prospects for the production of a number of important petrochemicals expected to come from the OAPEC area will now be examined.

Apart from ammonia and urea, and methanol in the Libyan Jamahiriya and Algeria, little else of export significance is currently being produced in OAPEC. Algeria's existing petrochemical facilities for ethylene and its derivatives, and BTX seem to be mainly directed to the domestic market, judging from their capacities. All OAPEC member states currently run significant production capacities of ammonia and urea, 4.052 and 2.66 million t/yr respectively. These chemicals are produced from natural gas, using mature technologies, with no important by-products to worry about and with ready and expanding domestic (fertilizers) and export (urea) markets. These factors explain the early start of their production. They are perhaps the least complicated direct industrial undertakings for valorizing flared associated natural gas in non-energy applications. The imputed price for natural gas in the production of ammonia and urea is crucial to their production economics. With a relatively efficient capacity utilization rate, the break-even imputed price for associated natural gas consumed at an ammonia smelter is $1.08/million Btus, or three times its gathering cost. This gives the producers a distinct comparative advantage at the current opportunity value for most associated gas produced. Finally, in addition to being a fertilizer, urea goes into the making of a few urea-resins. This provides opportunities for manufacturing other petrochemicals. (In Kuwait, for example, a urea melamin resin production facility is under construction with a capacity of 15,000 t/yr.)

As for the petrochemical output coming from projects either under construction or planned for completion by 1990, their market prospects are difficult to determine with any degree of accuracy. Significant production of additional quantities of ammonia and urea is expected from Iraq, Egypt, the Libyan Jamahiriya and Algeria, a total of 3.476 and 2.231 million t/yr respectively. Their production cost is favourable relative to Western Europe and presumably, Asia and the Far East. Deficits in major markets are predicted only for North America (780,000 t/yr of urea in 1982), but at persistently lower levels to 1987. By then a deficit is also predicted for Asia. A potential deficit for ammonia is also predicted for 1987. Deficits for both urea and ammonia are predicted to be even sharper in 1992. The competitiveness of anhydrous ammonia in deficit markets depends on the state of development of cryogenic tanker transport and complementary facilities (terminals and distribution networks at importing points).[1]

Being solid, urea poses no such problems of transport development. The recommended future markets for ammonia are North America and Western Europe by direct sale to major users. North America offers the best opportunity through negotiated contracts with existing suppliers. The Indian market may also be expanded through government-to-government deals.

The other major petrochemicals expected from projects now under construction are end-products such as LDPE, HDPE, PVC, and some intermediates such as MVC. These are expected to come from Iraq, Egypt and the Libyan Jamahiriya. Products would go mostly for domestic consumption, except in the Libyan Jamahiriya. The exportable quantities involved are not significant in terms of major markets volume and can be absorbed regionally. Here the marketing problems of performance chemicals referred to earlier are relevant.

Beyond 1985, Kuwait and Saudi Arabia will be the only OAPEC countries to face problems in petrochemical production marketing (other than for ammonia and urea), unless new projects are planned in the near term. Ethylene and its major derivatives (HDPE, LDPE and ethylene glycol) will be produced in the two countries, in addition to BTX (and styrene) in Kuwait and methanol in Saudi Arabia. The latter is reportedly proceeding on the basis of joint ventures, i.e. using the foreign partner's marketing network. Kuwait follows the direct or joint venture approach depending on the petrochemical output in question. However these are distant projects and one can now only discuss them in general terms. The following points on marketing strategies for some petrochemicals are relevant.

Ethylene: Except for South-east Asia and parts of Africa, where a deficit is predicted for the three target years 1982, 1987 and 1990, no natural target markets are recommended to the Gulf producers. Asia's deficit is put at 100,000, 220,000 and 120,000 t/yr for the years 1982, 1987 and 1992 respectively. The study advises OAPEC-type co-operation to establish contacts in target markets and look into the possibility of using the European pipeline system to distribute ethylene.

[1] Currently the freight rate for such materials as ammonia is 0.71 ¢/nautical mile/ton for a no return tanker of 15,000 m^3 capacity. Transport costs average 4 ¢/nautical mile/ton return freight for such materials as urea.

Low and High Density Polyethylenes (LDPE & HDPE): The analysis predicts continued over-capacity and intense competition for LDPE in the markets of Western Europe, North America and Japan. Therefore dismissing them as 'unattractive', it recommends greater attention to Arab states outside OAPEC (Syria, Jordan, and Lebanon), and Iran, East Africa, and Asia. The marketing strategy recommended is of the joint-venture type, owing to the OAPEC states' lack of commercial and technical expertise.

Vinyl Chloride Monomer, Ethylene Dichloride, and PVC: Target markets named are EDC for Japan, Taiwan and North America, and VMC for South-east Asia and in the medium-term Greece and Turkey. Longer-term markets could be Pakistan and Western Europe. As for PVC, the Arab region can absorb all output locally, as pointed out earlier.

The long-term contract is recommended for selling intermediates to Japan (also for spot sales to meet temporary imbalances). Because of the interrelated nature of EDC, VCM and PVC, they are best sold together, especially to Japan.

Methanol: The natural target markets are Western Europe, North America and Japan. The preferred strategy is the long-term contract with producers of the same materials in their markets. Cost-sharing (among OAPEC producers) at receiving terminals could cut marketing costs among the major potential methanol producers – Algeria, the Libyan Jamahiriya, Saudi Arabia and, perhaps, Kuwait.

BTX: Compared with their competitors in export markets, the OAPEC producers enjoy no significant cost advantage in producing these materials, except perhaps for paraxylene in the form of exports of DMT/TPA to South-east Asia and East Africa.

As for strategy, the study sees a future for benzene in making styrene for local needs. Only ortho-xylene for producing plasticisers locally and para-xylene hold promise for the export markets named.

Styrene: A deficit is predicted in the markets of Western Europe and Japan/South-east Asia, but at no cost advantage to OAPEC producers. Derived polystyrene could be targeted to the regional Arab/Middle East market.

Prospects to the Year 2000

Apart from ammonia, urea and methanol, an examination of expected production of all major petrochemical commodities and performance materials shows either surplus or balance in these markets. Newcomers in OAPEC states have to penetrate existing markets to carve out a share for themselves. In situations of excess capacity and slower growth in major markets, price competition can be intense. This would come, not from the traditional producers and merchants, but from other newcomers – Latin America aiming at the North American market and the East European aiming at Western Europe. Competition among producers for limited deficit spots could lead to price-cutting which would be loss-making for all.

Furthermore, the protectionism already being practised might get even worse. The problem between now and the year 1985 does not appear so drastic. Beyond that, and assuming all planned projects are carried out, the outcome lies in the area of international relations – in other words, whether there will be mutual accommodation or conflict. In 1974 OAPEC presented a paper to the European Petrochemical Association Conference. Its message then was simple enough: Let us co-operate. The suggested basis for that was (and remains) recognition of the shifts in petroleum prices and the new institutional arrangements that have emerged since 1973.

An even more pressing problem than that of protectionism is flare gas currently being wasted in huge quantities. Flare gas must be utilized to conserve other non-renewable natural resources as the volume of whatever can be used in economic activity will replace an equivalent quantity of crude oil in reserves. Relocating additional new capacities of a few bulk petrochemicals manufactured at favourable relative production costs may be a logical beginning. In the meantime, additional European speciality production continues feeding the growing chemical trade, including OAPEC's requirements. The question of co-operation has been taken up during intermittent meetings between OAPEC and the EEC on future development in the areas of refining and petrochemicals. But apart from the two sides conceding to the size and duration of current and expected over-capacity, little concrete progress has been made, although many public statements on both sides have urged consensus.

In October last year, during a seminar sponsored by OPEC on the problems and opportunities for downstream development, the views of the producers and those operating in the major markets in the West and Japan were put forward. It is worth trying to encapsulate the different positions.

The points favouring OPEC/OAPEC viewpoints were:

(a) Time is crucial. Industrialization of oil (downstream) must take place while petroleum reserves are still large enough to support a viable downstream industry. This is basically the exhaustible asset argument.

(b) The need to penetrate existing downstream industry in world markets with home-developed efficient-size refineries and petrochemical plants able to meet existing and potential Arab demand with some quantity left for export. They assert that this is how it was done initially in the industrial world.

(c) A comparative cost advantage exists in flare gas-based projects, such as ammonia, LPG and LNG, which justifies investment in export projects. World markets will grow tight as the slack in capacity is absorbed.

(d) Regional markets justifying economic-size projects (as in the case of OAPEC) provide inputs for individual country projects such as lube oil blending plants, catalysts, SBR and detergents, and others of the joint-venture type.

(e) The OAPEC producers would accept a low return in order to develop downstream industry as part of their economic development burden. Return on investment must be considered for major projects, but on the basis of a national cost/benefit analysis. A purely commercial approach to economic development would condemn these states to the status of raw material producers until such time as they run out of their raw material assets.

(f) Downstream investment provides the means and medium for technology acquisition, know-how and some employment. It helps in the restructuring and diversification of economies, whereas the production and export of raw materials means that wealth is transferred without adding real value in relation to cost replacing depletable wealth.

The developed countries argued that industry must be justified on the basis of profitability and a reasonable return on investment guided by market forces. They emphasised that the present market situation makes large exports from the OPEC/OAPEC area a losing commercial proposition fcr the near to medium terms, adding the following disadvantages for the oil producers: (a) Higher investment cost in a capital-intensive industry. (b) Technological change in the industry, i.e. fast obsolescence requires frequent modifications, hence adding further to costs. (c) Underdeveloped infrastructure; unstructured, inadequate labour force raising cost yet higher by the need to invest in new complementary facilities.

The results are: non-competitive, losing plants which ultimately have to close down; idle capacity in an OECD area calling for protection (tariffs and quotas); and unemployment with its consequent social problems.

In a more rigorous policy analysis, the UNIDO-sponsored study arrives at conclusions also favouring consensus as to the best type of relationship among actors on the scene. Their scenario analysis takes account of the objectives of various primary actors, the main economic gameboard areas, and their relative negotiating and staying powers.

The perceived aim of governments in developed market economies is to encourage investments and/or utilization under the constraint of increasing environmental restrictions. Their staying power lies in the management of their economies, their influence over the world economy – especially their ability to depress it – and in their bargaining power in dealing with other governments. The second actors, governments of CPEs, wish to accelerate development, raise living standards, and increase exports to obtain foreign exchange. Their staying power lies in their almost total control over their economic resources and priorities, and their bargaining power with other governments and MNCs through buy-back arrangements. The governments of the oil-exporting countries aim at the downstream integration of their oil activities and access to large markets for their future petrochemical output. Their staying power is based on control of most of the world's flare gas, their oil exports, their large financial resources, and hence their bargaining power in dealing with other governments and MNCs. Governments of non-oil exporting developing countries with large markets for petrochemicals aim at self-sufficiency first, and then exports, if possible. Their staying power lies in their control over markets, the management of their economies and their bargaining power.

The multi-national oil companies have the perceived aim to diversify into energy companies to find a profitable outlet for their cash flow into petrochemicals. Their staying power is based on their control of a variety of feedstocks, and advantages in the future processing (steam-cracking) of co-products. This staying power is reinforced by the ability to self-finance, their control of basic petrochemical markets, and their growing shares of several intermediates. They also have bargaining power in dealing with governments. The multi-national petrochemical companies aim at becoming speciality producers of end petrochemicals for higher growth, profits and cashflow. Their staying power is based on their control over technologies and a dominant market share of petrochemicals and in their co-operation and goodwill with processing industries; and they, too, have bargaining power.

The gameboard areas are: markets, production cost, raw materials, and economic and financial assets. After assessing 3 alternative scenarios[1] that explore the possibility of LDCs meeting the Lima Declaration's declared target of 25% of industrial output by the year 2000, the study concludes that 'the only way of attaining a 25% share or higher is through a high degree of co-operation among these countries'. The study then goes on to outline a four-point proposal for organizing co-operation, calling for the establishment of a working group whose task would be to validate information, work out an analysis of strategies for the actors in the petrochemical industry, elaborate the various scenarios, and then negotiate by choosing a reference scenario and analysing the conditions for the realization of this scenario to submit to the actors for discussion. The latter would include a proposed plan of action for co-operation in production and trading and transfers of various types, including financing, as well as a programme for the progressive development of the petrochemical industry in the developing countries.

Paradoxically, the study suggests that the global, and not the regional framework, is the proper dimension for market regulation. It explains this by saying that the contradictory points of view between free market adherents and advocates of CPE imply that consensus possibilities exist without the need to resort to control at the borders between countries. It adds that the borders of the petrochemical industry are on a world scale. Therefore, the petrochemical projects should be the object of negotiation. Those who reject the principle of global interdependence risk destabilizing the best established national or regional projects. The central idea of its proposed plan is an indicative contract for the development of the world petrochemical industry and negotiations among the interested actors on such an indicative medium and long-term reference framework for the development of the industry.

Concluding Remarks

The greatest opportunity for the development of refining and petrochemical industries in the major OAPEC states lies in their control of great reserves of oil and natural gas. The immediate pressure for early investment within the framework of an integrated refining/petrochemical industry is the unacceptable waste of associated natural gas, most of which is currently being flared. The economics of producing ammonia, urea, methanol, and ethylene from flare gas is decisively favourable. The economics of refining are more

[1] These are namely, mutual respect leading to lasting co-operation, agreement through contractual co-operation, or latent conflict leading to lasting conflict.

favourable when it is considered that flare gas as a co-product in the petroleum extraction process is, in fact, being upgraded from zero value. Being capital intensive, refining and petrochemical production provide absorption outlets for large financial flows whose opportunity cost in an inflationary world is less uncertain in the form of productive assets than in most forms of conventional financial claims. The availability of capital to OAPEC state enterprises reinforces the favourable factor of the energy intensity of downstream projects. A fast-growing Arab region provides the market potential for output in efficient-size production units equal or smaller than regional market size or even some individual domestic markets.

There are problems, however. The external ones may be summed up as the existence of refining and petrochemical industries in major markets that have surplus capacities but are likely to enjoy lower fixed capital costs for the next few years. Furthermore, the industry is in a state of shift tending to a lower growth rate than that of recent history, and economies of scale and technological improvement are no longer adequate to offset the inflationary rise in the cost of new investment. This is partly offset in the OAPEC area by the more favourable variable cost. But these states suffer from higher capital costs, partly because of the low level of their participation in the design, selection, procurement, installation and operation of downstream projects. Since these activities are labour intensive, the OAPEC states – owing to the unstructured state of their labour force and their organization deficiencies – incur maximum payment for their imported capital equipment. The opportunity cost in developing, structuring and substituting their skilled manpower is very high, but any improvement takes time. In the meantime, countries are losing a learning opportunity. Marketing is another major problem for OAPEC states because of the need to penetrate major existing markets which are imposing or are likely to impose protective measures. The situation to 1985 indeed seems like a zero-sum game.

In the longer run, policies and strategies aimed at developing OAPEC countries' domestic and regional markets on the one hand and negotiating long-term arrangements for indicative investment and production plans in major markets on the other, offer the best prospects for refining and petrochemical industries, which are recognizably global. This requires governments, MNCs and NOCs to adopt policies and attitudes that can accommodate differentiated growth and specialization in refining and petrochemical industries by OPEC/OAPEC states. Each one must concede the need to optimize scarce, non-renewable resources and to enable LDCs to participate in creating value-added capabilities in a fundamental manner, i.e. by contributing to productivity gains by the skills of their labour. This requires a policy of commitment to such aims and a plan of action to develop their technological base beyond project creation on a turn-key basis by international engineering conglomerates. The time left to achieve this level of indigenous self reliance is not long – perhaps years and not decades – for most major grassroots investments in OAPEC states will have to be made before the turn of this century.

25 THE ECONOMIC FUTURE OF OPEC COUNTRIES: CONSTRAINTS AND OPPORTUNITIES

George Corm

Oil-exporting countries are often discussed as if they are purely bonanza countries, having received a gift from the gods that permits them to live a never-ending life of ease, leisure and luxury. According to the annual *World Bank Atlas*, some of these countries head the list of the ten countries enjoying the highest per capita income in the world, and they are in this position without having undergone all the pains of an industrial revolution. This image is so well popularised in the West that not only does it create hostility towards oil-exporting countries, who have become an easy scapegoat for all the failures of the industrialised world to manage production and distribution of energy, but it has also prevented any analysis of the various social and economic constraints which the oil countries have faced since the early 1970s as a result of their dependence on oil exports.

In fact, when speaking of an energy crisis, one should not think exclusively in terms of problems created inside the economies of industrialised countries, or of non-oil developing countries, but equally in terms of oil-exporting countries. These countries are actually subject to a number of interrelated constraints, both local and international, to the extent that the opportunities offered by the disposal of oil have become negative factors antagonistic to their own political, social and economic aspirations.

As the economics of underdevelopment has largely demonstrated, specialisation in the production of one main export commodity is one of the key factors underlying most of the imbalances that characterise the underdeveloped economy and, furthermore, it paralyses all efforts to correct such imbalances. The experience of LDCs in the last 30 years has shown that reducing the dependency of the local economy on the export of raw materials, and correcting the structural imbalances created by such a dependency, is a very hard task that few countries have been in a position to perform successfully. More so, when the exported raw material is a vital input for industrialised countries, which possess several powerful tools to interfere in the local policies of the producing countries, ranging from military pressure, to currency manipulations that might affect export receipts.

In brief, this dependence on the export of one strategic raw material to provide a balanced social and economic growth, constitutes a tremendous challenge for oil-exporting countries, a challenge that is seldom analysed. The atmosphere of passion that surrounds all aspects of the so-called energy crisis in the West is such that any sound discussion of the economic future of oil-exporting countries is impossible. This is not without impact inside the exporting countries themselves. It encourages the governments' natural tendency to preserve the present status quo in their local and foreign policies in order to avoid disturbing what is portrayed as a very precarious equilibrium.

One should note, however, that great opportunities have been lost in the past few years, when the industrialised countries although reluctantly, began to accept the idea of linking oil problems to an improvement in the functioning of international economic relations. In this respect the North–South Dialogue did not perform its task successfully. The industrialised world succeeded in avoiding all important issues: a general agreement on stabilising the price of raw material that cannot be freely disposed of by the producers preserve the value of foreign assets of LDCs, and the real transfer of technology to LDCs. In the final analysis, the outcome of the North–South Dialogue was to open the way to disagreements between oil and non-oil LDCs. This was clearly demonstrated at the last UNCTAD meeting in Manila where opposition between oil and non-oil LDCs paralysed the discussions for several days. Today, oil-exporting countries are once again isolated and subject to strong international pressures. Oil, as a highly strategic product, is considered to be a raw material that cannot be freely disposed of by the procedures without endangering world peace.

It is in such a context that the economic challenge facing oil-exporting countries should be looked at. What our analysis will show is mainly that opportunities arising in an underdeveloped environment may rapidly become new constraints, aggravating imbalances in the local economy, while the traditional constraints to real economic growth are not effectively removed. In fact, in the case of oil-exporting countries, it is the opportunities created by the flow of exported oil that exacerbates the problems already existing in the local economies. The last increase in oil receipts will not fundamentally change this situation.

Let us now examine how this mechanism operates by looking at the various factors that are usually considered to be opportunities. We will examine three main points: the disposal of energy, the disposal of foreign exchange earnings either as an accumulation of financial surpluses or as a tool to accelerate technology transfer, and the increase in regional and international influence.

The Disposal of Energy

The modern industrialised world is based on energy. Without energy the present industrial machinery in the Western world and in industrialised socialist countries would come to a halt. In fact, not only would industry come to a stop without energy, but agricultural yields in Western countries would also fall dramatically, given the extreme importance of various energy inputs in modern mechanised agriculture. Thus, any country endowed with easily accessible energy resources in the present stage of technology, is considered to enjoy an important advantage towards economic growth. But a country at an early stage of industrialisation would not need large quantities of energy in the short term. The logical economic behaviour for such a country in the present context of a world shortage of energy, if it had at its disposal cheap energy resources, would be to conserve it for as long as possible so that it would still be available locally at a cheap cost when its industrial sector had matured and would need large quantities of energy in order to operate.

At present, oil-exporting countries are not, however, following this path. OPEC exports of oil, ranging from 27 to 30 million barrels a day, represent approximately 90% of total production. OPEC countries have a total population of 305 million, or 90 million more than the US. Assuming that this population will grow at an average rate of 2.3% per annum, it will reach 481 million after 20 years, or more than twice the present US population. If such a population should by then consume only half of what the US takes at present – which would mean less than ¼ of the US 1978 consumption level on a per capita basis – it would need more than 17 million barrels a day. It is doubtful that such future local needs are compatible with the present level of production and ratio of exports to production, given the size of proven reserves and the increasing costs incurred in expanding production capacities. In fact, producing at such a high rate today means considerable expenses tomorrow to make energy available for local industrial and agricultural needs. Thus OPEC countries might be losing one of their best opportunities for development in the next century, i.e. a cheap energy resource with which they could build up an efficient and competitive industrial and agricultural sector, and bridge the enormous economic gap currently existing between their level of development and that of the industrialised world. One should add here that the rational behaviour for OPEC countries would be to plan oil and gas exports not only in relation to their own local needs, but also in relation to the needs of their own regional neighbours.

Yet, it is difficult to conceive that OPEC countries would implement a policy of optimising available energy resources, locally and regionally, within the framework of a long-term strategy for the next 50 years of self-sustained, inward-looking industrialisation and agricultural modernisation. Although only the implementation of such a long-term strategy could avoid the continuing underdevelopment of OPEC countries in the next century, many factors militate against it. First, pressures from industrialised countries would become unbearable if there was a dramatic reduction in the rate of exports bringing about a more reasonable ratio of local consumption to exports. For instance, a ratio of 1 to 5, which is already quite high, would only allow exports at the level of 15 million barrels a day. Such a reduction would be considered as a *casuș belli* by the oil-importing countries and would provoke an explosive international situation. Secondly, many OPEC countries have implemented ambitious development plans. These plans have raised the spending capacities of these countries on imported technology and goods, so that any substantial reduction of oil exports might directly affect the local economy. Saudi Arabia, for example, was believed a few years ago to be unable to make use of most of its oil receipts, but has now reached a position in which it spends almost 70% of its huge oil receipts on fixed capital formation. The Saudi budget, less than $3 billion in 1972, is now approaching $43.5 billion, and while actual spending was lower than allocation before 1973 it now tends to exceed it.

The Disposal of Foreign Exchange Earnings

The disposal of foreign exchange earnings should be the second opportunity enjoyed by the OPEC countries, in varying degrees according to production capacity, population and territory. For the purpose of our analysis we could divide OPEC members into two groups. One which accumulates foreign currency surpluses and is condemned to do so to satisfy the energy needs of the industrialised countries: these are the OPEC countries whose population and territory are small. They have no alternative but to accumulate foreign financial assets and try to manage them in the best possible way. Qatar, Abu-Dhabi and Kuwait are in this situation. The other group consists of countries capable of increasing expenditure on foreign goods and services to meet the needs of their large population, so that an accumulation of surpluses, if any, would only be a temporary phenomenon. In view of what has already been said of Saudi Arabia, it is difficult to classify this country in either of these two categories, especially if we consider its great potential capacity to adapt its oil production rate. It is interesting to note that between 1973 and 1977 Saudi Arabia accumulated huge amounts of foreign assets, standing at $60 billion by the end of 1978, although presently and until the last increase in oil prices, its expenditures were growing nearer to its oil receipts.

Let us now study the opportunities and constraints of each group of OPEC countries; on the one hand those having no other alternative than to accumulate financial surpluses and, on the other hand, those using their oil receipts to promote an accelerated transfer of technology.

The Accumulation of Financial Surpluses. Since the international monetary system has lost its gold standard, the value of any accumulation of international liquidity is no longer linked to any real value. As is known, OPEC countries, and more broadly LDCs, do not possess important quantities of gold, in fact 3.55% and 9.12% respectively of official gold holdings (socialist countries excluded). As the price of gold goes up, industrialised countries enjoy a tremendous increase in the value of the gold content of their foreign assets, while those OPEC countries compelled to maintain large balances in US dollars incur severe losses. The more so because any weakness in the US dollar will tend to boost the price of gold.

It is true that balances maintained in US dollars since the early 1970s have generated interest varying between 6 and 12%, but these rates of interest have always been lower than the respective rates of inflation. In this respect it is interesting to note that an amount of $35, earning an average compound interest rate of 8.5% per annum since 1970 would have only reached $67.22 in 1978, while one ounce of gold bought at $35 in 1970 would have been worth $300; an average compound annual interest rate of 31% on the dollar would have been needed to match this performance.

These few examples show that the accumulation of financial surpluses within the present international monetary system does not represent a real opportunity. The exchange of a real asset available in limited quantities, having a strategic importance and a growing international value, against financial assets with fluctuating values and a rate of return inferior to the rate of inflation, in no way represents an economic opportunity. It represents rather a net loss, a depletion of the national wealth of the oil-exporting countries.

In addition, the concentration of these financial assets in a few industrialised countries, has its disadvantages. In the case of a political or economic conflict of interest with one or several of the recipient countries, this accumulation of financial assets and its concentration would prove to be a very stringent constraint on the freedom of oil exporters. There is no doubt that this constraint is already in operation in the relations between the West, particularly the US, and the Gulf countries.

For some of the OPEC countries there is no alternative to the accumulation of foreign financial assets, given the energy needs of industrial countries and the rising price of oil on the one hand, and the absence of integrated economic development at the regional level on the other hand. Such an integrated plan would allow regional distribution of the oil wealth, thus maximizing its local effects. The absence of regional economic integration stems from the state of underdevelopment of whole areas surrounding OPEC countries, so that investing foreign assets in the industrialised world looks safer. This situation contributes to the perpetuation of the state of dependency characterising underdeveloped countries.

OPEC countries with large populations and/or territory, are not accumulating huge amounts of foreign assets; some of them are even net borrowers on the international financial markets. Oil receipts of these countries are devoted to accelerating the path of modernisation, i.e. building infrastructures, coping with the high rate of urbanisation, and promoting industrialisation in order to secure employment for the growing number of active members of the population. In brief, we could say that financial assets are used as a tool to promote rapid transfer of technology, as most of the oil receipts are used to import foreign technology in the form of goods and services.

Oil Receipts and Transfer of Technology. The present mechanism of technology transfer from industrialised countries to LDCs has been subject to growing criticism in recent developmental analysis. First, several UNCTAD studies have shown the importance of additional costs that LDCs have to bear when dealing with transnational firms.[1] The practice of overpricing imports of intermediate products and equipment is becoming wider and wider. According to UNCTAD's findings which were, incidentally, based on a sample of the few LDCs that gave sufficient data on their procedures for the importation of technology, direct costs involved in technology transfer (i.e. payments of patents, licences, know-how and other technical services) amounted in 1968 to $1.5 billion annually. This represented 0.5% of the LDC's GDP and 4.5% of their exports. However, the study showed that, according to the different countries surveyed, the rate of increase in these payments was between 1 and 6 times higher than the rate of increase of GDP. UNCTAD has also estimated that by the end of the 1970s these payments could reach $9 billion, representing 15% of the value of LDC's exports. This estimate was made before the present inflation pushed the price of equipment and the salaries of technical assistants to a very high level.[2]

In addition to these costs, one should take into account specific engineering and R & D costs linked with solving unexpected problems and adapting or modifying the technology, pre-start-up training costs and excess manufacturing costs during the initial phase to achieve the specification performances. All these are transfer costs unembodied in the price of equipment, the cost of the detailed engineering design of the plant or the project.[3] Data based on a sample of 26 fairly recent cases of international transfers of technology in petro-chemicals and machinery showed that on average these extra costs amounted to 19% of total project costs (from 2% to 59% according to projects).[4]

Foreign payments under technical assistance agreements of individual employment contracts with foreign expatriates is the last item to be included in the financial outflows involved in technology transfer. No data are available, for such costs are not always recorded in the balance of payments, and when recorded they are aggregated with various other items.[5]

[1] For a few of the relevant studies see: UNCTAD (1971/1972) *Transfer of Technology* (TD/106); UNCTAD (1972) *Guidelines for the Study of the Transfer of Technology to Developing Countries: A Study by the UNCTAD Secretariat* (TD/B/AC 11/9), UNCTAD (1975) *Major Issues Arising from the Transfer of Technology to Developing Countries: Report by the UNCTAD Secretariat* (TD/B/AC 11/10). Some recent case studies on LDCs include UNCTAD (1974) *Major Issues Arising from the Transfer of Technology: A Case Study of Spain* (TD/B/AC 11/17); of Hungary (TD/B/AC 11/18); of Chile (TD/B/AC 11/20); of Ethiopia (TD/B/AC 11/21); and a 1975 case study of Sri Lanka (TD/B/C 6/6).

[2] See note 1 TD/106 and TD/B/AC 11/10.

[3] D. J. Teece: 'Technology Transfer by Multinational Firms: The Resource Cost of Transferring Technological Know-how', *The Economic Journal*, 87 (1977), 242–261.

[4] Ibid.

[5] On all these points, see G. Corm, 'Finance and Technology Transfer', in *Technology Transfer and Change in the Arab World*, ed. by A. B. Zahlan, Pergamon Press, 1973.

All these extra costs are reflected in the dramatic increase in the burden of the import of services of most LDCs, and in particular of OPEC countries. In effect, between 1970 and 1977, 26 LDCs – including OPEC countries – have seen their import bill for services rocket from $15 billion to $73.3 billion, almost a fivefold increase. For OPEC countries this has represented an increase from $7.9 billion in 1970 to $44 billion in 1977, or a foreign services bill increasing at an average annual rate of 28% (see Table 1).

Secondly, the present mechanism of technology transfer – either in the form of turnkey contract or in the form of joint venture – is not only a source of increased technical dependency on transnational firms, but is also paralysing the emergence of local engineering capacities. Is it not strange that some OPEC countries in which oil has now been flowing for approximately half a century have still not succeeded in building satisfactory engineering capacities in the field of oil and related industries? Without such capacities the disposal of energy remains a captive enclave in the local economy, and the best that could be achieved is a maximisation of the financial resources generated by the oil sector. These financial resources, when spent massively on technology transfer – as it now operates – tend to aggravate the dependency of the countries concerned on imported goods and services. In fact the relatively easy access to foreign exchange encourages this dependency on borrowed and immediately available technology. As one scholar wrote, 'learning-by-doing becomes learning-by-watching'.[1] Such easy access constitutes a disincentive to the promotion of local capacities in the field of technology and research and development, and develops a technological consumerism based on imported ready-made and unpackable technological hardware. These negative aspects of the oil revenues have recently been emphasised at the First Arab Energy Conference held in Abu-Dhabi from March 4 to March 8 1979.[2]

The dramatic decline in agricultural productivity has also been a common feature of most OPEC countries since the increase in oil revenues, in spite of the big potential existing in countries like Iraq, Nigeria, Iran, Indonesia, and Venezuela. This decline is due to the concentration of investments in the urban areas through importing sophisticated and expensive technological hardware. Investments in

[1] R. I. McKinnon, *Money and Capital in Economic Development*, The Brookings Institution, Washington, 1973.

[2] See in particular, Yusif A-Sayegh, 'The Social Cost of Oil Revenues', in *Arab Report and Memo*, Vol. 3, No. 20, May 14, 1979; and R. Mabro, 'Oil receipts and the Cost of Economic and Social Development', in *Al Mustakbal Al-Arabi*, No. 6, 1979.

Table 1 Services imports of developing countries with highest growth rates in these imports (in millions of US dollars)

OPEC	*Average annual growth rate 70/77*	*1970*	*1972*	*1974*	*1976*	*1977*
Algeria	31%	414	419	1,017	2,107	2,705
Saudi Arabia	41%	1,208	1,619	4,394	11,425	13,674
Iraq	20%	679	335	1,927	2,000*	2,400*
Indonesia	34%	449	777	2,281	2,881	3,525
Iran	25%	1,466	2,086	3,156	6,677	7,071
Libya	13%	1,059	1,012	1,864	2,497	2,489
Venezuela	18%	1,132	1,202	2,115	2,326	3,596
Nigeria	26%	857	1,310	2,643	3,884	4,266
Gabon	29%	117	177	391	687	687
Ecuador	75%	139	176	451	415	645
United Arab Emirates*	31%	100	200	400	600	650
Qatar*	29%	100	200	350	550	600
Kuwait	36%	200*	450*	700*	1,167	1,682
Total OPEC countries	28%	7,920	9,963	21,689	37,216	43,990
Other Countries						
Brazil	27%	1,192	2,011	4,371	5,499	6,454
South Korea	33%	376	516	1,147	1,718	2,761
Ivory Coast	23%	260	364	528	836	1,080
Egypt	24%	369	417	829	1,260	1,681
Spain	20%	1,123	1,714	2,979	3,685	4,019
Mexico	14%	1,794	2,187	3,574	4,867	4,581
Turkey	28%	331	532	924	1,360	1,841
Philippines	19%	399	401	867	1,128	1,335
Singapore	29%	265	515	1,139	1,353	1,564
Sudan	16%	99	113	200	268	274
Tunisia	20%	149	202	369	495	521
Jugoslavia	22%	681	939	1,610	2,211	2,753
Syria	30%	73	126	197	248	451
Total OPEC and other countries	25%	15,031	20,000	40,423	62,144	73,305

*Estimates.
Source: IMF, *International Financial Statistics*. July 1977 and June 1979.

agriculture are targetting towards creating agro-business with the help of transnational firms, an unrealistic goal in relation to the various basic unsatisfied needs of small landowners and agricultural workers.

The conclusion of this analysis is that the disposal by OPEC countries of a relatively abundant flow of foreign exchange earnings[1]

[1] It should be remembered here that this flow of foreign exchange is not greater than

cannot be considered, given the present world economic system, as a pure opportunity to secure balanced, inward-looking economic and social development.

No doubt, the situation would have been different if OPEC countries were already on the road to mature industrialisation in a positive regional and international environment. But OPEC countries have suffered between 1950 and 1968 from stagnant oil prices, and were not in a position to promote vigorous industrial policies and to develop the necessary network of infrastructure. As Hollis B. Chenery remarked in his article 'Restructuring the World Economy' published in *Foreign Affairs* in January 1975, it is the suddenness rather than the magnitude of the oil price increase that created the challenge for the industrialised world. This suddenness, in my opinion, was as much a challenge to OPEC countries, given the past stagnancy of their underdeveloped economies and the highly unfavourable international context. In this same article, Chenery remarks that a 3% annual increase in oil prices between 1950 and 1975 would have allowed gradual adjustments and much better use of energy resources in industrialised countries. Continuing his analysis, the author states: 'Instead, the progressive cheapening of oil for 20 years led to its wasteful use – particularly in the US – and postponed the development of other energy resources.' In fact this would have also completely changed the picture for OPEC countries in terms of their local development which is now experiencing, after a period of complete stagnancy of foreign resources, a relative glut of foreign financial assets in a local environment not yet prepared to absorb them.

The mis-management of world energy resources to which Chenery alludes is a responsibility that the industrialised world bears, and constitutes the core of the so-called oil-crisis and the challenge it now creates to the various partners involved. It is in this context that opportunities and constraints facing OPEC countries should be dealt with, especially since there is no doubt that the fragile capacities of OPEC countries, still largely underdeveloped countries, cannot be compared to the powerful economic structures with which the industrialised countries can face the present challenge of world energy markets.

But before examining this last point, let us turn to the third and last opportunity oil has created for OPEC countries.

the amount earned by one industrialised country like Germany from its merchandise exports, standing at $138 billion in 1979.

The Increase in the Regional and International Role of OPEC Countries

At first sight, one feels in a much safer position to describe the opportunities provided by oil to OPEC countries in relation to their increased influence in world affairs. Here again, however, one should be very cautious. No doubt countries like Saudi Arabia, Iran, Iraq and Venezuela have enjoyed a substantial influence in world affairs since the rise in oil prices. Their international role is enhanced now by the regional influence each of these countries has developed. They also play an important role in most international institutions concerned with Third World Development, due to their increasing financial contribution to international aid. In addition, Arab banks and financial institutions have become a force to be reckoned with on the international financial markets. But looking in more detail at what has been effectively achieved in terms of changing the present international world order, one can realise that very few concrete changes have taken place. We have already seen that the North-South Dialogue did not achieve any positive results, and LDCs, including OPEC countries, are still subject to the same economic international constraints paralysing their local development efforts. The only concrete achievement was the permanent seat given to Saudi Arabia on the board of directors of the IMF. However, all important international economic and financial decisions are still taken in isolation by the closed club of the 5 to 10 largest industrial countries.

On the political front on the other hand, one can see that the vital issue of the Palestinian right to self-determination, which is of great concern for the future stability of Arab OPEC countries, has not progressed since the 1973 war and since the partial and temporary oil embargo that was implemented for a few months but then waived before any real solution of the Palestinian plight had been achieved.

In fact, as stated in our introduction, the nature of oil as a strategic raw material for the continued growth of the industrialised countries, and the formidable pressures to which OPEC countries are thus exposed by these countries, tends to favour the present political and economic world status quo. OPEC countries, given their fragile economic situation as largely underdeveloped countries, are in no position to challenge in depth the world status quo. Their successful action on oil prices is attributable, in addition to their solidarity, to a clear market trend in which easily available oil resources are becoming more and more scarce, while alternative energy sources are not developed sufficiently to cope with the world demand.

Towards a New Co-operation between Oil-exporting Countries and Consumer Countries

There is no doubt that oil markets should be organised in such a way as to preserve both the interests of OPEC countries and consumer countries. However, this could not take place if the industrialised countries continue to behave as in the last few years, their main preoccupation being to dislocate OPEC solidarity on the one hand and to develop antagonism and contradictions between oil-exporting and non-oil-exporting LDCs on the other hand. It is time to face the fact that world energy distribution has to be organised so that the long-term interests of the oil exporters will be safeguarded. After all, oil is a strategic raw material not only for industrialised countries, but also for the exporters themselves because no real industrialisation could take place in these countries without the future availability of easily accessible energy resources. The present exchange of energy assets against financial assets or technological hardware is without doubt a losing trade for the oil exporters and will badly hurt the well-being of future generations. Financial assets, in the present context of high inflation and absence of real standard of value, are a constraint more than an opportunity. Importing technological hardware without the prerequisite of a successful transfer and mastering of such technology, is also a constraint bringing dramatic social, economic and financial local wastage.

These are issues that must be dealt with in order to establish the foundation of a fruitful dialogue between oil exporters and industrialised oil consumers. In this context guarantees could be supplied by the industrialised countries concerning the future supplies of energy to OPEC countries, still largely underdeveloped and currently depleting their oil reserves for the sake of rich countries. What form such guarantees should take is something needing detailed studies, but given that OPEC countries are depleting precious non-renewable energy resources against their future energy needs – one should consider the present oil exports from OPEC countries as a form of energy lending to industrialised countries, to be reimbursed at a later stage.

There is no doubt also that the time has come for the industrialised countries and their transnational firms to devote more serious efforts and research into supplying OPEC countries and other LDCs with technology in a form that will seriously promote the local engineering capacities of these countries.

In any case, it appears clear that the only alternative to a renewed and fruitful dialogue between North and South will be for OPEC countries to continue to try to maximize the financial benefits reaped from growing imbalances in the world energy market, in a game in which all partners involved will ultimately be losing. Let us hope that some wisdom will soon prevail and that OPEC countries and more generally LDCs, will be permitted to free themselves from the constraints of the present world order which is paralysing their efforts towards the development of their countries and the well-being of their population.

26 OIL AND REGIONAL CO-OPERATION AMONG THE ARAB COUNTRIES

Ali Attiga

The 1973–74 oil price increase initiated a series of fundamental changes in the world energy markets. Although many of these changes represented basic adjustments to the hard realities of the world energy balance, the sudden and sharp increase in oil prices became the cause for harsh attacks on OPEC and its member countries. The fact that the price of oil had declined in both monetary and real terms during the 25 years following World War II was generally ignored.

Fortunately, as the debate over oil prices gained world-wide importance and became a central issue in the search for adequate national economic growth and a viable world economic order, it became increasingly clear that the availability of oil supply was perhaps more important in the long run than the level of oil prices. The world began to realize for the first time that known oil reserves could not be relied upon as a cheap and convenient source of energy for an indefinite period.

Regardless of cost or price, the physical availability of oil is not sufficient to sustain present and future world consumption. This highly significant and rather belated realization means that known oil resources can at best be only a transitional source of energy and industrial raw materials. It also means that higher oil prices are essential for the efficient exploitation and use of scarce oil resources, whether in the OPEC area or elsewhere.

Oil Provides a Common Bridge

In spite of the basic differences between oil exporting and importing countries over the price and supply of oil, there is a wide consensus that existing oil resources provide the only energy bridge over which the world may be able to pass from its present state of energy mix to a new and much desired era of renewable energy resources, with coal and nuclear power providing the main substitutes for oil during the transitional period.

Although the transitional period has significant consequences for both oil importing and exporting countries, only those affecting the latter will be discussed in this paper. The main focus of discussion will deal with the essential need for regional co-operation during the transitional period.

Foreign-exchange revenues from the export of oil provide the OPEC countries with the purchasing power for current consumption and the necessary capital for investment to provide for future consumption. Thus, oil revenues provide the only bridge over which the oil-exporting countries can hope to pass from their present state of heavy dependence on oil resources to an era of sustained economic growth without oil. In this regard, oil-exporting and importing countries have a fundamental interest in gradually decreasing their dependence on depletable oil resources by making effective use of the transitional period to develop alternative sources of energy for the world at large, and alternative sources of wealth and income for the oil-exporting countries in particular. But, unfortunately, vested interests in the advanced oil-importing countries are constantly attempting to manage the transitional period in such a way as to. relegate the long-term interests of the oil-exporting countries to the short-term economic and political problems associated with business fluctuations and social affluence in the advanced countries. This tendency has so far greatly limited the scope for effective co-operation between oil-exporting and importing countries.

Faced with constant and often vicious attacks on their inherent rights as producers to determine the production and prices of their oil, the OPEC countries have had no alternative but to continue to defend their rights, and to hope to use their oil resources to attain sustainable economic growth for the post-oil era.

The real question is, can they realize such an objective? What are the necessary conditions for reaching it? Can they reasonably achieve their long-term economic development goals by working separately or even in competition with one another? What are the opportunities and limitations of regional co-operation in assisting them in this task?

Co-operation Within the Oil Sector

Any attempt to answer these questions should, in my opinion, begin with the need for co-operation within the oil sector itself. Prior to the 1973–74 oil price adjustment and the take-over of oil resources by the producing countries, the international oil industry was largely vertically and horizontally integrated. Thus, control and management of upstream and downstream operations were in the hands of a few transnational corporations, which planned their activities on a regional basis, regardless of national borders or political differences between sovereign states.

This structure of the international oil industry was radically altered when the ownership and control of upstream operations became the responsibility of the oil-exporting countries. A sharp division was brought about in jurisdiction and authority concerning upstream and downstream integration, co-ordination, planning and management. Yet much of the technology and skills needed for the operation of upstream activities remained with the former concession holders, which also remained in almost total control of downstream operations.

The basic control and management of the oil industry was an still is in the hands of those who control the downstream operations and outlets. These are largely owned and managed by transnational oil corporations belonging to the advanced oil-importing countries. They generally manage their operations on regional and integrated bases. Their commercial interest for co-ordination has been greatly reinforced by increasing co-operation among their home governments in the field of energy.

While the take-over of upstream operations by the oil-exporting countries has undoubtedly weakened the control of the foreign oil companies, it has also created a new situation which could, if left unchecked, greatly work to the advantage of these companies and their native countries. The present fragmentation of ownerships, management and decision-making of upstream operations between independent sovereign states can have serious consequences for the efficient development and management of their oil resources. Oil exploration, production, transportation and marketing need a certain degree of planning and co-ordination which takes into account the individual and collective interests of the oil exporters. It is common knowledge that at present there is little or no co-operation among oil exporters in these areas. OPEC performance has so far been largely limited to collective agreement on the determination of a basic price for market crude. At times, even that limited objective has been very difficult to achieve.

Common ownership of oil facilities and industries (except for what was inherited from past concessions), or the co-ordination of development policies of these industries, are largely non-existent among the oil-exporting countries. In fact, even the exchange of technical and economic data is encountering increasing resistance for bureaucratic and political considerations. Without information concerning the individual and overall positions of their technical needs, resource potentials and development policies, it is relatively easy for the transnational oil companies and others to gain substantial power over the policy options of any individual oil exporter. This kind of situation also makes it easy for the advanced oil-importing countries to maintain the status quo with regard to the producers' share in downstream operations.

Fortunately, there is growing awareness of the danger of working alone in the planning and management of the oil sector. OPEC is beginning to extend its activities to include such important areas as co-operation among the national oil companies of its member countries and the promotion of international relations with the developed countries. In order to contribute to the economic and social development of energy-deficit developing countries, it has established a special institution affiliated with OPEC.

The OPEC Special Fund represents a new dimension of co-operation between OPEC and other developing countries. Although its function has so far been limited to the collective assistance of developing countries, it should be stressed that such an institution can also be used to promote economic co-operation and policy considerations among its members. Indeed, it can be argued that the OPEC countries need this kind of co-operation as a prerequisite for their success in the management of their oil industries and the development of their economies, both of which are essential for maintaining OPEC assistance to developing countries.

The Case of OAPEC and Regional Co-operation

While there is a wide scope for co-operation within OPEC, its chief limitation is the diversity of its members in terms of geographic location and socio-economic conditions and political affiliations. But within OPEC there are seven Arab countries whose conditions are theoretically more favourable to regional co-operation. From only three Arab members of OPEC that OAPEC established in January 1968, the organization has grown to ten Arab countries. As a regional institution it includes countries with contiguous borders (except where broken up by the Israeli occupation of Palestine) and stretches from Algeria in the West to Iraq in the East. It has more than 70% of the population of the Arab world and generates more than 80% of its GNP. OAPEC members constitute a region within the Arab world which shares common values and social institutions with other Arab countries that are not members.

The Arab countries as a group have many historical, cultural, religious and linguistic ties, besides their complementary natural and human resources.

OAPEC deals only with hydrocarbons and its main objectives are as follows:

(a) The promotion of the co-ordination of the petroleum economic policies of its members and their co-operation in working out solutions to problems facing them in the petroleum industry.
(b) Exchange of information, harmonization of their relevant legal systems in the oil sector, and the promotion of training and exchange of expertise and employment opportunities in the petroleum industries of its member countries.
(c) The use of the resources and common potentialities of its member countries in establishing joint projects in various phases of the oil industry in which all or some of OAPEC members may participate. Such joint projects take the form of commercial companies aimed at creating among other things, a common Arab capability in both upstream and downstream activities of the oil industry.

OAPEC has created four joint companies, one training institute and it is working towards the formal establishment of a Judicial Board. It is also undertaking studies for the establishment of an engineering consultant firm and a number of projects in the petrochemical industry such as lube oils, detergents, synthetic rubber and carbon black for domestic consumption. OAPEC is active in promoting regional co-operation in the field of energy for domestic consumption in the Arab countries. It is also engaged in promoting better international relations between oil exporters and importers through the dissemination of information and the organization of direct contacts concerning the development needs of less developed countries, including the oil exporters.

The main objectives of the OAPEC companies are to enable their member countries to acquire common capabilities and experience in the management of their oil resources. Two of the four companies are concerned with oil transportation and ship-repair facilities and skills. The third company deals with the promotion of investment of petroleum industries within the OAPEC member countries and the Arab world as a whole. Its main objective is to promote the development of hydrocarbon resources, especially downstream activities on a regional basis. For this purpose, it is required to give priority to equity participation in joint oil and gas projects. It can also provide such projects with direct loans and help to mobilize additional financial resources for their initial establishment or expansion.

The fourth OAPEC-sponsored company operates in the general field of petroleum services associated with exploration and production. It is a holding company with plans to establish a number of subsidiary companies either wholly owned or in partnership with foreign companies possessing well-established technology. It has started with the establishment of a joint drilling company, which should be in operation before the end of 1979.

OAPEC-sponsored companies are legally incorporated as independent companies operating under the supervision and direction of their shareholders and boards of directors. OAPEC as such has no direct or indirect control or supervision over the policy directions or administrative performance of these companies. Theoretically, they are expected to follow the overall objectives of OAPEC as specified in their basic statutes. In practice, however, it is rather difficult to ensure this kind of policy direction without regular consultative meetings and policy guidelines from the OAPEC Council of Ministers. So far these arrangements have not been institutionalized, except for an annual routine meeting between the OAPEC Secretariat and the company managers.

While these companies taken together with their sponsoring organization provide an adequate institutional framework for regional co-operation in the development of hydrocarbon resources, their actual performances are greatly limited by the tendency of their member countries to plan and manage their oil industry individually and, at times, even in competition with each other. This tendency has resulted in the neglect of joint projects and common policy formulation in the petroleum sector. In fact, it is common for a member country to promote its own separate projects and facilities, while not doing the same for the joint ventures. Yet we have already noted that real and lasting success in the management and development of oil resources requires regional co-operation among the producers for exports and for domestic markets.

The Arab Petroleum Training Institute was formally established in 1979. Its main objective is the training of prospective instructors for teaching in the training institutes of their respective countries. It is also expected to promote the production of teaching materials and various aids, as well as the use of Arabic as the language of instruction. The Institute is expected to become a regional centre for technical meetings, seminars and special short courses designed to promote professional contacts and efficiency among the member countries. It is open to the nationals of all Arab countries.

Regional Co-operation in the Non-Oil Sectors

Although the non-oil sectors are outside the jurisdiction of OAPEC, oil revenues play a decisive role in promoting co-operation in these sectors. Prior to 1973, oil wealth tended to have a divisive impact on relations between oil and non-oil Arab countries. Beginning with the colonial history of oil concessions and the world power struggle, associated with that history, the Arab world became divided into separate political and legal entities with strong local and foreign vested interests protecting the new political and social situation. At that time oil revenues were just enough to create divisive trends, but not enough to provide for significant aid and assistance designed to initiate trends favouring regional co-operation within the Arab world. But after the 1973–74 oil price adjustment it became possible for the oil states to provide bilateral and multilateral assistance to other Arab countries. Many national and regional development funds, banks and companies have been established during the last 5 years. Billions of dollars have been invested in these institutions and in other national and bilateral development projects. However, the regional aspect of intra-Arab aid has so far been rather limited.

What are the areas most suitable for regional co-operation within the Arab world? Theoretically, the scope is very wide. It can cover many areas of a regional nature and mutual interest where a common approach is essential for significant results. Perhaps the following areas offer the best opportunities for regional co-operation.

Transportation and Communications. There is a great and pressing need to develop and maintain regional networks for modern land, sea and air transportation within and between the Arab countries. The Arab Fund for Economic and Social Development has estimated that a minimum of 20,000 km in new inter-Arab roads and highways are required to meet prospective needs over the next two decades, at a cost of about $10 billion. The development of regional railways, air transportation, and national and regional telecommunications also needs billions of dollars which can only be mobilized through regional co-operation.

Agricultural Development. It is well-known that the Arab world is heavily dependent on food imports. With its high population growth and increasing levels of income, its future food requirements are increasing rapidly and will continue to do so over the next 10 years. At present no Arab country is self-sufficient in food production, and all of the Arab oil-exporting countries are heavily dependent on food imports. The present value of net food imports of the Arab world is around $8–10 billion. Yet some Arab countries – Sudan, Iraq, Syria, Algeria, Morocco, Tunisia and even Libyan Jamahiria – have considerable agricultural potential, while others (the oil-exporting and food deficit countries) have substantial financial resources which can contribute to regional co-operation in the field of agriculture and livestock production and trade. Thus, only through regional co-operation can the food deficit problems of the Arab world be reasonably tackled.

Development of Domestic Energy Resources. Although the Arab world contributes about 50% of world oil exports, the majority of the Arab countries are either energy-deficit or barely energy balanced in terms of their domestic resources. Only one-third of the population of the Arab world, may be energy secure for the next 50 years. Even this security is based on depletable oil and gas, which are also required for export. Thus, despite the present state of energy surplus in the oil-exporting countries, the future energy needs of the Arab region as a whole are very serious. It is true that there are some Arab countries that will remain energy balanced for the rest of this century, but these countries – Egypt, Syria and Tunisia – are exporting much of their limited reserves. As for the energy-deficit countries, they urgently need assistance through regional co-operation for present consumption and for the development of their domestic energy resources. At present the Arab countries as a group possess only 9% of world energy resources, but they provide 16% of global energy consumption.

There is a wide scope for regional co-operation in the intensification of exploration for oil, gas and solid fuels. The same is true for the production and distribution of electric power supplies which can utilize flared gas in the oil-exporting countries and hydropower in some others (Sudan, Morocco and Iraq). But this kind of co-operation will require, among other things, many billions of dollars in capital investment, which can again only be mobilized through regional co-operation. There is also scope for regional co-operation in the applied research and technology of alternative sources of energy, especially solar. The First Arab Energy Conference held in Abu Dhabi in March 1979 provided a suitable forum for discussing the problems and opportunities for regional co-operation in the field of energy production, distribution and consumption. One of the important recommendations of that Conference was the establishment of a Joint Arab Energy Committee to promote and co-ordinate regional co-operation in the energy field. OAPEC and the Arab

Fund are currently studying in consultation with their member governments ways and means to implement the establishment of such a committee.

Banking, Finance and Investment. There is a great need for increasing regional co-operation in the mobilization and investment of capital funds within the Arab world. Although a number of joint banks and companies have been established during the last 5 years, the scope is still wide for regional investments. There are still many barriers to the movement of capital across national borders. Yet the Arab world is today one of the main saving centres for financing the investment needs of the advanced industrial countries through loans and bank deposits whose purchasing power is constantly being diminished by inflation and currency fluctuations.

Education, Research and Training. General and specialized education, research and training are still quite backward in all or most of the Arab countries. Although considerable progress has been achieved in these areas during the last 20 years, present and future needs for qualified professionals and skilled labour are so great that individual country efforts are not adequate to cope with the situation. There are great opportunities for co-operation among the Arab countries in the promotion of adult literacy, technical training and specialized education, such as engineering and medicine. Countries with substantial financial resources can help to finance common institutions in these fields. Such an approach would greatly accelerate the development of human resources in the Arab world. The existence of a common language, history and culture among the Arab countries offer considerable advantages for the success of this kind of regional co-operation. In terms of investment needs, this area can absorb substantial capital for the construction of schools and other teaching facilities.

Applied research and technology in various development fields are badly needed in all of the Arab countries. Here again individual efforts on a country basis are not sufficient. There is a wide scope for regional research institutions in agriculture and fishery, public health, energy, minerals and other areas of common interest. Present oil revenues can mobilize resources for joint projects in these fields. The present level of expenditure for research and education is not adequate to cope with the great development problems of the OAPEC region.

Problems and Limitations

Political Division and Differences. There are now 21 Arab countries all with full sovereign rights. The basic division took place during periods of foreign occupation, especially European competition and domination. In Libya, Lebanon, Iraq, Sudan and other Arab countries, there were attempts to create more than one state. After independence political differences concerning who should govern developed into major conflicts and competitions, both within and between different states.

Excessive Economic Fragmentation. The national economies of most Arab countries are too small for sustained and balanced economic growth on an individual basis. Each is heavily dependent on one sector for domestic needs and exports. But vested interests have been built around these fragmented economies which resist serious attempts for regional co-operation, not to mention integration.

Each country has developed a system of isolating its economy from its neighbours by a set of economic and administrative barriers to the free movement of natural and human resources. The limited movements which take place at present are on a temporary basis and highly regulated.

Foreign Conquest and Foreign Penetration and Influences. From 1832 until the end of World War I the Arab world was taken over by military conquest from the Ottoman Empire which by that time ceased to exist. The three Western Colonial powers which competed for territorial possessions in the Arab world were France, Britain and Italy, while Spain was trying hard to maintain its minor possessions in Morocco and the Western Sahara desert. Each colonial power divided its possessions into separate administrative units with formal borders and other barriers between these units. France and Italy regarded their possessions as part of the motherland and Britain handed over Palestine to the Zionist movement.

With the end of direct colonialism and the emergence of independent states the Arab world became politically divided in accordance with the border arrangements that were established by the colonial powers. Strong economic and cultural relations were also established between these new states and their former colonial powers. This kind of inherited relationship created a new kind of dependence on the former colonial countries, which was detrimental to regional co-operation within the Arab world. With the emergence of oil revenues the oil-exporting countries became more and more dependent on foreign markets for their exports and imports. The consumption sectors were and still are becoming integrated with the productive sectors of the advanced industrial countries, especially their former colonial powers. Thus international economic dependence on imports in the oil countries and excessive external economic fragmentation within the Arab world as a whole make it very difficult to formulate or implement regional schemes based on the resource complementary of the area. Today the prevailing world power structure (East vs. West, US vs. Western Europe and Japan, the North–South conflicts) are reinforcing the existing political and economic fragmentation in the Arab world.

Conclusion

Regional co-operation is highly essential for the viable development of the oil-exporting countries. It is also essential for the most obvious needs of the other Arab countries. Even the proper development and management of the hydrocarbon resources are dependent upon adequate regional co-operation and policy co-ordination among oil-exporting countries. For the region as a whole there are sufficient complementary resources for sustained economic and social development. For individual Arab countries working alone there are serious limitations on their balanced economic and social development regardless of how much capital or manpower or other natural resources they may possess.

27 OPEC AND THE ENERGY CRISIS

Humberto Calderon Berti

In this paper I intend to refer briefly to the world energy situation and the resulting energy crisis, as it has come to be known. I will also talk about OPEC, about its real significance and importance; about the problems of energy and development which follow from the international situation, and about possible modes of co-operation towards the solution of such problems.

The energy problem as we know it today can be explained almost totally on the basis of three factors: proven reserves of oil, production of oil and consumption of oil on a worldwide basis. In fact, this is so because the energy problem is, more than anything else, a problem of the relative scarcity of oil or liquid energy. Some people might add that a fourth factor is needed to complete the picture of the energy crisis, that is the price of oil. Price, however, is a result of supply and demand conditions, and the low price of oil probably caused a rapid increase in demand during three decades of this century, and the consequences are only now being realized.

At present the world is producing and consuming petroleum at a rate of some 22 billion barrels per year, which is more than the entire proven reserves of Venezuela. If we accept an estimate of total world proven reserves of 640 billion barrels of oil, this would yield a reserve to production ratio of the order of 29 years. This indicates the theoretical duration of world reserves at present rates of production and consumption. Supposing this figure of 29 years to be correct, it would be necessary to discover more than one 'Venezuela' each year, just to be able to maintain the world reserve to production ratio at that level. I think you will all agree that this is a fairly difficult if not impossible task, and that the figure of 29 years is not necessarily a desirable one, nor even close to what could be considered as optimum. On the contrary, it indicates what appears to be a rather short duration for present world reserves, and this gives an idea of the magnitude of the problem we are facing.

Considering it from another point of view, the figure of 29 years could be indicative of what I would like to call the transition period from an oil-based world economy, to whatever other energy form is to be used principally in the future. This transition period may turn out to be longer than 29 years, but it is unlikely that it will be any shorter. Like almost all transitions in history, this one too will be accompanied by difficulties, maladjustments, problems, and – if nothing is done – by crisis. I am personally convinced that the real energy crisis does not yet exist; what we have experienced during this decade are only preludes to the future energy crisis, which have been the result of political cries. Each political crisis has been accompanied by oil price increases and temporary price volatility throughout the world, both of which are indicative of the precarious balance of oil supply and demand on a worldwide basis. I expect, however, that during the coming decade of the 1980s such a precarious balance will turn into a shortage; world demand will overtake world supply potential and we will confront the makings of a real crisis.

It is in this context of an impending energy crisis that OPEC has acquired the necessary strength to make its voice heard on the international scene. OPEC existed for over 10 years before anyone paid any attention to its ideas. The international oil companies had existed for several decades and manipulated supply and prices in their own short-sighted interests, without regard to the future needs of the world or the interests of the oil-producing countries. In the particular case of Venezuela, there occurred a transfer of very real resources, real wealth in the form of oil, for over 50 years at prices which, both in nominal and real terms, were actually declining during the greater part of that period. The same can be said of other oil-exporting countries. What we observed during that period was a direct transfer of wealth from the oil producers to the industrialized nations, who built up their economies to a large degree on the basis of our cheap energy, while we were left in a position of economic indigence.

What we are seeing today is the opposite side of the same coin; the producing countries are now in a position to obtain a transfer of real resources from the industrialized countries which will enable us to build up our own economies. Not often has history provided such an example of international social and economic justice.

Our position is, however, not that of simple exploiters of an oligopolistic situation. Our position is much stronger than that, both economically and ethically, since it is derived from the real scarcity of an exhaustible resource, which constitutes the wealth of our countries and which should be used in the future only in its most valuable applications. The world would hardly be thinking in terms of conservation and developing alternative energy sources, were it not for the significant increases in oil prices that have occurred during this decade; and prices have increased significantly and have been supported by the marketplace simply because oil is a scarce and exhaustible resource. If it had not been for these price increases, I venture to say that the energy crisis to be faced by the world would be of a catastrophic magnitude, and perhaps not as manageable as it may yet turn out to be.

But OPEC is more than just oil and prices. As I mentioned previously, OPEC is an example of international economic justice which can serve as a model of what can be achieved through the solidary action of developing countries. OPEC is well aware of its role as one of the most important instruments for the development of Third World countries. Through diverse individual and collective measures of financial assistance OPEC has tried to alleviate some of the economic problems faced by these countries, particularly those related to increases of oil prices.

What I am trying to point out is that the OPEC member countries are an integral part of the developing world and will continue in this position in the future. There is no 'Fourth World' as some would like us to believe, and the manoeuvres by some industrialized countries to try and break the position of solidarity of all developing countries in diverse international forums have not succeeded, nor will they succeed in the future. Why? Because both OPEC member countries and other developing countries recognize their fundamentally common interests in international economic matters.

This is one of the reasons why OPEC established its Long-run Strategy Committee which is about to conclude its work. One of the main topics under consideration by this committee is the question of defining a long-run OPEC policy *vis-à-vis* the other developing countries of the world, and I trust that during the next ordinary OPEC conference to be held in Caracas in December 1979, we will be able to announce to the world the main thrust of such a policy of co-operation with the rest of the developing countries. Fully aware that OPEC member countries form part of the developing world and that the energy problem is part of a broader context which includes all international economic and development problems, we recognize that important efforts toward international co-operation must be undertaken with a view to establishing a stable and solid basis for economic growth and development.

Undoubtedly, the world economy finds itself at present in a difficult situation with high inflation rates, an unstable monetary system, increasing energy prices, unemployment and underdevelopment on a world-wide basis. These circumstances are certainly not conducive to the stability necessary for economic growth and development, but rather create an atmosphere of tension between developing and industrialized countries. These problems are not only related to oil and energy but include the whole gamut of international economic relations; oil only represents a small proportion of the total value of international trade. In spite of this, OPEC is aware of its important presence in the international economic system and realizes that there is much to be gained from a new concerted effort toward co-operation between developed and developing countries. But I must stress, this is on the understanding that the energy problem is not an isolated issue but something that belongs to a broader context of economic problems currently affecting the world. This has been the position of OPEC and of the rest of the developing countries, as far back as the nearly forgotten North–South Dialogue which took place a few years ago. This is a firm position that cannot and will not change, because as I previously stated, OPEC is quite conscious of its role as a promoter of the economic development of the developing countries as a whole.

Thus, any new effort towards a dialogue must cover a large variety of subjects beyond the energy question, such as the international monetary problems, the question of opening up markets for the manufactured goods of the developing countries, the need to facilitate transfer of technology under favourable conditions, etc. As far as the energy problem itself is concerned, we could discuss questions related to security and stability of supplies, the relation of oil prices to the prices of goods imported by the oil-producing countries, the development of additional conventional and non-conventional energy sources, etc.

As regards the price of oil specifically, I believe that the experience of the last few years indicates the convenience of having oil prices increase in the future at stable and predictable rates; this would introduce a necessary element of stability in the interntional petroleum markets and, hence, in the international economy. However, the oil price problem is not the only one. There is also the problem of general inflation rates on a world-wide scale which have offset, to a high degree, some of the increases in oil prices which occurred in the past. This type of vicious circle is undesirable from all points of view and I think it could constitute an important subject for a new dialogue. In my opinion, the important points here are twofold; first to guarantee the real purchasing power of a barrel of oil and, second, to allow predictable increases in the real price of oil, in accordance with the relative scarcity and exhaustible nature of this resource. If an understanding can be reached on these points, I am sure there will be adequate stability in the markets, conservation efforts will continue to be implemented, and the development of alternative energy sources will be fostered.

There are many other areas in the energy field where international co-operation could be very productive, and at the same time would be of great help in the solution of problems of development and of energy supplies. I am thinking, in particular, of a concerted international effort for the development of indigenous energy resources in many developing countries in the world. Many countries have large potential resources of oil, coal, geothermal and hydroelectric energy — to mention only the more obvious ones — but have not been able to tap these resources simply because of a lack of know-how, technology and financing. A concerted international effort in this direction would serve, not only to increase world energy resources, but also to liberate significant quantities of the ever more scarce oil supplies, which could then be used for those purposes where today it is truly irreplaceable — mainly in the transportation sectors. At the same time, such an endeavour would foster the development of many developing countries and would help alleviate their balance-of-payments problems by diminishing their oil imports. In the long run, the economic development of all countries is in the interest of the entire international community, since it would provide incentives and opportunities for additional international trade and industrial development.

I would like to conclude these remarks by saying that the OPEC member countries are prepared to reinitiate an international economic dialogue along the lines I have discussed and others that could be suggested. We are well aware of the fact that international problems are troublesome for all members of the world community, and that much is to be gained from a concerted effort toward international co-operation. We hope that the industrialized countries will also view the situation from this perspective, and that concrete efforts will be made to achieve the objectives of development and economic stability.

28 CONCEPTUAL PERSPECTIVE FOR A LONG-RANGE OIL PRODUCTION POLICY

Ali Khalifa Al Sabah

The views expressed in this paper are those of a citizen of an oil-producing state, concerned with the finite nature of oil reserves and, at the same time, with the vision of the unrepeatable opportunity this natural resource provides to improve the quality of life, economic, social and ecological, for citizens at present, in the future, and after its depletion.

It is within this specific framework, that I will present a long-range oil production scenario for an oil producer which takes into consideration the depletable nature of oil, though utilizing it as a conveyor belt for economic and social transformation and not just a crutch.

A Perspective on Oil and Energy

I wish to emphasize the words 'A Perspective', for, I believe that oil and energy are topics on which we need many informed views, a lot of dialogue, and fewer manifestos that sound as though the speaker is pronouncing the one and only word on this subject.

The proven reserves of the world are limited in nature, and their life span is not far out of sight. This self-evident fact is applicable whether a modest increase in oil consumption is assumed, or alternatively, even in the event of no increase in oil consumption. In the event that new oil explorations lead to new oil discoveries, and even with higher oil prices, and assuming an optimistic scenario for developing new alternative energy sources, still the possibility of extending the time horizon for the proved oil reserves is at best limited.

On the basis of present proved oil reserves (which is a more realistic concept than unproved), the world outside the Communist areas had oil reserves of 85 billion tons at the end of 1978. (I have intentionally taken the proved oil reserves in the world outside the Communist areas as they have been and will continue to represent the mainstream of internationally traded oil.) Assuming that annual economic growth could be maintained at 3% per annum, with only 1.5% annual increase in oil consumption (a highly suspect relationship ratio), and assuming 1 billion tons of oil discoveries per year over the next 10 years, this would mean a cumulative consumption of 100 billion tons of oil between 1979 and 2008 inclusive. In other words, all present known oil reserves plus the additional 10 billion tons to be discovered outside the Communist areas will have been exhausted. Assuming, on the other hand, no increase in oil consumption outside the Communist areas (as of 1978 approximately 2.5 billion tons per annum), the proved reserves of 85 billion tons plus the 10 billion tons to be discovered will be exhausted by the year 2018.

Recent oil discoveries have been running at just over 10 billion barrels a year. If we are to reach the end of the second quarter of the 21st century (a period of 70 years from 1980), the amount of oil which has to be discovered to meet this level of consumption has to be approximately five times the present known reserves. This is assuming a 1.5% growth rate in energy (and oil).

I believe that this is a tall order, since what has been discovered so far are 'easiest' reserves and future oil finds will prove more difficult to recover. This is because future finds will be located in smaller fields and/or harsher environments than previously and, therefore, their technical cost will be higher. Even if another Middle East is discovered, this would only mean that oil supplies would suffice for about 10 years of oil consumption; a North Sea can satisfy the world appetite for oil only for about a year and a half, and a North Slope for about six months. An oil discovery of 10 billion barrels per annum will only meet the needs of oil consumption outside the Communist areas for about six months.

Alternative energy sources, a subject which is currently being viewed as the panacea to redress the oil imbalance in the short and medium terms, will not substantially prolong the limited life span of oil reserves. This is because of the enormously long lead times involved in implementing energy technology – even proved technology, i.e. 10 to 12 years lead time to bring a nuclear plant on stream in the US. A significant volume of production from the US oil shale deposits will probably require more than 15 years, and a similar time frame would apply to gas and liquids from coal, etc.

Looking ahead from the vantage point of the early 1970s, an observer of the oil scene would have perceived clear, unmistakable and obtrusive smoke signals that the rate of oil discoveries was not sufficient to sustain the growth in oil consumption, as was the case in the 1950s and 1960s. The conspicuous sign of this structural change in the oil balance has been the steady increase in the prices of oil since the early 1970s (whether in real or monetary terms), and these increases came into being irrespective of OPEC policies, or, the wish of OPEC countries.

The lesson to be drawn from the limited nature of oil reserves, is the constraint this factor imposes on OPEC in its efforts to administer oil prices in the future. In other words, a fact of life which we shall have to recognize and accept in the future is that the limited life span of oil reserves will be reflected in the prices of oil in the market-place, irrespective of policies adopted by individual OPEC countries, or collectively, through OPEC. If we wish to engage ourselves in an exercise in self-deception, however, with the comforting illusion that things will get better, either through new oil discoveries or the development of new alternative sources of energy, we will only be helping to exacerbate the problem rather than attempting to solve it.

Scenario for Oil Production Policies

In the context of the foregoing oil/energy perspective, I would like to tackle the question of what the likely OPEC oil production policies are going to be, not in the medium or short run, but in the long run.

I envisage that OPEC countries' production policies will be attuned to the requirements of their economic and social development. We must remember that all OPEC Member States are in fact under-developed countries, dependent for their present and future well-being on a finite commodity, namely oil, a natural resource which is depletable. This raw material provides the cash-crop as in all developing countries, but, in our case, unfortunately, it is a non-renewable one. Further, as we all know, there are no short cuts to economic and social development. As such, the process of economic development cannot conceivably be completed in a short period during one, two or three five-year plans. In my opinion, this process will take the better part of a century at the very least. For, it involves not only the physical changes in the landscape, but, more basically, the fundamental transformation of the human being and the social fabric of a society.

This transformation should, in my opinion, be made at a measured pace. By a measured pace, I mean a not too slow and not too fast mobilization of available resources; natural, human and financial. For, if it is too slow, the critical mass necessary for the so-called 'take-off' stage of development may not be achieved; on the other hand, if it is too fast, it will cause inflation, bottlenecks, scarcities, market distortions, social dislocations and maldistribution of benefits that do not serve to maximize the social welfare. So, it has to be a 'measured' pace.

If the economic and social development of a third world country is perceived in terms of a long process of change and at a measured pace, and if this process is dependent on a finite oil resource as a conveyor belt, then it is reasonable to expect that the oil production policy of such a state should be tailored to fit within this broad context.

Within the foregoing framework, it is not unreasonable that an oil producer should have an oil policy based on a ratio of oil reserves to production around 100:1. (Some may argue: why not production rates implying reserves/output rates of say, 25 to 30 years? This ratio, in my opinion, is not acceptable for the simple reason that these states are underdeveloped, lacking a diversified and flexible productive structure.) Commercial oil companies, on the other hand, have traditionally maintained a ratio between 10 and 15:1, as they tended to maximize the present value of the future income of their shareholders at a time when they expected a fall in the future real price of oil. In other words, oil companies as commercial entities in the market-place, and acting according to their private interest, are not concerned with long-term economic and social objectives in the host countries. The contrast is clear: the oil companies tend to use a commercial criterion towards their shareholders and to minimize risk, while an oil producer has a long-term view of oil reserves as a vehicle for economic and social development.

The notion of maintaining a high oil reserve to production ratio will, in my opinion, take some time to be accepted and articulated by individual OPEC Member Countries. The resistance to this trend of thought will come in the short to medium term from a number of factors, not mutually exclusive, which are as follows:

First. There is an association in the minds of politicians between the level of oil production and political power. This is because politicians, by nature, tend to have a much shorter time span than the nations they govern. Even in the case of the individual man in the street, there is a psychological association between oil production and security. I have been asked on so many occasions by individuals not ordinarily concerned with oil policies, what ranking Kuwait would have as an oil producer if it reduced its oil production ceiling to 1.5 M b/d. I believe that the prevalent association between the level of oil production and political power will tend to decrease in the future in OPEC Member Countries, as both the general level of education and political awareness increase.

Second. The existing confusion between ends and means in the oil policies of individual OPEC countries. There are those who still think that it is important to grant production incentives to the oil companies in order to encourage exploration in the hope of finding more oil (so they end up producing initially much more oil than they require) and to indicate that future oil production policy will be more accommodating if an investment in exploration and development is made and large new reserves are found. But they will bear in mind that this production policy can be changed freely in the future if it outlives its purpose. In this connection, I wish to stress two points: (a) This oil policy, in my opinion, is the most inefficient and expensive mode of incentive to explore and find more oil. In fact, such a policy is not dissimilar to the behaviour of a man who gambles his family food budget on the pools in the hope that if he wins not only will he be able to feed his family but take them on a Mediterranean Cruise; and (b) Changing production policies of countries can and should be constrained by legal and international obligations once contracted. Time is needed before the individual OPEC Member Countries develop more efficient incentives to explore for oil than the incentive of overproduction.

Third. There is a misconceived and perhaps unconscious belief amongst some oil producers that once alternative sources of energy have been developed, say, in the first quarter of the next century, the prices of oil and gas would fall in real terms.

Again, I feel that this perception is highly dubious in view of the high cost of the alternative sources of energy. And when I speak of the cost of alternative sources of energy, I do not mean the cost of additional energy equivalent to one barrel of oil that can be produced from an alternative source today, but the cost of substituting large quantities of oil to satisfy energy demand by the alternative sources.

The difference in dimension between the above two concepts is readily apparent. The first would wrongly assume a constant or horizontal cost curve, while the second takes into account the fact that to produce a sizeable percentage to satisfy a large demand for energy, would change and dramatically affect all the factors that go into the production of energy from the alternative sources.

This is to say, the cost curve of energy from an alternative source is likely to be steeply and positively sloped, even if it remains constant or falls slightly in real terms over time.

Thus, while certain individuals may equate in their minds the price of oil to the marginal cost of the first equivalent barrel of alternative energy produced, it is more correct to be concerned with the cost of the last equivalent alternative barrel required to make a substantial shift from oil.

Let me also add, that even if we do not take into consideration the dynamic nature of costs nor the past experience of always underestimating the cost of alternatives, still the price of oil compares favourably with that of alternatives. On the basis of current estimates, shale oil is 1.3 times more expensive than conventional crude oil, syncrude from coal is 1.4 times more expensive and syngas 1.8 times more expensive. Moreover, exotic alternative energy sources such as nuclear, solar or wind, basically compete with oil solely in the field of power generation. Oil, conventional and non-conventional, will continue to be required for those end-uses for which it has a special value, such as automotive power, petrochemical and fertilizer feedstocks, and for speciality products, uses which will, even by the end of the century, continue to account for more than 50% of energy demand currently met by oil.

Also, in this instance, some time is needed for a re-orientation to this new perception of things.

In short, the fundamental theme in the scenario I have discussed is the call for a shift in the production policies of oil producers, from being essentially oriented to world markets and influenced by developments in the energy demand patterns of the industrialized countries (usually reflected in the actual and anticipated demand for imported oil into those countries); to production policies based on a long-range vision for the economic and social transformation of their societies. Further, as this scenario constitutes an essential feature of this transformation, its inevitability, in my opinion, is beyond question.

Conclusion

The long-range oil policy scenario which I have sketched out, is not envisaged to take place immediately, that is, in the next 2 to 5 years. It implicitly entails, however, a long-run reduction in the oil production by OPEC, almost to half its existing level. This perspective, in turn, necessitates drastic adjustment problems for the economies of the industrialized countries. The question that comes to mind is: does this, by necessity, imply a conflict of interest?

If the oil reserves were in the consuming areas where demand pressures on production policy are great, or, if the OPEC countries were willing to produce more (sacrificing their long-term interest for the benefit of the short-term interest of the industrialized countries), the rate of depletion would be higher, and the day of final reckoning would come sooner.

Therefore, given this context, is it not to the mutual benefit of both the industrialized and the producing countries that the reserves fall outside the main areas of consumption, with the oil producers need for rational development serving as a rationing device. This will bring forward the period of adjustment, but, at the same time will stretch out the adjustment process.

Is not this perception similar to Adam Smith's concept of the 'Invisible Hand' where the pursuit of self-interest brings optimality to the market? Similarly, the long-range target of development in the producing countries helps to bring optimality to consumption.

29 CONCLUSIONS AND SUMMING-UP OF THE SEMINAR DEBATES

Robert Mabro

A broad consensus was achieved on the following features of the energy situation. Temporary, none the less acute, imbalances between supply and desired demand for crude oil are likely to recur at irregular intervals in the short and medium term. This possibility underlines the need to improve or indeed develop methods for the management of short-term crisis by all parties concerned – companies and Governments of both producing and consuming countries. The nexus between the performance of the world economy and the energy problem is widely recognized, and no consolation is to be sought when the supply/demand imbalance for oil is corrected through recession and growing unemployment. The long-term energy problem relates to the difficulties and to the extended time lags involved in effecting a smooth transition from oil to other forms of energy.

Participants from oil-producing countries (including non-OPEC members such as Mexico and Malaysia) expressed concern about finding themselves in the situation of 'residual suppliers', called upon to absorb the fluctuations of total primary demand, without regard for their preferred rates of production. These rates are consistent, at the bottom of the range, with short-term revenue needs, and at the top of the range, with long-term 'conservation of the resource' objectives. They expressed also concern about 'being accommodated with the minimum possible concessions' in the next 10 or 15 years until industrialized countries succeed in reducing significantly the volume of oil imports, and about being left thereafter to solve alone their long-term development problems without support from their trade in the strong commodity, oil.

Participants from all backgrounds seem to have reached a fuller understanding of the significance of long-term development objectives, the vital priority for all producing countries.

Participants from industrial countries expressed concern about the security of oil supplies and price movements. There was no clear agreement about the concept of 'security of supplies'. On prices, the need for gradual rises (ensuring, at least compensatory increases against inflation), and the disruptive effect of sudden changes were recognized. Participants from industrial countries displayed various degrees of faith or scepticism in the scope for energy conservation in use. The numerous constraints (financial, socio-political, environmental) on the rapid development of alternative sources of energy were also identified. Hence, an enhanced awareness of the difficulties of the transition, of the length of the lead-in times (not all technological) involved, and of the need to design a number of different bridges to facilitate the passage from the present to a different energy world. A feeling of increased dissatisfaction with the energy policies of consuming countries, and with their apparent lack of initiative in the field of international co-operation with oil-producing and developing countries, began to manifest itself towards the end of the Seminar among some participants from industrial countries.

Progress towards mutual understanding was noticeable on the following points. Participants from Europe, North America and Japan saw more clearly that the worries of producing countries about depletion rates relate to vital interests because oil is often the sole resource and the development horizon is very long. That oil should be traded for assets of more perennial value than paper dollars and inappropriate industrial projects is the central issue. Participants from oil-producing countries gained a clearer idea of the factors – political processes, public opinion, independence of the media, etc. – which interfere with the formulation and implementation of policy in Western countries.

Yet, neither side was fully satisfied by the case put by the other. Participants from consuming countries were constantly seeking a more concrete formulation of the solutions envisaged by producing countries for their problems and objectives, especially on issues such as the 'transfer of technology', the 'store of value' to be acquired in exchange for depletable oil, the development strategy, etc. Participants from producing countries felt that public opinion, democratic institutions and the media, though constraining Governments in consuming countries, can also provide convenient excuses for inaction. What is certain is that they cannot excuse the failure of Governments to inform and educate. It was also felt that the weight of public opinion is always invoked by industrial countries as something specific to the West. Governments in oil-producing countries also have to contend with public opinion, constituencies and pressure groups, the universal features of any polity.

The urgent need for a dialogue was almost unanimously recognized.

There was no general agreement on whether the dialogue should be pursued at the informal level by *ad hoc* groups, like the Seminar itself, or whether it should be structured officially. In the latter case, the question of the institutional framework arises. The oil-producing side sees the dialogue on oil as part and parcel of a wider agenda involving all the main issues of the third world/industrialized countries relationship. The industrialized countries side seemed divided, some favouring the wider agenda and others preferring a piecemeal approach focusing on distinct issues one at a time.

The realization that a fundamental asymmetry characterises the relationship between oil-producing and industrialized countries seemed to crystallize towards the end of the seminar. Oil-producing countries have a source of power in the medium-term because of oil. They do not enjoy power or strength in other areas, economic, military, demographic, not even financial (despite certain appearances to the contrary in surplus countries). Unless they translate the oil wealth of today into economic development for tomorrow, they may find themselves in a very weak position in 15 or 20 years from hence. And they are dependent to a great extent on the outside world for this successful translation of oil into economic wealth for the future. The industrialized countries enjoy economic, political, technological, military, and, despite the shambles of currency markets, financial power. They are dependent in one area only, that is energy, on the oil-producing countries. They hope to reduce this dependence in perhaps 15 or 20 years. The resources needed to remedy their perceived weakness in energy – finance, technology, skills – are, potentially at least, available to them from within. The irresistible temptation is to ask the other side for a period of grace, 'until we sort ourselves out'. But the oil-producing countries themselves need oil to buy the long period of time necessary for their own development. To give it away, fast and relatively cheaply, as a grace to others, would leave them at too early a stage of their development, stripped of assets and leverage, dependent and weak. The real challenge for all those who seek a successful and fair agreement between the two sides is to come to grips with the implications of this asymmetry in the relationship.

Discussion groups among participants attempted to identify major issues for a hypothetical dialogue between producers and consumers and to sketch approaches to possible solutions. Apart from the issue of the dialogue itself (the main questions having been raised above) the following items tended to recur in most reports.

(a) The demand for a store of value. For countries with surplus funds, a possible 'store of value' is a new financial asset protected against the erosion of inflationary forces and the vagaries of exchange fluctuations. Energy bonds and world unit trusts were mentioned. But the concept of a 'store of value' has wider dimensions. Some think that a reform of the world monetary system and actions by industrial countries to control inflation are essential prerequisites. Since a large proportion of oil revenues is spent on development projects, several groups emphasized the need for new forms of co-operation with industrialized countries through which the latter would provide technological and managerial assistance.

(b) An investigation of the concept of 'transfer of technology'. This issue has many facets. Most developing countries feel that they are charged monopoly prices for technological packages and that barriers to access are often erected. There is some ambiguity in many discussions about absorption by the recipient country. If the transfer fails, blame tends to be put by one side on problems of absorption, and by the other side on restrictive practices and on inappropriate packages offered without consideration for the specific situation of the developing country.

(c) The removal of trade barriers against exports from developing countries. Oil-producing countries feel that their industrialization involves exports of petrochemicals, refined petroleum products, metals, etc. Access to markets, which means absence of tariffs, import prohibition and quotas in the industrialized countries is essential. But access to markets also means expertise and introduction to distribution channels, all of which may require specific assistance.

(d) Non-oil producing countries were considered in all the reports. Suggestions for complementary efforts by OPEC and OECD countries were made. One proposal is a joint energy fund with contributions from both sides for the development of energy resources. Several groups clearly indicated that the opening-up of mrkets for exports, the transfer of technology, etc., are issues which concern oil-producing and other developing countries as a single group.

(e) On the price of oil, though recognizing that the sovereignty of OPEC is not open to negotiation, all syndicates emphasized the need to explore alternative strategies. The main objective is to avoid sudden rises. The difficulties of reconciling the principle of gradual and regular increases with that of sovereignty and with the need to respond flexibly to varying market conditions have not been properly explored. The issue of oil supply is more complex. Clearly security of supply cannot be interpreted as 'satisfaction of consumers' desired demand irrespective of its level'. The problem, rather, is how to make the rate of production preferred by producing countries (given their development objectives) match a rate of consumption that is not so low as to jeopardise a healthy growth performance of the world economy. A further problem is that the desire of consuming countries to reduce their dependence on imported oil within, say 10 or 15 years is difficult to reconcile with that of producing countries to stretch the useful (and economically profitable) life of their reserves for a period of say 30 or 40 years. The reconciliation may involve a radical change of attitude towards conservation policies and development of alternatives in consuming countries. The mutuality of interest would be better served if the aims of energy policy were not stated in terms of reducing dependence (interpreted as 'putting the producers out of business') but were rather defined in terms of a long-run transition of the interdependent economies of producers and consumers from one state of the world to another. This may mean more conservation in use immediately, more economic growth than contemplated in the medium-term by pessimistic forecasters, and a carefully phased programme of investment and development of alternatives for the long-run.

(f) Preoccupations about a possible supply crisis in the short-term led some groups to discuss ways of managing crises. It is understood that relatively small, often purely accidental, short falls in supply tend to send market prices of oil rocketing high. The question is whether a joint intervention on pre-agreed principles by producing and consuming countries is possible during such a crisis? The problems of agreement seem formidable. But the notion that the willingness to provide the market from stocks and excess productive capacity in the event of a short-term supply imbalance is the best antidote to speculative fever, is not devoid of commonsense.

This summing-up is by no means exhaustive. The Seminar did not exhaust all the issues, and may not have identified them all. But the consensus among participants, if I interpreted it correctly, was that some of the problems listed here deserve urgent and profound consideration. The short-term issues can be serious because a crisis tends to fall upon us without sufficient warning and there is no strong certainty that existing institutions will always be able to cope effectively. The long-run issues cannot be ignored or postponed because the time required to provide an adequate solution is itself very long. To postpone action on these latter issues simply means that after the slippage of a few idle years we shall have to face them again as the short-term crisis of the late 1980s or the 1990s.

121102